GW01237843

Heterogeneous Catalysts

Heterogeneous Catalysts

Advanced Design, Characterization and Applications

Edited by
Wey Yang Teoh, Atsushi Urakawa, Yun Hau Ng, and Patrick Sit

Volume 2

Editors

Prof. Wey Yang Teoh
University of Malaya
Department of Chemical Engineering
50603 Kuala Lumpur
Malaysia

Prof. Atsushi Urakawa
Delft University of Technology
Faculty of Applied Sciences
Building 58 E2 100
Van der Maasweg 9
2629 Delft
The Netherlands

Prof. Yun Hau Ng
City University of Hong Kong
School of Energy and Environment
Tat Chee Avenue
Kowloon
Hong Kong, S.A.R.

Prof. Patrick Sit
City University of Hong Kong
School of Energy and Environment
Tat Chee Avenue
Kowloon
Hong Kong, S.A.R.

Cover
Cover image: Courtesy of Wey Yang Teoh and Nat Phongprueksathat

Library of Congress Card No.:
applied for

British Library Cataloguing-in-Publication Data
A catalogue record for this book is available from the British Library.

Bibliographic information published by the Deutsche Nationalbibliothek
The Deutsche Nationalbibliothek lists this publication in the Deutsche Nationalbibliografie; detailed bibliographic data are available on the Internet at <http://dnb.d-nb.de>.

Print ISBN: 978-3-527-34415-4
ePDF ISBN: 978-3-527-81356-8
ePub ISBN: 978-3-527-81358-2
oBook ISBN: 978-3-527-81359-9

Typesetting SPi Global, Chennai, India
Printing and Binding CPI Group (UK) Ltd, Croydon, CR0 4YY

Printed on acid-free paper

10 9 8 7 6 5 4 3 2 1

Contents

Preface

Heterogeneous catalyst to a chemical reaction is akin to a microscope to microbiology or a sail to a yacht. It is a necessary tool that helps to speed up reactions and at the same time steers the reactions in such a way that maximum selectivity of the desirable product can be attained. In that sense, heterogeneous catalysts will always be relevant as far as chemical reactions are of interest, whether at large industrial scales (e.g. commodity chemicals, petrochemical refineries), small scales (e.g. devices with catalytic functions such as automobile catalytic converters), or even microscales (e.g. microfluidic devices, catalytic nanomachines). The dedication of scientists and engineers working in the field of heterogeneous catalysis throughout the twentieth century was instrumental in solving of some of the most important problems facing humanity, including the nitrogen food chain issue (the so-called Nitrogen problem), production of high-quality automobile fuels (e.g. gasoline and diesel), abatement of noxious airborne pollutants, and the manufacturing of methanol as well as other building block chemicals.

Entering the twenty-first century, the two immediate and overarching challenges in heterogeneous catalysis are (i) to address the issues related to global warming and climate change and (ii) to align with the United Nations sustainable development goals. Catalytic reactions such as water splitting, reduction of carbon dioxide, waste biomass conversion, removal of emerging aqueous micropollutants, and the abatement of NO_x that are focused on enhancing renewable energy security and environmental sustainability will take the center stage against a backdrop of swelling population and increasing urbanization.

In adapting to these grand challenges, new and sophisticated emerging techniques in heterogeneous catalysis are continuously being developed to overcome the various limitations in catalyst design and to understand the mechanism of the molecular reactions occurring on the catalytic surface (information that can feed back to the catalyst design). At the same time, catalysts with different modes of activation are increasingly being appreciated, which besides the conventional thermal catalysts now include electrocatalysts and photocatalysts and their underlying physics. For newcomers entering the field, acquiring such vast amount of knowledge, although essential, can be overwhelming. That is not to mention the tenacity in mastering the basic fundamentals in heterogeneous catalysis, itself a century worth of knowledge, prior to the appreciation of these state-of-the-art advancements. With this in hindsight, the book is geared toward attracting and assisting beginners who wish for a head start and quick overview on some of the most important emerging tools for catalyst design, techniques for *operando/in situ* characterization and ab initio computation, as well as a glimpse of the advancements in heterogeneous catalysis toward some of the grand challenges. Undergraduates with some prior exposure to reaction engineering/heterogeneous catalysis and analytical chemistry/spectroscopy or early postgraduates pursuing research on heterogeneous catalysis but only with some primitive background of the field shall find the book useful. Because the aim is to bridge the gap between amateur readers and expert knowledge, each chapter provides a brief description of the required basic fundamentals that lead to the appreciation of state-of-the-art advancements.

It should be mentioned that the Contributors of the different chapters in the book are themselves among the most promising Emerging and Pioneering Researchers in the field of heterogeneous catalysis. We capitalized on that point in our book design to allow each Contributor to articulate the advancement of his/her own technique in a semitutorial manner that can be appreciated by the target readers. We strive to preserve a delicate balance between readability and articulating the complexity of the advanced techniques. In that sense, we present the content in a less mathematical (in a semiquantitative form, as much as we could) but comprehensible setting as the first step to inculcate interest and inspiration among beginners. With some patience, self-learning is highly possible, following which readers should have the ability to pursue more quantitative references of specific techniques. With the heightened expectation of "cross skills" among the new generation of catalyst researchers, this book shall come in handy for readers to gain appreciation on some of the most advanced techniques before deciding to specialize in some of them. In fact, we hope that the book would serve as a platform to inspire readers to potentially develop their own original or hybrid techniques in a wider effort to tackling the grand challenges using heterogeneous catalysts.

Finally, we take the opportunity to thank Emerging and Pioneering Researchers who have contributed to this book, its vision and purpose. It has been a massive effort that took us more than three years to put together this book, and we thank the Contributors for their patience.

Wey Yang Teoh
University of Malaya, Malaysia
23 November 2020
Atsushi Urakawa
Delft University of Technology, The Netherlands
23 November 2020
Yun Hau Ng
City University of Hong Kong, S.A.R.
23 November 2020
Patrick Sit
City University of Hong Kong, S.A.R.
23 November 2020

Section III

Ab Initio Techniques in Heterogeneous Catalysis

22

Quantum Approaches to Predicting Molecular Reactions on Catalytic Surfaces

Patrick Sit

City University of Hong Kong, School of Energy and Environment, Kowloon, Hong Kong SAR, PR China

22.1 Heterogeneous Catalysis and Computer Simulations

Heterogeneous catalysis has important applications in many different areas in chemistry and chemical engineering. The increasing demand from these applications necessitates the development of high-performance, robust catalysts, as well as design of better catalytic processes. Such development needs to be guided by the fundamental understanding of the physiochemical properties of the catalysts and the mechanistic details of the reaction processes. Computer simulations are powerful tools to provide this understanding from the macroscopic to the atomic and quantum scale that are often difficult to obtain by experiments alone due to the limitations in experimental techniques.

The "equipment" needed to carry out simulation studies are computers. Nowadays, simulations are commonly run in high-performance computing (HPC) clusters that consist of a large number of computers linked together with high-speed connection. Through aggregation of computing power, studies can be carried out much more efficiently when the simulation programs are well parallelized over the CPUs with fast communication. The performance of an HPC cluster is typically measured by the number floating-point operations it carries out per second (FLOPS). As of 2019, the fastest HPC cluster or supercomputer can perform over a hundred petaFLOPS [1], and this number is expected to continue to increase. With the ever-increasing computing power, better parallelization in programming, and, most importantly, development of more efficient and accurate computational techniques, simulation studies can reveal much more information in heterogeneous catalysis study than previously could.

Different computational approaches have been developed to simulate reaction processes and material properties over a wide range of length scales and time scales. Figure 22.1 shows the scales of physical problems and examples of computational techniques applicable in the different regimes. At the macroscopic scale that involves reactor study and design, continuum approaches are typically

Heterogeneous Catalysts: Advanced Design, Characterization and Applications, First Edition.
Edited by Wey Yang Teoh, Atsushi Urakawa, Yun Hau Ng, and Patrick Sit.

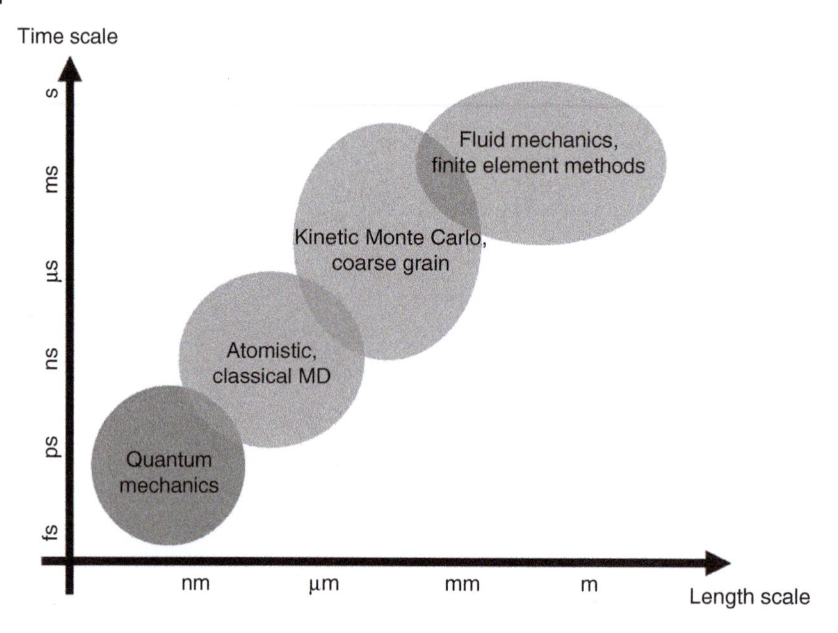

Figure 22.1 Different length scales and time scales of simulations.

used to model the heat, mass, and momentum transfer processes. For example, the flow of the reactants in the reactor can be modeled by solving differential equations like the Navier–Stokes equations using numerical techniques.

On the other hand, at the mesoscopic scale, different techniques can be used depending on the results of interest. For example, reaction rates are often modeled using kinetic Monte Carlo techniques. The coarse-grain approach has also been adopted to model systems in the micrometer to millimeter length scale. In the coarse-grain approach, "pseudoatoms" are used to approximately represent groups of atoms. The interactions between these coarse-grained particles are described by potentials with predetermined parameters. By decreasing the degrees of freedom of the systems, the simulations can afford much larger systems and longer simulation times.

Going further down in the length scale and time scale, we have atomic-scale simulations in the nanometer regime. For example, classical molecular dynamics (CMD) simulations are typically used to study the structures of nanoscale materials or enzymes. Similar to the coarse-grain simulations, the interactions between atoms in CMD are described by the predetermined force field parameters. Since CMD does not consider electronic properties, it typically cannot study chemical reactions like electron transfer, bond breaking, and bond formation. Although there have been reactive force field potentials [2] developed for chemical reaction study, their application and transferability can be limited.

In order to study catalytic reactions, we need to properly describe the electronic properties using the theory of quantum mechanics. Quantum mechanics is a groundbreaking theory developed in the early twentieth century that deals with properties of matters at the microscopic scale. Through proper description of the electronic properties, quantum mechanical simulations can provide energetic and mechanistic details of chemical reactions, as well as the physiochemical properties of the catalyst surface.

22.2 Theory of Quantum Mechanics

At the turn of the twentieth century, two groundbreaking theories appeared that revolutionized our view of the physical world. The first one is the theory of relativity introduced by Albert Einstein, which completely changed our understanding of the properties of matters at high speed and our understanding of gravity [3].

Another groundbreaking theory of modern physics is the theory of quantum mechanics. Unlike the theory of relativity, the development of quantum mechanics was a collective effort of many prominent scientists in the first decades of the twentieth century. One important concept in quantum mechanics is wave–particle duality, which states that all matters and light exhibit properties of both wave and particles. Light, traditionally considered as wave, can also behave like particles known as photons. On the other hand, elementary particles like electrons also exhibit wavelike behaviors and should be described by the wavefunctions. The postulates in quantum mechanics state that the wavefunctions determine everything that can be known about the quantum mechanical system. To obtain the electronic wavefunctions, one needs to solve a differential equation called the Schrödinger equation. As will be discussed in detail in Chapters 23–28, quantum mechanical simulation techniques center around solving the Schrödinger equation of quantum mechanical systems for their ground-state and excited-state properties.

22.3 Quantum Mechanical Techniques in the Study of Heterogeneous Catalysis

In Chapters 23–28, we will discuss state-of-the-art quantum mechanical approaches for the study of heterogeneous catalysis. In Chapter 23 we focus on the density functional theory (DFT). DFT has been the most widely used quantum mechanical technique due to the best balance between accuracy and efficiency in the study of ground-state properties. However, the accuracy of DFT calculations depends on the choice of the functionals and the development of corrections. In Chapter 24, the DFT-based ab initio (first principles) molecular dynamics technique will be discussed for the study of dynamical and finite-temperature properties.

In heterogeneous catalysis study, the physiochemical properties at the interface are important. In Chapter 25, we will discuss the recently developed technique to simulate the interfacial electrochemistry. On the other hand, as DFT is a ground-state theory, there have been ongoing efforts to develop techniques for the study of excited-state properties. Two important techniques in this direction are the time-dependent DFT (Chapter 26) and the GW method (Chapter 27). Lastly, Chapter 28 discusses the design of novel catalytic materials using high-throughput simulations and data analysis that has been an area of intense research in recent years.

References

1 https://www.top500.org/lists/2019/11/ (retrieved December 2019).
2 Senftle, T.P., Hong, S., Islam, M.M. et al. (2016). The ReaxFF reactive force-field: development, applications and future directions. *npj Comput. Mater.* 2: 15011.
3 Einstein, A. (2017). *Relativity: The Special and General Theory: Original Version*. CreateSpace Independent Publishing Platform.

23

Density Functional Theory in Heterogeneous Catalysis

Patrick Sit and Linghai Zhang

City University of Hong Kong, School of Energy and Environment, Kowloon, Hong Kong SAR, PR China

23.1 Introduction

This chapter focuses on the first principles approach to study the properties of heterogeneous catalysis. A simulation is called first principles if it relies on the established physical laws without empirical parameters. In the study of electronic properties, the Schrödinger equation is the central physical law. Although the form of the Schrödinger equation is known, getting the exact solution in realistic systems with multiple electrons is usually impossible due to the coulomb interaction between the electrons. Different approximated approaches have been introduced to circumvent this problem. Among them, density functional theory (DFT) [1, 2] is the most widely used approach to solve the Schrödinger equation with the best balance between accuracy and efficiency.

The DFT approach is particularly useful in the study of heterogeneous catalysis to reveal the atomic-scale details of the catalytic reactions. For example, the energetic and structural properties of the reaction intermediates and the energy barriers of the reaction steps can be determined. With such information, the possible reaction mechanisms can be identified. DFT calculations also allow us to examine the ion-transfer processes and electron flow during the reactions in heterogeneous catalysis.

In this chapter, we will first provide an overview of DFT in terms of its basic concepts and practical implementation. Although DFT is an exact theory, how well it describes the material electronic properties relies on the accuracy of the exchange–correlation energy functional. We will discuss the common functionals used nowadays and the efforts toward improving the accuracy of the results. Lastly, we will highlight some examples of the application of DFT simulations in the study of heterogeneous catalysis and surface chemistry.

Heterogeneous Catalysts: Advanced Design, Characterization and Applications, First Edition.
Edited by Wey Yang Teoh, Atsushi Urakawa, Yun Hau Ng, and Patrick Sit.

23.2 Basics of Density Functional Theory Calculations

23.2.1 Born–Oppenheimer Approximation

As stated earlier, the central equation in quantum mechanics is the Schrödinger equation. The full Schrödinger equation considers both the electrons and the nuclei as quantum mechanical particles. However, the intrinsic speed of the motion of the nuclei is usually significantly lower than that of the motion of the electrons due to the larger nuclei masses. We can therefore simplify the full Schrödinger equation by separating the nuclei and electronic degrees of freedom and treat only the electrons as quantum particles. This is called the Born–Oppenheimer approximation. Under this approximation, the electrons respond instantaneously to the motion of the nuclei. For a system with M atoms and N electrons, the electronic wavefunctions are obtained by solving the electronic Schrödinger equation:

$$\hat{H}\psi(\bar{r}_1, \ldots, \bar{r}_N) = E\psi(\bar{r}_1, \ldots, \bar{r}_N)$$

$$\hat{H} = \hat{T} + \hat{V}_{\mathrm{ne}} + \hat{V}_{\mathrm{ee}} = -\frac{\hbar^2}{2m_e}\sum_{i=1}^{N}\nabla_i^2 + \sum_{i=1}^{N}\sum_{I=1}^{M}\frac{-e^2 Z_I}{|\bar{r}_i - \bar{R}_I|} + \sum_{i=1}^{N}\sum_{i<j}\frac{e^2}{|\bar{r}_i - \bar{r}_j|} \tag{23.1}$$

Here $\hat{H}$ is the multi-electron Hamiltonian, $\psi(\bar{r}_1, \ldots, \bar{r}_N)$ is the multi-electron wavefunction that is a function of the coordinates of all the electrons, and E is the total energy. The three terms in the Hamiltonian are, from left to right, the kinetic energy operator, the nuclei–electron Coulomb potential energy operator, and the electron–electron Coulomb potential energy operator. Despite the relatively simple form of the Schrödinger equation, solving the equation for the wavefunction is impossible in multi-electron systems due to the electron–electron interaction term. DFT was therefore introduced to tackle this problem.

23.2.2 The Hohenberg–Kohn Theorems and the Kohn–Sham Approach

DFT is based on two fundamental theorems developed by Hohenberg and Kohn [1]. Here, we only discuss the consequences of the two theorems. For interested readers, refer to review papers, books, and the original papers for the detailed discussion and proofs [1, 3, 4].

The first Hohenberg–Kohn theorem states that given a ground-state electronic density, $n_0(\bar{r})$, the external potential (i.e. nuclei–electron Coulomb potential) is uniquely defined. Then when the external potential is defined, the ground-state many-body wavefunction, $\psi_0(\bar{r}_1, \ldots, \bar{r}_N)$, can be determined by solving the Schrödinger equation. In other words, the ground-state multi-electron wavefunction is uniquely defined with a given $n_0(\bar{r})$. Therefore, the total energy, besides being a functional of the ground-state wavefunction, can be rewritten as a functional of the electronic density only. In this regard, Hohenberg and Kohn introduced

$$E[n(\bar{r})] \equiv \langle\psi|\hat{T}|\psi\rangle + \langle\psi|\hat{V}_{\text{ee}}|\psi\rangle + \int v_{\text{ext}} n(\bar{r}) \tag{23.2}$$

where $E[n(\bar{r})]$ is the total energy functional that includes (from left to right on the right-hand side) the kinetic energy term, the electron–electron interaction term, and the external potential energy term.

The second Hohenberg–Kohn theorem states that the electronic density that minimizes the total energy is the exact ground-state electronic density and the corresponding energy is the ground-state energy. It provides a scheme to search for the electronic ground state via the variational principle:

$$E_0 \leq E[n(\bar{r})] \tag{23.3}$$

In this regard, determining the ground state of the Schrödinger equation is simplified significantly because, instead of minimizing the energy functional with respect to the multi-electron wavefunction, minimization of the energy functional can now be performed with respect to the electronic density. However, the exact form of the energy functional, especially the kinetic energy term and the electron–electron interaction term, is not known in practice.

Kohn and Sham [2] presented a practical approach to solve this problem by introducing a noninteracting electron reference system:

$$\hat{H}_s\phi_i(\bar{r}) = \left[-\frac{\hbar^2}{2m_e}\sum_{i=1}^{N}\nabla_i^2 + \sum_{i=1}^{N} v_s(\bar{r}_i)\right]\phi_i(\bar{r}) \tag{23.4}$$

where the potential $v_s(\bar{r}_i)$ is chosen in such a way that the ground-state electronic density of this noninteracting system is the same as that of the interacting system. Equation (23.4) is called the Kohn–Sham equation. $\phi_i(\bar{r})$ are the noninteracting electronic wavefunctions called the Kohn–Sham wavefunctions. The electronic density of an N-electron system is therefore given by

$$n(\bar{r}) = \sum_{i}^{N} \phi_i^*(\bar{r})\phi_i(\bar{r}) = \sum_{i}^{N} |\phi_i(\bar{r})|^2 \tag{23.5}$$

where the summation is over the N lowest eigenstates of the noninteracting Hamiltonian.

With the Kohn–Sham approach, we have transformed the problem from finding the ground-state electronic density, $n(\bar{r})$, to finding the set of noninteracting electronic wavefunctions (Kohn–Sham wavefunctions) of the Kohn–Sham equation. The total energy functional can be written as

$$E[n(\bar{r})] = T_s[n(\bar{r})] + E_H[n(\bar{r})] + E_{\text{xc}}[n(\bar{r})] + \int v_{\text{ext}} n(\bar{r}) \tag{23.6}$$

where

$$E_{\text{xc}}[n(\bar{r})] \equiv T[n(\bar{r})] - T_s[n(\bar{r})] + V_{\text{ee}}[n(\bar{r})] - E_H[n(\bar{r})] \tag{23.7}$$

$T_s[n(\bar{r})]$ is the kinetic energy, and $E_H[n(\bar{r})]$ is the Hartree energy of the noninteracting electronic system. $E_{xc}[n(\bar{r})]$ is the exchange–correlation energy functional that contains the difference between $T[n(\bar{r})]$ and $T_s[n(\bar{r})]$, as well as the nonclassical part of $V_{ee}[n(\bar{r})]$. With the energy functional established, we can obtain the corresponding Hamiltonian:

$$\begin{aligned}\hat{H}_s &= \frac{\delta E[n(\bar{r})]}{\delta n(r)} = \frac{\delta \left\{T_s[n(\bar{r})] + E_H[n(\bar{r})] + E_{xc}[n(\bar{r})] + \int v_{ext} n(\bar{r})\right\}}{\delta n(r)} \\ &= -\frac{\hbar^2}{2m_e}\sum_{i=1}^{N} \nabla_i^2 + v_{ext}(\bar{r}) + \int \frac{n(r')}{|\bar{r} - \bar{r}'|} dr' + v_{xc}(\bar{r})\end{aligned} \tag{23.8}$$

and with the exchange–correlation potential

$$v_{xc} = \frac{\delta E_{xc}[n(\bar{r})]}{\delta n(r)} \tag{23.9}$$

Up to this point, the Kohn–Sham approach is still exact, but it requires the knowledge of the exact exchange–correlation functional, $E_{xc}[n(\bar{r})]$, which is not known. Practical calculations involve the use of approximated exchange–correlation functionals. As will be discussed later, the search for more accurate $E_{xc}[n(\bar{r})]$ is an active field of research.

Beside the approximations in the exchange–correlation energy functional, another issue is that the Kohn–Sham Hamiltonian is now a functional of the electronic density. Before we can determine the Kohn–Sham wavefunctions, we should first know the electronic density. Therefore, the input to the Kohn–Sham equations relies on the output via the single-electron Kohn–Sham wavefunctions $\phi_i(\bar{r})$. To tackle this issue, the solutions of the Kohn–Sham equations are obtained by the self-consistent field (SCF) procedure. The details of the SCF cycle are shown in Figure 23.1: (i) The initial guess of the electron density is made. One way to construct such initial guess is to use the electron densities that correspond to the individual atoms in the system. (ii) Using the given electron

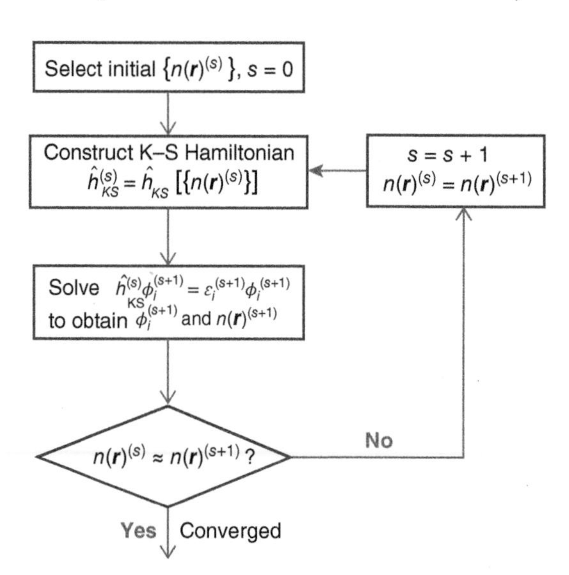

Figure 23.1 Schematic diagram for the self-consistent field calculations when solving the Kohn–Sham equation. The "≈" sign indicates the check whether the difference between the new electronic density and the old electronic density is less than a target tolerance.

density, we construct the Hamiltonian of the Kohn–Sham equations. (iii) We solve the Kohn–Sham equations to get the occupied single-particle orbitals $\phi_i(\overline{r})$, and then we obtain the updated electronic density. (iv) If the difference between the updated electronic density and the old electronic density is less than a target tolerance, the calculation is converged. If not, the SCF calculation continues until convergence is achieved. After the SCF calculation, we will be able to obtain the ground-state energy, charge density, and spin density of the system. Such information is important for evaluation of the molecular adsorption energy on the surface and examining possible charge and spin transfer upon adsorption.

23.2.3 Basis Sets

Due to the limit in computer's memory, the electronic wavefunctions need to be expanded by a finite set of basis functions. One type of basis functions is the localized Gaussian-type basis. The localized basis sets are better suited to study localized electronic systems such as molecules or molecular solids like metal–organic frameworks (MOFs). Examples of DFT packages that use localized basis sets are Gaussian [5] and NWChem [6]. Another type of popular basis sets is the plane-wave basis set, which is particularly useful for the study of extended systems with the periodic boundary conditions as in the study of heterogeneous catalysis. Some common packages that use the plane-wave basis are VASP [7], Quantum ESPRESSO [8], and CPMD [9].

The plane-wave basis has some advantages over other basis. These include the simple form of the basis functions with no pre-assumption of the form of the solution. Unlike the localized basis, the plane-wave basis has no basis-set superposition error. The plane-wave basis also allows efficient calculation of the forces on ions with no Pulay forces as the basis set is independent of the atomic positions. Moreover, the convergence of the results with respect to the basis set can be easily tested and monitored by adjusting the plane-wave cutoff energy. On the other hand, there are also simulation packages (e.g. CP2K [10]) that use mixed plane-wave/Gaussian basis benefiting from the advantages of both to enhance computational efficiency.

23.2.4 Forces on the Ions

In the study of heterogeneous catalysis, we often need to obtain the optimized structure of molecular adsorption on the surface, which requires the calculation of the forces on the ions. Forces are also needed to perform first principles molecular dynamics simulations. Under the Born–Oppenheimer approximation, the ions propagate on the potential energy surface of the electronic ground state $V = \langle\psi|\hat{H}|\psi\rangle$, where $\hat{H}$ is the Hamiltonian. The forces on the ions thus become

$$\begin{aligned} F_I &= -\frac{dV}{dR_I} = -\frac{d\langle\psi|\hat{H}|\psi\rangle}{dR_I} = -\frac{\partial\langle\psi|\hat{H}|\psi\rangle}{dR_I} - \langle\psi|\frac{\partial\hat{H}}{\partial R_I}|\psi\rangle - \left\langle\psi|\hat{H}\left|\frac{\partial\psi}{\partial R_I}\right.\right\rangle \\ &= -E\frac{\partial\langle\psi\mid\psi\rangle}{\partial R_I} - \langle\psi|\frac{\partial\hat{H}}{\partial R_I}|\psi\rangle = -\langle\psi|\frac{\partial\hat{V}}{\partial R_I}|\psi\rangle \end{aligned} \tag{23.10}$$

where $\hat{V}$ is the external potential operator, which is the only term that depends on R_I in $\hat{H}$. This is called the Hellmann–Feynman theorem, which simplifies the determination of the forces on the ions as the calculation now does not involve the computation of the derivatives of the electronic wavefunctions with respect to the positions of ions. Instead, only the analytical form of $\frac{\partial \hat{V}}{\partial R_I}$ is needed. Besides getting the optimized structures and their energetics, knowing the forces also allows us to search for the transition state, the transition path, and the activation energy in a reaction. Two commonly used techniques for this purpose are the nudged elastic band (NEB) [11] and the dimer [12] methods.

23.3 The Search for Better Energy Functionals

23.3.1 Energy Functional Development

Throughout the years, many exchange–correlation density functionals have been developed aiming at more accurate study of quantum mechanical problems [13]. The functionals can generally be categorized into the semiempirical and nonempirical functionals. The semiempirical functionals typically have flexible functional forms of the electronic density in which the coefficients were obtained by fitting to the reference values from high-level quantum chemistry calculations, such as the coupled-cluster method, or the experimental data. The B3LYP functional [14] that is the most popular functional in the chemical community is one example. For the nonempirical functionals, the functional form and the parameters were derived from the many-body Schrödinger equation, which means that these functionals are truly free from the empirical parameters. The Perdew–Burke-Ernzerhof (PBE) functional [15] that is the most widely used functional in computational materials science is an example.

Current functionals usually face accuracy problems like poor description of the van der Waals interactions and of the electronic correlation [16, 17]. The spurious self-interaction of the electrons in most exchange–correlation functionals also causes poor results. The self-interaction error (SIE) arises as the energy functional is a functional of the electronic density. Individual electron can therefore feel the repulsion with itself without the accurate exchange–correlation functional. The SIE is particularly serious in systems with localized electronic density like the d-electrons in transition metals.

Proposed by Perdew et al. [18], the development of energy functionals follows a "Jacob's ladder"-like hierarchy as shown in Figure 23.2. Building on the foundations in the "Hartree world," where the exchange–correlation energy is zero, additional ingredients are added at each ladder up for better accuracy. The first addition is the exchange and correlation under the local density approximation (LDA). In this case, the exchange–correlation functional is dependent only on the electronic density. In other words, the exchange and correlation are treated as local in character:

$$E_{\mathrm{xc}}^{\mathrm{LDA}}[n(r)] = \int n(\bar{r}) v_{\mathrm{xc}}^{\mathrm{LDA}}[n(\bar{r})] d^3 r \tag{23.11}$$

Despite showing improvements, simulations using LDA functionals tend to overestimate the strength of interaction leading to overbinding of molecules on the surface and shorter bond lengths.

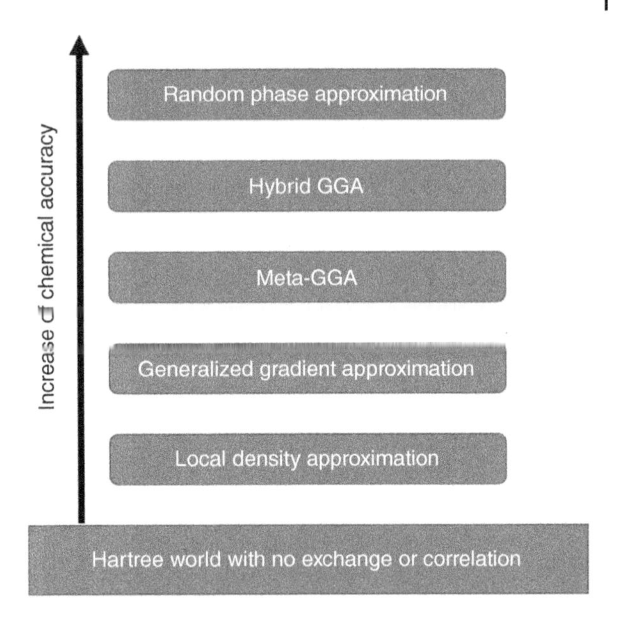

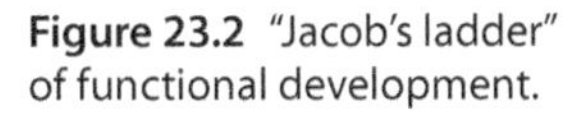
Figure 23.2 "Jacob's ladder" of functional development.

The generalized gradient approximation (GGA) is next in the level, which includes the dependence of the magnitude of the gradient of the electronic density $|\nabla n(\overline{r})|$. The functional form of the GGA functional is

$$E_{\mathrm{xc}}^{\mathrm{LDA}}[n(r)] = \int n(\overline{r})v_{\mathrm{xc}}^{\mathrm{GGA}}[n(\overline{r}), |\nabla n(\overline{r})|]d^3r \tag{23.12}$$

Numerous GGA exchange–correlation functionals have been developed, and among them, the examples are B88 [19], PW91 [20], BLYP [21], and PBE [15].

Next step in the development is meta-GGA. A meta-GGA functional involves the additional dependence of the exchange–correlation functional on the kinetic energy density ($\tau = [\nabla n(\overline{r})]^2$). Typical ones are B88C [22], B95 [23], Becke Roussel [24], and TPSS [25]. The next level of development involves introducing functionals of nonlocal nature like the inclusion of a fraction of the Hartree–Fock exchange term constructed from the Kohn–Sham occupied orbitals. Lastly, one can include all the information of the occupied and unoccupied orbitals through the random phase approximation and the second-order many-body perturbation theory.

23.3.2 Other Corrections and Approaches

Besides the hierarchy of improvement of the exchange correlation functional, correction terms have also been introduced to improve the accuracy. While the LDA and GGA functionals can reasonably describe the covalent and ionic bonds, they are known to not be able to properly treat the weak van der Waals forces. This problem leads to prediction of incorrect interlayer distance of the graphene sheets in graphite and poor description of molecular adsorption structures and energetics. Different correction schemes have been introduced to the existing functional to fix this problem. Some of these schemes are free of empirical input [26], while others involved empirical corrections [27–29].

Another common problem of current functionals is that the strongly correlated systems are poorly treated. The Coulomb repulsion between individual electrons is not properly described in standard LDA or GGA functionals. These functional also contain serious SIE. One notable consequence is that the electrons have the tendency to over-delocalize. This could also lead to incorrectly predicting metallic ground states for insulators. This problem is particularly prominent in systems containing transition metals with strongly localized d-electrons. One way to tackle this problem is the Hubbard U corrections [30] to properly describe the on-site repulsion of the d-electrons. Key in this correction is the Hubbard U parameter, which quantifies the interaction between the d-electrons. The Hubbard U values can be determined experimentally or obtained from DFT calculations [31, 32]. This correction has been shown to adequately reproduce the correct electronic localization, the insulating behavior, and the spin distribution without introducing additional computational cost. For interested readers, you can refer to these works [31, 32] for more details.

Besides, an approach called constrained density functional theory (CDFT) [33–35] was also introduced as a workaround to the SIE problem. It also allows the study of the excited-state properties that is not assessable in DFT calculations. In CDFT, a constraint term is added to the Kohn–Sham energy functional so that the electronic density or magnetization around the target atom or a group of atoms is constrained to a predefined value. With the appropriate choice of the constraint, the over-delocalization of the electronic density due to SIE can be corrected, and the excited-state properties related to photoexcitation and electron transfer can be studied. Recently, there is also a new approach of CDFT that targets specifically the oxidation states of the transition metal (TM) atoms [36]. Interested readers can refer to these works [33–36] for more details.

23.4 DFT Applications in Heterogeneous Catalysis

DFT calculations have been extensively used to study heterogeneous catalysis. They allow us to obtain the energetic and structural properties of materials. Energy barriers for the chemical reactions can also be determined, which enable the identification of the possible mechanisms. The details of the ion-transfer processes and electron flow during the reactions can also be examined. Here, we highlight some of these examples.

In the study of the reactions in heterogeneous catalysis, it is of important to gain atomistic understanding of the adsorption of reactants on the surface. DFT calculations are the ideal tool to provide such structural and energetic information. Figure 23.3 shows the results of the adsorption study of different molecules on the Pt(111) surface [37]. Such atomic-scale understanding of the adsorption structures is crucial to understand the subsequent reaction mechanisms.

After knowing the adsorption structures, DFT calculations can be applied to study the reaction energetics and kinetics through the determination intermediate energies and the barriers. Figure 23.4 shows the study of the reaction of the

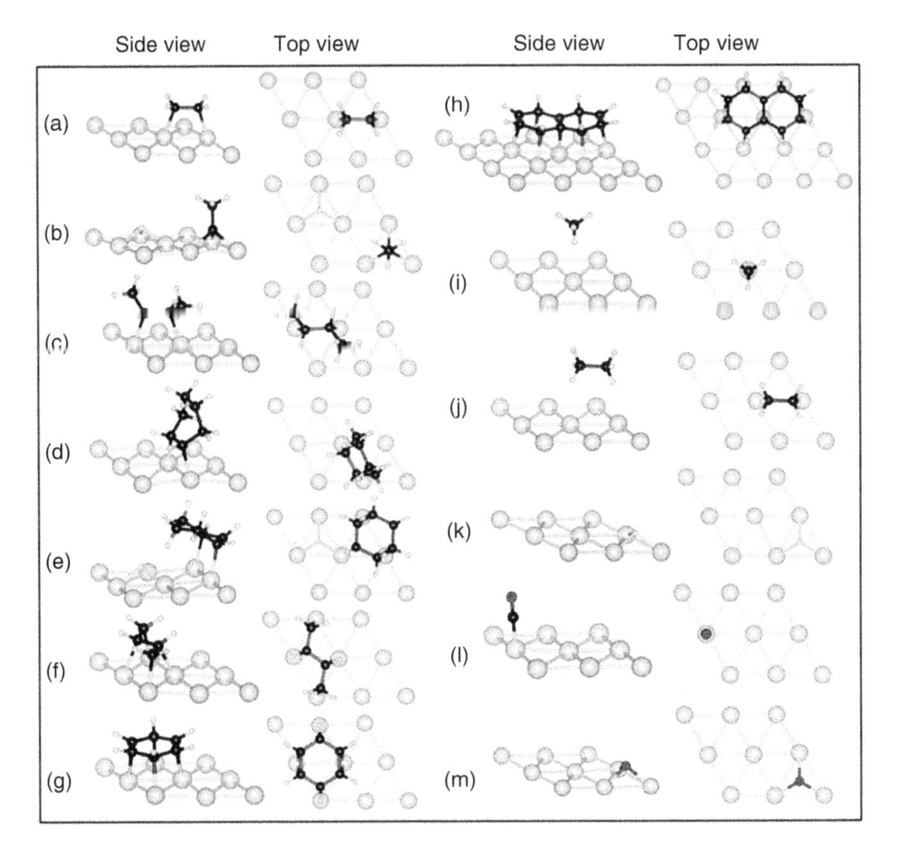

Figure 23.3 Side view and top view of the adsorption geometry on Pt(111) for (a) ethylene C_2H_4, (b) ethylidyne CCH_3 and one H, (c) butene C_4H_8, (d) cyclohexene C_6H_{10}, (e) C_6H_9 and one H, (f) butadiene C_4H_6, (g) benzene C_6H_6, (h) naphthalene $C_{10}H_8$, (i) methane CH_4, (j) ethane C_2H_6, (k) H atom, in face-centered cubic (fcc) hollow position, (l) CO, and (m) O atom. Source: Gautier et al. 2015 [37]. Adapted with permission of Royal Society of Chemistry.

pyrite (100) surface with water and oxygen [38]. Through DFT calculations coupled with the NEB approach, the reaction intermediates, their energetics, and the barriers have been identified from the initial state of a surface sulfur to the final state of the formation of a surface sulfate.

Besides the structural and energetic information, the Kohn–Sham wavefunctions also contain valuable information of the bonding configurations and electron flow along the reaction pathway. For example, the Kohn–Sham wavefunctions can be transformed to a set of localized orbital functions called the maximally localized Wannier functions (MLWFs) [39], which allows us to track the change in the electronic structures during the reaction. Figure 23.5 shows the MLWFs of four representative frames obtained from the first principles molecular dynamics simulations in the study of the first protonation processes of a bio-inspired hydrogen-producing catalyst [40]. The flow of the electrons and the qualitative change in the bonding during the first protonation processes can be visualized through these orbitals. One can see that the proton is first attracted

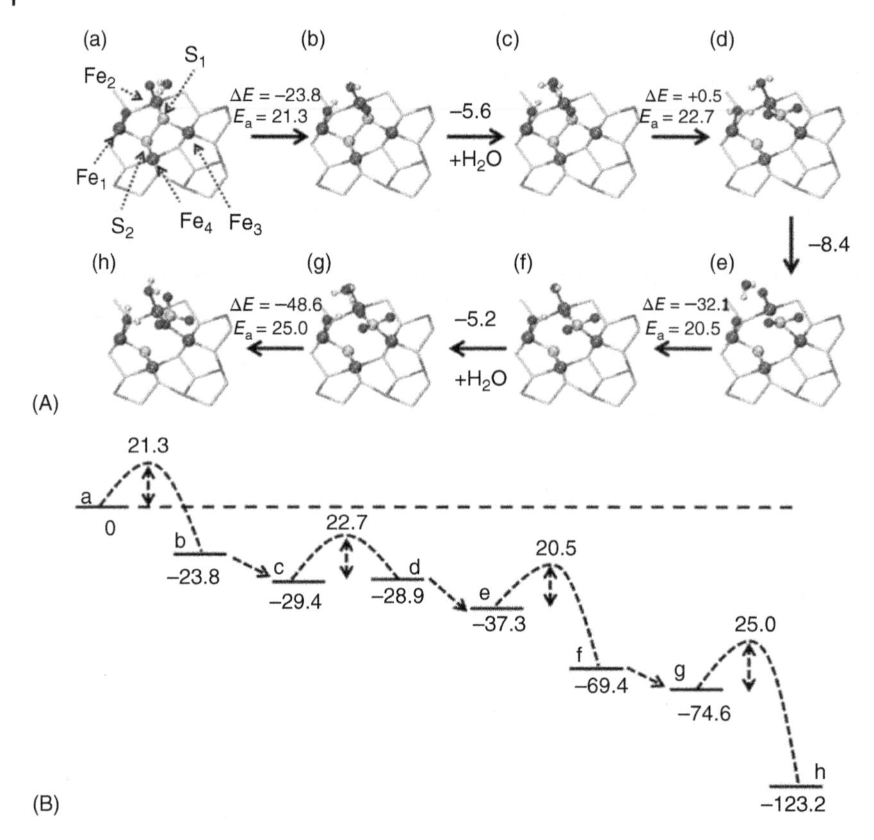

Figure 23.4 (A) Possible mechanism of the oxidation of a surface S to SO_4^{2-}. Energy changes along the reaction pathway and energy barriers of some of the key steps are summarized in panel (B) in kcal/mol. Source: Sit et al. 2012 [38]. Adapted with permission of American Chemical Society.

by the bond pair electrons between the Fe and the P before moving over to bond to the Fe active center. The electron trajectories can be more sensitive probe of change than the nuclear trajectories. Such information is useful in the computational study of catalytic reactions and provide useful guide for catalyst design.

23.5 Conclusions and Perspective

In this chapter, we provide an overview of the fundamentals of the DFT. Considering the trade-off between accuracy and efficiency, DFT is currently the most widely used quantum mechanical techniques to study realistic problems. Although DFT is in principle an exact ground-state theory, the accuracy of the results depends strongly on the accuracy of the exchange–correlation functional used. The search for more accurate functionals up the "Jacob's ladder" is inevitably one key direction of research in DFT. Moreover, inclusion of different corrections like the van der Waals interaction and the Hubbard U term helps in improving the calculations.

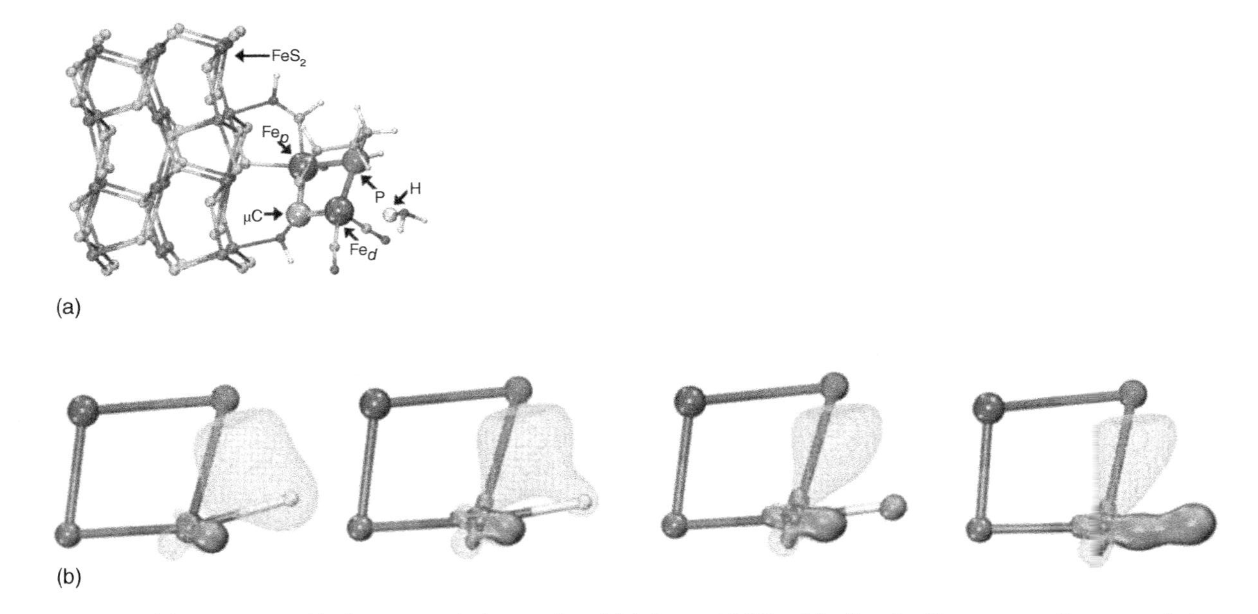

Figure 23.5 (a) Side view of the bio-inspired hydrogen-producing catalyst. (b) Relevant MLWFs of the five significant atoms, C gray, two Fe brown, P violet, and H white during the first protonation process from left to right. Source: Sit et al. 2011 [40]. Adapted with permission of John Wiley & Sons. (See online version for color figure).

On the other hand, being a ground-state theory, DFT cannot provide accurate excited-state properties. The use of CDFT discussed above can partly circumvent this problem in the study of electron transfer problems. DFT can also be coupled with excited-state theories like the time-dependent density functional theory (TDDFT) and the GW approach to have accurate description of the excited states. These approaches will be discussed in detail in the later chapters.

References

1 Hohenberg, P. and Kohn, W. (1964). Inhomogeneous electron gas. *Phys. Rev.* 136: B864–B871.

2 Kohn, W. and Sham, L.J. (1965). Self-consistent equations including exchange and correlation effects. *Phys. Rev.* 140: A1133–A1138.

3 Burke, K. and Wagner, L.O. (2013). DFT in a nutshell. *Int. J. Quantum Chem.* 113: 96–101.

4 Parr, R.G. and Yang, W. (1994). *Density Functional Theory of Atoms and Molecules*. Oxford University Press.

5 Frisch, M.J., Trucks, G.W., Schlegel, H.B. et al. (2009). *Gaussian 09, Revision A.02*. Wallingford, CT: Gaussian, Inc.

6 Valiev, M., Bylaska, E.J., Govind, N. et al. (2010). NWChem: a comprehensive and scalable open-source solution for large scale molecular simulations. *Comput. Phys. Commun.* 181: 1477–1489.

7 Kresse, G. and Furthmüller, J. (1996). Efficiency of ab-initio total energy calculations for metals and semiconductors using a plane-wave basis set. *Comput. Mater. Sci.* 6: 15–50.

8 Giannozzi, P., Andreussi, O., Brumme, T. et al. (2017). Advanced capabilities for materials modelling with Quantum ESPRESSO. *J. Phys. Condens. Matter* 29: 465901.

9 CPMD, http://www.cpmd.org/, Copyright IBM Corp 1990–2019, C. M. für F. S. 1997–2001.

10 Hutter, J., Iannuzzi, M., Schiffmann, F., and VandeVondele, J. (2014). CP2K: atomistic simulations of condensed matter systems. *WIREs Comput. Mol. Sci.* 4: 15–25.

11 Henkelman, G., Uberuaga, B.P., and Jónsson, H. (2000). A climbing image nudged elastic band method for finding saddle points and minimum energy paths. *J. Chem. Phys.* 113: 9901–9904.

12 Henkelman, G. and Jónsson, H. (1999). A dimer method for finding saddle points on high dimensional potential surfaces using only first derivatives. *J. Chem. Phys.* 111: 7010–7022.

13 Mardirossian, N. and Head-Gordon, M. (2017). Thirty years of density functional theory in computational chemistry: an overview and extensive assessment of 200 density functionals. *Mol. Phys.* 115: 2315–2372.

14 Becke, A.D. (1993). A new mixing of Hartree–Fock and local density-functional theories. *J. Chem. Phys.* 98: 1372–1377.

15 Perdew, J.P., Burke, K., and Ernzerhof, M. (1996). Generalized gradient approximation made simple. *Phys. Rev. Lett.* 77: 3865–3868.

16 Cohen, A.J., Mori-Sánchez, P., and Yang, W. (2012). Challenges for density functional theory. *Chem. Rev.* 112: 289–320.
17 Jones, R.O. (2015). Density functional theory: its origins, rise to prominence, and future. *Rev. Mod. Phys.* 87: 897–923.
18 Perdew, J.P., Ruzsinszky, A., Tao, J. et al. (2005). Prescription for the design and selection of density functional approximations: more constraint satisfaction with fewer fits. *J. Chem. Phys.* 123: 62201.
19 Becke, A.D. (1988). Density-functional exchange-energy approximation with correct asymptotic behavior. *Phys. Rev. A: At. Mol. Opt. Phys.* 38: 3098–3100.
20 Perdew, J.P., Chevary, J.A., Vosko, S.H. et al. (1992). Atoms, molecules, solids, and surfaces: applications of the generalized gradient approximation for exchange and correlation. *Phys. Rev. B: Condens. Matter* 46: 6671–6687.
21 Lee, C., Yang, W., and Parr, R.G. (1988). Development of the Colle-Salvetti correlation-energy formula into a functional of the electron density. *Phys. Rev. B: Condens. Matter* 37: 785–789.
22 Becke, A.D. (1988). Correlation energy of an inhomogeneous electron gas: a coordinate-space model. *J. Chem. Phys.* 88: 1053–1062.
23 Becke, A.D. (1994). Thermochemical tests of a kinetic-energy dependent exchange-correlation approximation. *Int. J. Quantum Chem.* 52: 625–632.
24 Becke, A.D. and Roussel, M.R. (1989). Exchange holes in inhomogeneous systems: a coordinate-space model. *Phys. Rev. A: At. Mol. Opt. Phys.* 39: 3761–3767.
25 Tao, J., Perdew, J.P., Staroverov, V.N., and Scuseria, G.E. (2003). Climbing the density functional ladder: nonempirical meta–generalized gradient approximation designed for molecules and solids. *Phys. Rev. Lett.* 91: 146401.
26 Dion, M., Rydberg, H., Schröder, E. et al. (2004). van der Waals density functional for general geometries. *Phys. Rev. Lett.* 92: 246401.
27 Grimme, S. (2006). Semiempirical GGA-type density functional constructed with a long-range dispersion correction. *J. Comput. Chem.* 27: 1787–1799.
28 Grimme, S., Antony, J., Ehrlich, S., and Krieg, H. (2010). A consistent and accurate ab initio parametrization of density functional dispersion correction (DFT-D) for the 94 elements H-Pu. *J. Chem. Phys.* 132: 154104.
29 Tkatchenko, A. and Scheffler, M. (2009). Accurate molecular van der Waals interactions from ground-state electron density and free-atom reference data. *Phys. Rev. Lett.* 102: 73005.
30 Anisimov, V.I., Zaanen, J., and Andersen, O.K. (1991). Band theory and Mott insulators: Hubbard *U* instead of Stoner *I*. *Phys. Rev. B: Condens. Matter* 44: 943–954.
31 Cococcioni, M. and de Gironcoli, S. (2005). Linear response approach to the calculation of the effective interaction parameters in the **LDA + U** method. *Phys. Rev. B: Condens. Matter* 71: 35105.
32 Kulik, H.J., Cococcioni, M., Scherlis, D.A., and Marzari, N. (2006). Density functional theory in transition-metal chemistry: a self-consistent Hubbard ***U*** approach. *Phys. Rev. Lett.* 97: 103001.
33 Dederichs, P.H., Blügel, S., Zeller, R., and Akai, H. (1984). Ground states of constrained systems: application to cerium impurities. *Phys. Rev. Lett.* 53: 2512–2515.

34 Wu, Q. and Van Voorhis, T. (2006). Constrained density functional theory and its application in long-range electron transfer. *J. Chem. Theory Comput.* 2: 765–774.

35 Kaduk, B., Kowalczyk, T., and Van Voorhis, T. (2012). Constrained density functional theory. *Chem. Rev.* 112: 321–370.

36 Ku, C. and Sit, P.H.-L. (2019). Oxidation-state constrained density functional theory for the study of electron-transfer reactions. *J. Chem. Theory Comput.* 15: 4781–4789.

37 Gautier, S., Steinmann, S.N., Michel, C. et al. (2015). Molecular adsorption at Pt(111). How accurate are DFT functionals? *Phys. Chem. Chem. Phys.* 17: 28921–28930.

38 Sit, P.H.-L., Cohen, M.H., and Selloni, A. (2012). Interaction of oxygen and water with the (100) surface of pyrite: mechanism of sulfur oxidation. *J. Phys. Chem. Lett.* 3: 2409–2414.

39 Marzari, N. and Vanderbilt, D. (1997). Maximally localized generalized Wannier functions for composite energy bands. *Phys. Rev. B: Condens. Matter* 56: 12847–12865.

40 Sit, P.H.-L., Zipoli, F., Chen, J. et al. (2011). Oxidation state changes and electron flow in enzymatic catalysis and electrocatalysis through Wannier-function analysis. *Chem. Eur. J.* 17: 12136–12143.

24

Ab Initio Molecular Dynamics in Heterogeneous Catalysis

Ye-Fei Li

Fudan University, Collaborative Innovation Center of Chemistry for Energy Material, Department of Chemistry, Key Laboratory of Computational Physical Science (Ministry of Education), Shanghai Key Laboratory of Molecular Catalysis and Innovative Materials, 2005 Songhu Rd, Shanghai 200433, PR China

24.1 Introduction

Molecular dynamics (MD) is a powerful sampling technique to estimate the free energy landscape of chemical reactions in complex environments such as solid–liquid interface, protein. The power of this technique lies in its ability to reproduce experimental observable quantities accurately and, at the same time, giving the mechanistic details of chemical reactions at the atomic level. In many aspects, MD is very similar to real experiments. As a result, MD is often used to complement experimental investigations and to help in interpreting experiments and designing new ones.

Most classical MD using empirical force field cannot deal with the chemical reactions involving breaking and forming of chemical bonds. Besides, the transferability of parameters in force field is poor. The parameters are closely related to the chemical environments, such as oxidation state (OS), hybridization (i.e. sp, sp^2, sp^3). For example, let us consider two systems, $Fe(OH)_2$ and FeOOH with the different OS of the Fe. Although both systems consist of the same elements, one still needs two sets of force field parameters for Fe^{2+} and Fe^{3+}, respectively. As for ab initio MD, one does not need to worry much about these issues. The whole system is described by a unified formalism, namely, Schrödinger equation. Therefore, the ability to study chemical reactions in complex catalytic systems is a significant motivation to use ab initio MD in heterocatalysis.

Although great success has been achieved for MD simulations, the major challenge in this field is the short simulation time due to the high cost of calculations. In the ab initio method, this simulation time is usually within tens of picoseconds (ps). In unbiased MD simulations, such time scale is too short to sample the phase space around the transition state (TS) sufficiently, which hinders us from obtaining the full free energy landscapes of chemical reactions. In order to resolve this problem, various biased MD simulations have been developed to make the sampling of phase space around the transition state possible.

In this chapter, we focus on the MD methods that are related to heterogeneous catalysis. In Sections 24.2 and 24.3, we introduce the basic algorithm for MD

Heterogeneous Catalysts: Advanced Design, Characterization and Applications, First Edition.
Edited by Wey Yang Teoh, Atsushi Urakawa, Yun Hau Ng, and Patrick Sit.

simulation in microcanonical (NVE) ensemble and show how this MD can be extended to unconventional canonical (NVT) ensemble. Next, in Section 24.4, we introduce the transition state theory (TST). In Section 24.5, we demonstrate three techniques of biased MD simulations to calculate free energy barriers, e.g. thermodynamic integration, umbrella sampling, and metadynamics. In Section 24.6, we briefly introduce the applications of neural networks (NNs) in MD, which is an appealing technology that emerged in recent years and can significantly speed up the calculations of MD simulation. Lastly, in Section 24.7, we give some examples for MD simulations in heterogeneous catalysis.

Here, we assume that you are familiar with the Lagrangian formulation of classical mechanisms, as well as with the concepts of ergodicity, ensemble, free energy, and partition function in statistical mechanics. If you are not familiar with these concepts, we recommend referring to the textbooks of classical mechanics [1] and statistical mechanics [2].

24.2 Basic Algorithm of Molecular Dynamics

In most MD simulations of heterogeneous catalysis, the motion of nucleus is based on Born–Oppenheimer (BO) approximation, which states that the motion of nucleus is decoupled with that of the electron due to the difference between nuclear and electronic masses. This approximation is essential for MD. With this approximation, the motion of the nucleus can be calculated by the classical mechanism.

In the classical mechanism, the conformation of the system evolves through Newton's laws of motion. The trajectory is obtained by solving the differential equations of Newton's second law:

$$\boldsymbol{a} = \frac{d^2\boldsymbol{r}}{dt^2} = \frac{\boldsymbol{F}}{m} = -\frac{\nabla \boldsymbol{U}}{m} \tag{24.1}$$

The acceleration $\boldsymbol{a}$ is determined by force ($\boldsymbol{F}$) and mass (m). To perform MD simulation, the key step is to calculate the force for a given atomic position ($\boldsymbol{r}$). Mathematically, the force is the negative gradient of potential energy (U) (i.e. $-\nabla \boldsymbol{U}$). In ab initio approaches or empirical force field, the force can be calculated analytically. Therefore, MD simulations can be easily combined with different levels of theoretical calculations as long as the force can be easily obtained.

In MD, the forces on the particles will continuously change whenever the particles change their positions. The motion of the particles is a nonlinear multibody problem, which cannot be resolved analytically. The numerical method of finite difference has been applied to solve this problem. In such an approach, the continuous evolution of the atomic positions is approximated by a discretized sampling-point sequence $t_n = n\Delta t$, where a discretized time step Δt is chosen. In such a discretization scheme, the position of particles is approximated as Taylor series expansions:

$$\boldsymbol{r}_{t+\Delta t} = \boldsymbol{r}_t + \boldsymbol{v}_t\Delta t + \frac{1}{2}\boldsymbol{a}_t\Delta t^2 + \mathcal{O}(\Delta t^3) \tag{24.2}$$

where $\boldsymbol{v}_t$ is the velocity, $\boldsymbol{a}_t$ is the acceleration, and $\mathcal{O}(\Delta t^3)$ is the error in a single time step (in the order of Δt^3). If we truncate the expansion beyond the term in Δt^2, we obtain the Euler algorithm as shown in Eq. (24.2). In this approach, the truncation error accumulates with the order of Δt^3. During long-time MD simulations, this error grows large, which eventually fails the simulations. As a result, one key task for MD simulations is to build an algorithm that slows the accumulation of errors as much as possible.

24.2.1 Verlet Algorithm

In MD, the most common and usually best time integration algorithm is the Verlet algorithm. The basic idea is to write two third-order Taylor expansions for two positions, forward and backward in time as in Eqs. (24.3) and (24.4):

$$\boldsymbol{r}_{t+\Delta t} = \boldsymbol{r}_t + \boldsymbol{v}_t\Delta t + \frac{1}{2}\boldsymbol{a}_t\Delta t^2 + \frac{1}{6}\boldsymbol{b}_t\Delta t^3 + \mathcal{O}(\Delta t^4) \tag{24.3}$$

$$\boldsymbol{r}_{t-\Delta t} = \boldsymbol{r}_t - \boldsymbol{v}_t\Delta t + \frac{1}{2}\boldsymbol{a}_t\Delta t^2 - \frac{1}{6}\boldsymbol{b}_t\Delta t^3 + \mathcal{O}(\Delta t^4) \tag{24.4}$$

where $\boldsymbol{b}_t$ is the jerk. By adding Eqs. (24.3) and (24.4), we get

$$\boldsymbol{r}_{t+\Delta t} = 2\boldsymbol{r}_t - \boldsymbol{r}_{t-\Delta t} + \boldsymbol{a}_t\Delta t^2 + \mathcal{O}(\Delta t^4) \tag{24.5}$$

Immediately, we can see that the first- (Δt) and third-order terms (Δt^3) cancel out, and the error is in the order of Δt^4, even if jerk ($\boldsymbol{b}_t$) does not appear explicitly. This formula is more precise than the direct use of the Taylor series expansions to second order as in Eq. (24.2), which results in a truncation error in the order of Δt^3. Equation (24.5) is the basic formula for the time evolution of particles in MD simulations.

A problem of this version of the Verlet algorithm is that the velocities cannot be directly obtained. Although the velocities are not needed for calculating the trajectory, they are necessary to compute many physical quantities such as kinetic energy (K) and temperature (T). One could compute the velocities from the trajectory by using

$$v_t = \frac{x_{t+\Delta t} - x_{t-\Delta t}}{2\Delta t} + \mathcal{O}(\Delta t^2) \tag{24.6}$$

where the error for this expression for velocity is only in the order of Δt^2.

24.2.2 Velocity Verlet Algorithm

It should be mentioned that in the Verlet algorithm, to calculate the velocity at time t, we need the position at $t + \Delta t$. As a result, the position and velocity cannot be calculated simultaneously, which is not convenient in practical simulations. Alternatively, there is an equivalent form to the original Verlet algorithm, which follows the formulas of

$$\boldsymbol{r}_{t+\Delta t} = \boldsymbol{r}_t + \boldsymbol{v}_t\Delta t + \frac{1}{2}\boldsymbol{a}_t\Delta t^2 + \mathcal{O}(\Delta t^4) \tag{24.7}$$

$$\boldsymbol{v}_{t+\Delta t} = \boldsymbol{v}_t + \frac{1}{2}(\boldsymbol{a}_{t+\Delta t} + \boldsymbol{a}_t)\Delta t + \mathcal{O}(\Delta t^2) \tag{24.8}$$

The formulation of velocity Verlet can be derived from the basic Verlet algorithm. The velocity Verlet algorithmgives the same precision as in the Verlet algorithm, and the advantage is that both the position and velocity can be resolved in the same time step.

24.2.3 Conserved Quantity

According to the law of conservation of energy, the total energy $H = K + U$ (involving kinetic energy K and potential energy U) should be conserved during MD simulations. The conservation of total energy is one of the most important tests to verify that an MD simulation proceeds correctly. If a drift of the total energy exists during simulations, one should decrease the time step Δt or check the precision of the force.

24.3 Molecular Dynamics in Canonical Ensembles

The MD technique discussed in Section 24.2 can simulate the time evolution of a system where the number of particles (N), volume (V), and total energy (E) are fixed variables. In statistical mechanics, such a system is called a microcanonical (NVE) ensemble. However, in the experimental conditions of heterogeneous catalysis, the reactions are usually performed under "constant temperature" or "constant temperature and constant pressure" conditions. In statistical mechanics, the phase space of a system under "constant temperature" or "constant temperature and constant pressure" corresponds to the canonical (NVT) and isothermal–isobaric (NPT) ensembles, respectively. Initially, it seems impossible to perform MD simulations other than the NVE ensemble. Fortunately, by introducing an additional thermostat, it has been proved that NVT and NPT ensembles can be achieved in MD simulations. In this section, we focus on the NVT ensemble. The NPT ensemble can be attained by MD simulations with a similar approach to that in the NVT ensemble [3].

To perform MD simulations in the NVT ensemble, Nosé first constructed an extended Lagrangian formulation by adding a degree of freedom s in a classical N-body system: [4]

$$L_{\text{Nosé}} = \sum_{i=1}^{N} \frac{m_i}{2} s^2 \dot{r}_i^2 - U + \frac{Q}{2}\dot{s}^2 - k_B TL \ln(s) \tag{24.9}$$

where s is a "time-scale" (or "mass-scale") variable. L is a parameter that will be fixed later. Q is the virtual "mass" associated with s, while $k_B TL \ln(s)$ is the potential energy of additional variable s. With this extended Lagrangian, Nosé showed that the average of NVT ensemble is generated by conventional MD simulations. Later, this approach was reformulated by Hoover [5] to be more convenient for numerical simulation. So, the final version of this method is called the Nosé–Hoover thermostat.

Here we skip the details of the derivation, directly going to the final equations of motion, which help us to understand how the Nosé–Hoover thermostat works in practice. If readers are interested in the details of the derivation, they can refer to the textbook (section 6.1.2) by Frenkel and Smit [5]. The formulas of motion are as follows:

$$\dot{\boldsymbol{r}}_i = \boldsymbol{p}_i / m_i \tag{24.10}$$

$$\dot{\boldsymbol{p}}_i = \boldsymbol{F} - \dot{\xi}\boldsymbol{p}_i \tag{24.11}$$

$$\dot{\xi} = p_\xi / Q \tag{24.12}$$

$$\ddot{\xi} = \left(\sum_i \frac{\boldsymbol{p}_i^{\,2}}{m_i} - k_B T L \right) / Q \tag{24.13}$$

The L is a parameter that is equal to the degree of freedom of the system, involving vibration, rotation, and translation. In an isolated system containing N particles, L is $3N$. While in periodic boundary condition (which is commonly used in many simulation codes), L is decreased to $3N - 3$, since three additional constraints exist by the boundary condition. It should be noticed that a parameter ξ rather than s appears in the equations of motion, which is related to s by the formula

$$\xi = \ln(s) \tag{24.14}$$

The conserved quantity $H_{\text{Nosé}}$ then becomes

$$H_{\text{Nosé}} = H + \frac{p_\xi^2}{2Q} + k_B T L \xi \tag{24.15}$$

The most significant feature of the equations of motion is that in Eq. (24.11), the $\dot{\boldsymbol{p}}_i$ is no longer proportional to the force of the particle. One additional term $\dot{\xi}\boldsymbol{p}_i$ exists, which is proportional to the momentum of the particle. Therefore, the dynamics in the Nosé–Hoover thermostat can be considered the motion coupled with a special type of damping. In such a thermostat, the parameter $\dot{\xi}$ can either be positive or negative, while the velocities of particles can either decrease or increase.

Next, let us discuss the physical meaning of Eq. (24.13), which determines the evolution of ξ during simulations. According to statistical mechanics, the kinetic energy of a system is related to its temperature with the following formula:

$$\frac{1}{2} \boldsymbol{k}_B \boldsymbol{T}' \boldsymbol{L} = \sum_i \frac{\boldsymbol{p}_i'^2}{2\boldsymbol{m}_i} \tag{24.16}$$

Therefore, Eq. (24.13) can be understood as the difference in kinetic energy between the real system and the virtual thermostat. Note that in Eq. (24.13), the term $\sum_i \frac{p_i^2}{m_i}$ involves $\dot{\xi}$ implicitly. As a result, this equation needs to be solved iteratively.

24.4 Transition State Theory

For a chemical reaction, no matter how complicated, it consists of a series of elementary reactions. For each elementary reaction, the reaction state can be

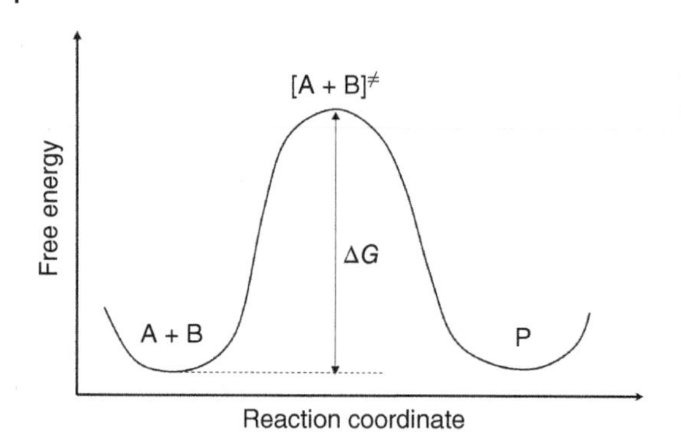

Figure 24.1 Schematic illustration of the free energy profile for an elementary reaction.

estimated using the TST. Therefore, in the simulations of heterogeneous catalysis, the essential task is to determine the reaction mechanism and calculate the reaction rate. Consider an elementary reaction of $A+B \rightarrow [AB]^{\neq} \rightarrow P$, where $[AB]^{\neq}$ is the transition state (or activated complex). Figure 24.1 shows the free energy profile for this reaction. The TST states that the reaction rate constant follows the formula of [6]

$$k = \frac{k_B T}{h} \exp(-\Delta G/RT) \tag{24.17}$$

where $\frac{k_B T}{h}$ is the crossing frequency ($\sim 10^{13}\,s^{-1}$ in 300 K) and ΔG is the free energy difference between the ground state (A + B) and transition state ($[AB]^{\neq}$). In heterogeneous catalysis, the most important issue is to calculate the free energy barrier ΔG of each elemental step.

There are two approaches to calculate ΔG. One is to directly locate the transition state using transition state search methods (e.g. nudged elastic band [NEB] and dimer methods), while the other is to estimate the free energy barrier using MD-based methods (e.g. thermodynamic integral, umbrella sampling, and metadynamics). In the first approach, the atom moves on the potential energy surface (PES) of 0 K. The problem of this approach is that the contribution due to finite temperature, notably entropy, is missing.

For MD-based methods, the major problem is that MD simulations are difficult to surmount free energy barrier, and even when it is tiny. In ab initio MD, solving the Schrödinger equation is very expensive, and the time scale of simulation is typically limited to be within 100 ps. To observe a chemical reaction in the time scale of MD simulation, it is required that this process occur with a rate larger than $\sim 1/100\,ps^{-1} = 10^{10}\,s^{-1}$. According to the TST, such a reaction rate corresponds to a barrier of ~0.1 eV. But for almost all chemical reactions, the barrier is significantly higher than 0.1 eV. As a result, no chemical reactions can occur in standard MD described in Section 24.3. Methods to enhance the sampling around TS are necessary to calculate the free energy barrier, which are introduced in Section 24.5.

24.5 Free Energy Calculations

To calculate the free energy profile, we first need to define a reaction coordinate. MD simulations perform sampling in the $6N$-dimensional phase space, which is a huge space. To distinguish all the peculiar states of interest during the process under investigation, we often analyze the heterogeneous catalysis regarding the reaction coordinate $\zeta(\boldsymbol{r})$, a low-dimensional function of the atomic coordinates. As an example, for proton transfer, the reaction coordinate could be the distances between the hydrogen and each of the two involved heavy atoms. In short, the reaction coordinate represents a sort of coarse description of the system, which projects the reaction pathway in low dimensionality. It should be mentioned that choosing the reaction coordinate is not trivial, which requires good chemical intuition. In the following, we assume that the reaction coordinate is already known for the process of interest.

24.5.1 Thermodynamic Integration and Constrained MD

The basic idea of thermodynamic integration is shown in Figure 24.2. Here we consider that the reaction of $A + B \rightarrow P$ occurs in the NVT ensemble, while the Helmholtz free energy (F) is used to describe this system. In the approach of thermodynamic integration, it performs a series of constrained MD simulations with fixed reaction coordinates ζ_i. In each time step of MD simulations, we obtain a general force, which is the derivative of the potential energy (U) with respect to the reaction coordinate ζ at $\zeta = \zeta_i$. Averaging this general force in MD simulation, we then obtain the potential mean force (PMF), which is the derivative of the Helmholtz free energy (F) with respect to the reaction coordinate ζ at $\zeta = \zeta_i$:

$$\left.\frac{\partial F(\zeta)}{\partial \zeta}\right|_{(\zeta=\zeta_i)} = \left\langle \frac{\partial U(\zeta)}{\partial \zeta} \right\rangle_{(\zeta=\zeta_i)} = \lim_{t\to\infty} \frac{1}{t} \int_0^t \frac{\partial U(\zeta, t_i)}{\partial \zeta} dt' \tag{24.18}$$

where $\langle \cdots \rangle$ indicates an ensemble average. The second equation is the so-called ergodicity assumption, which states that an ensemble average is equal to the time

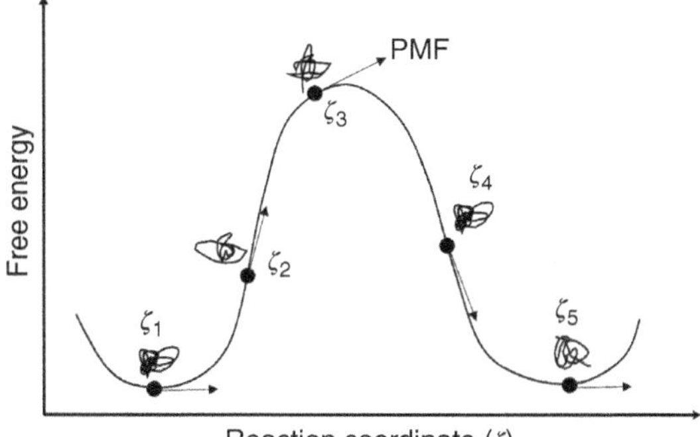

Figure 24.2 Schematic illustration of using thermodynamic integration to calculate the free energy barrier. The scratches represent MD sampling at a few points of the reaction coordinate.

average. In MD simulations, the time is discretized with Δt, and the simulation time is finite. So, we use the following approximation:

$$\lim_{t\to\infty} \frac{1}{t} \int_0^t \frac{\partial U(\zeta, t_i')}{\partial \zeta} dt' \approx \frac{1}{N} \sum_{i=1}^{N} \frac{\partial U(\zeta, t')}{\partial \zeta} \tag{24.19}$$

where N is the total number of steps in MD simulations. Next, integration of PMF in Eq. (24.18) gives the free energy profile of this process.

Obviously, the essential challenge in this method is how to perform constrained MD and then to calculate the general force $\frac{\partial U(\zeta,t_i)}{\partial \zeta}$. Equation (24.9) cannot solve the classical particle motion problems with constraints. Alternatively, this problem can be solved by the Lagrange formalism, which is a more general expression of Newton's second law, as shown in Eq. (24.20):

$$\frac{\partial L}{\partial \boldsymbol{q}_i} - \frac{d}{dt}\frac{\partial L}{\partial \dot{\boldsymbol{q}}_i} = 0 \tag{24.20}$$

where L is called the Lagrangian, which is equal to $K - U$, while $\boldsymbol{q}_i$ is the generalized coordinate. Notice that in MD simulations, it is easier to handle the Cartesian coordinates rather than the generalized coordinates. So, in the following text, we replace $\boldsymbol{q}_i$ by Cartesian coordinates $\boldsymbol{r}_i$. In this formalism, the constraints can be added into L by the method of Lagrange multipliers, which generate a new Lagrangian L' as

$$L' = L - \sum_{k=1}^{C} \lambda_k \sigma_k \tag{24.21}$$

where σ_k define the constraints, λ is Lagrange multipliers, and C is the number of constraints. The forces on the particles can then be calculated as

$$m_i \ddot{\boldsymbol{r}}_i = \frac{\partial L'}{\partial \boldsymbol{r}_i} = -\frac{\partial U}{\partial \boldsymbol{r}_i} - \sum_{k=1}^{C} \lambda_k \frac{\partial \sigma_k}{\partial \boldsymbol{r}_i} = \boldsymbol{F}_i + \boldsymbol{G}_i \tag{24.22}$$

where $\boldsymbol{F}_i$ and $\boldsymbol{G}_i$ are the force and constraint force, respectively.

In heterogeneous catalysis, the reaction coordinate is usually one dimensional. Thus, only one degree of freedom should be constrained during MD simulations. Here, we consider the simple condition with a single constraint and set $C = 1$. To demonstrate how constrained MD is performed, we choose the bond length as the reaction coordinate, which is commonly used in heterogeneous catalysis. In this case, the constraint σ is defined as

$$\sigma = |\boldsymbol{r}_i - \boldsymbol{r}_j| - d_{ij} = \boldsymbol{r}_{ij} - d_{ij} \tag{24.23}$$

where $|\boldsymbol{r}_i - \boldsymbol{r}_j|$is the definition of the reaction coordinate, while d_{ij} is the target bond length between atom i and atom j. The constraint force ($\boldsymbol{G}$) is then

$$\boldsymbol{G}_i = -\lambda \frac{\partial \sigma}{\partial \boldsymbol{r}_i} = -\lambda \frac{\boldsymbol{r}_{ij}}{|\boldsymbol{r}_{ij}|} \tag{24.24}$$

$$\boldsymbol{G}_j = -\lambda \frac{\partial \sigma}{\partial \boldsymbol{r}_j} = \lambda \frac{\boldsymbol{r}_{ij}}{|\boldsymbol{r}_{ij}|} \tag{24.25}$$

From Eqs. (24.24) and (24.25), we notice that the key to obtaining constraint force G is to calculate the Lagrange multiplier λ, which can be determined by the condition that the constraint is exactly satisfied after one time step in the Verlet algorithm. We performed a Taylor expansion of constraint σ and truncated it in the first term:

$$\sigma^c(t+\Delta t) = \sigma^u(t+\Delta t) + \sum_{i=1}^{N} \frac{\partial \sigma^u(t+\Delta t)}{\partial \boldsymbol{r}_i}(r_i^c(t+\Delta t) - r_i^u(t+\Delta t)) + \mathcal{O}(\Delta t^4) = 0 \tag{24.26}$$

According to the Verlet algorithm, we get

$$r_i^c(t+\Delta t) - r_i^u(t+\Delta t) = -\lambda \frac{\Delta t^2}{m_i} \frac{\partial \sigma^u(t)}{\partial \boldsymbol{r}_i} \tag{24.27}$$

Substituting Eq. (24.27) into Eq. (24.26)

$$\sigma^c(t+\Delta t) = \sigma^u(t+\Delta t) - \lambda \frac{\Delta t^2}{m_i} \sum_{i=1}^{N} \frac{\partial \sigma^u(t+\Delta t)}{\partial \boldsymbol{r}_i} \frac{\partial \sigma^u(t)}{\partial \boldsymbol{r}_i} + \mathcal{O}(\Delta t^4) = 0 \tag{24.28}$$

After reorganizing Eq. (24.28) and using $\sum_{i=1}^{N} \frac{\partial \sigma^u(t)}{\partial r_i} = \nabla \sigma^u(t)$, we get

$$\lambda = \frac{m_i}{\Delta t^2} \frac{\sigma^u(t+\Delta t)}{\nabla \sigma^u(t+\Delta t) \nabla \sigma^u(t)} \tag{24.29}$$

With Eq. (24.29), we then can solve the Lagrange multiplier λ. On the other hand, since the Taylor expansion of σ^c is truncated, Eq. (24.29) should be solved iteratively. As for the condition that more than one constraint exists during MD simulations, Eq. (24.29) can be applied to each constraint in succession. Equation (24.29) is called the SHAKE algorithm [7]. There are several other algorithms, such as M-SHAKE [8] and SHAPE [9]. The difference between these algorithms lies only in the methods to solve the system of equations.

At this point, we can perform the constrained MD simulation. Next, we want to explain the physical meaning of the Lagrange multiplier λ. As we will show in the following text, the Lagrange multiplier λ is exactly the general force in Eq. (24.18). To demonstrate this, let us consider a simple example that external force $\boldsymbol{F}$ in Eq. (24.22) is zero. According to Eqs. (24.24) and (24.25), the equation of motion is then

$$\ddot{\boldsymbol{r}}_i = -\frac{\lambda}{m_i} \frac{\boldsymbol{r}_{ij}}{|\boldsymbol{r}_{ij}|} \tag{24.30}$$

To solve for λ, we impose $\ddot{\sigma} = 0$:

$$\ddot{\sigma} = \frac{\frac{d}{dt}(\dot{\boldsymbol{r}}_{ij} \cdot \boldsymbol{r}_{ij})}{|\boldsymbol{r}_{ij}|} = \frac{\dot{\boldsymbol{r}}_{ij}^2 + \ddot{\boldsymbol{r}} \cdot \boldsymbol{r}_{ij}}{|\boldsymbol{r}_{ij}|} = 0 \tag{24.31}$$

Substituting Eq. (24.30) into Eq. (24.31), we get

$$\dot{\boldsymbol{r}}_{ij}^2 - \frac{\lambda}{m_i} \frac{\boldsymbol{r}_{ij}^2}{|\boldsymbol{r}_{ij}|} = 0 \tag{24.32}$$

Hence

$$\lambda = \frac{m_i \dot{r}_{ij}^2}{r_{ij}^2} |r_{ij}| \tag{24.33}$$

As a result, λ is the centripetal force in this simple example. In general form, the Lagrange multiplier λ is a general force $\frac{\partial U(\zeta)}{\partial \zeta}$, and PMF can be obtained by the ensemble average $\langle \lambda \rangle$:

$$\left. \frac{\partial F(\zeta)}{\partial \zeta} \right|_{(\zeta=\zeta_i)} = \langle \lambda \rangle_{(\zeta=\zeta_i)} \tag{24.34}$$

In thermodynamic integration, we need to predefine a series of sampling points along the reaction coordinate, and the calculated free energy profile is discrete. As a result, the choice of sampling points is important for the precision of the calculated free energy profile. There is another approach called umbrella sampling that can generate continuous free energy profiles. We introduce it in Section 24.5.2, which follows next.

24.5.2 Umbrella Sampling

In umbrella sampling [10], the reaction coordinate is not constrained but only restrained by a biased potential. Therefore, the full momentum space is sampled:

$$U^b(\boldsymbol{r}) = U^u(\boldsymbol{r}) + \omega(\zeta(\boldsymbol{r})) \tag{24.35}$$

where $\omega(\zeta(\boldsymbol{r}))$ is the applied biased potential. The probability distribution of this biased potential is

$$\begin{aligned} P^b(s) &= \frac{\int \exp\{-\beta[U^u(\boldsymbol{r}) + \omega(\zeta(\boldsymbol{r}))]\}\delta[\omega(\zeta(\boldsymbol{r})) - s]\, \boldsymbol{d}^N\boldsymbol{r}}{\int \exp\{-\beta[U^u(\boldsymbol{r}) + \omega(\zeta(\boldsymbol{r}))]\}\, \boldsymbol{d}^N\boldsymbol{r}} \\ &= \exp[-\beta\omega(\zeta(\boldsymbol{r}))] \frac{\int \exp\{-\beta[U^u(\boldsymbol{r})]\}\delta[\omega(\zeta(\boldsymbol{r})) - s]\, \boldsymbol{d}^N\boldsymbol{r}}{\int \exp\{-\beta[U^u(\boldsymbol{r}) + \omega(\zeta(\boldsymbol{r}))]\}\, \boldsymbol{d}^N\boldsymbol{r}} \\ &= \exp[-\beta\omega(\zeta(\boldsymbol{r}))] \frac{\int \exp(-\beta U^u(\boldsymbol{r}))\delta[\omega(\zeta(\boldsymbol{r})) - s]\, \boldsymbol{d}^N\boldsymbol{r}}{\int \exp(-\beta U^u(\boldsymbol{r}))\, \boldsymbol{d}^N\boldsymbol{r}} \\ &\quad \times \frac{\int \exp(-\beta U^u(\boldsymbol{r}))\, \boldsymbol{d}^N\boldsymbol{r}}{\int \exp(-\beta U^u(\boldsymbol{r})) \exp[-\beta\omega(\zeta(\boldsymbol{r}))]\, \boldsymbol{d}^N\boldsymbol{r}} \\ &= \frac{\exp[-\beta\omega(\zeta(\boldsymbol{r}))] P^u(s)}{\langle \exp[-\beta\omega(\zeta(\boldsymbol{r}))] \rangle} \end{aligned} \tag{24.36}$$

Free energy profile can then be calculated from the probability distribution by

$$A(s) = -k_B T \ln(P(s)) \tag{24.37}$$

Therefore

$$A^b(s) = \omega(\zeta(\boldsymbol{r})) + A^u(s) + k_B T \ln\langle \exp[-\beta\omega(\zeta(\boldsymbol{r}))] \rangle \tag{24.38}$$

Then the unbiased free energy profile is

$$A^u(s) = A^b(s) - \omega(\zeta(\boldsymbol{r})) + \boldsymbol{F}_i \tag{24.39}$$

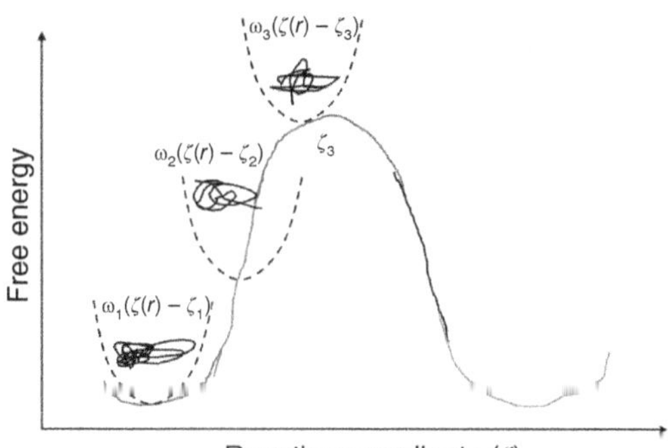

Figure 24.3 Schematic illustration of using umbrella sampling to calculate the free energy barrier. The dotted curves represent the biased potentials at different reaction coordinates. The solid curves represent the calculated free energy profiles in each window. (See online version for color figure).

where

$$\boldsymbol{F}_i = -k_B T \ln\langle \exp[-\beta\omega(\zeta(\boldsymbol{r}))]\rangle \tag{24.40}$$

which is a constant for a fixed biased potential $\omega(\zeta(\boldsymbol{r}))$. If the free energy profile of the whole range of reaction coordinate ζ of interest can be calculated with a single MD simulation, $\boldsymbol{F}_i$ need not to be calculated explicitly. In this case, the free energy profile of the biased potential can be easily transformed to the free energy profile for the unbiased potential.

However, the free energy profile is, of course, not known *a priori*. Therefore, one single simulation usually cannot cover the whole range of the reaction coordinate. Hence, we need to divide the reaction coordinate into several windows, represented by different colors in Figure 24.3. In principle, the biased potential can be any function of the reaction coordinate. In simulations, it is usually a harmonic potential as

$$\omega_i(\zeta(\boldsymbol{r}) - \zeta_i) = \frac{1}{2}k(\zeta(\boldsymbol{r}) - \zeta_i)^2 \tag{24.41}$$

where ζ_i is the target value of restraint $\omega_i(\zeta(\boldsymbol{r}) - \zeta_i)$. In each window, the MD simulation should sample the phase space sufficiently. The total free energy can be generated by combining the free energy profile in different windows. Notice that there exist undetermined constants $\boldsymbol{F}_i$ for each window. To combine the free energy profile in different windows together, we need to estimate these constants $\boldsymbol{F}_i$.

Let us consider the equation

$$\begin{aligned}\exp(-\beta \boldsymbol{F}_i) &= \langle \exp[-\beta\omega_i(\zeta(\boldsymbol{r}))]\rangle \\ &= \int P^u(\zeta)\exp[-\beta\omega_i(\zeta(\boldsymbol{r}))]\, d\zeta\end{aligned} \tag{24.42}$$

To determine the specific value of $\boldsymbol{F}_i$, we need to know the global probability distribution $P^u(\zeta)$, which is what we want to know. As a result, solving the constant $\boldsymbol{F}_i$ is coupled with probability distribution $P^u(\zeta)$ and should be resolved iteratively.

In practice, aligning the free energy profiles of different sampling windows can be achieved by the weighted histogram analysis method (WHAM) [11], which minimizes the variance of global distribution $\sigma[P^u(\zeta)]$:

$$P^u(\zeta) = \frac{\sum_{i=1}^{N_{\text{window}}} n_i(\zeta)}{\sum_{i=1}^{N_{\text{window}}} N_i \exp\{\beta[F_i - \omega_i(\zeta)]\}} \quad (24.43)$$

$$F_i = -k_B T \ln \int P^u(\zeta) \exp[-\beta\omega_i(\zeta)]\, d\zeta \quad (24.44)$$

where $n_i(\zeta)$ is the number of counts in a given reaction coordinate ζ, N_i is the total number of steps sampled for window i, and N_{window} is the number of windows. The final global probability distribution is obtained by resolving Eqs. (24.43) and (24.44) iteratively.

It should be mentioned that MD simulations in different windows can be performed in parallel. The validity of WHAM requires that there exists a sufficient overlap between the tails of free energy profiles in adjacent windows. If the overlap is not sufficient in some regions of the reaction coordinate, one needs to add a window in this area.

24.5.3 Metadynamics

The metadynamics aims to generate an adaptive biased potential to make a single MD simulation cover the whole range of reaction coordinate. Particularly, when the biased potential $\omega(\zeta) = -A(\zeta)$, this precisely flattens the biased free energy profile and leads to a uniform sampling along ζ. In this case, $\omega(\zeta)$ is exactly the free energy profile in the whole reaction coordinate. This approach overcomes the difficulty in combining the free energy profiles in different windows of umbrella sampling.

Laio and Parrinello [12] proposed an approach that the biased potential $\omega(\zeta, t)$ evolves within the MD simulation with the formula of

$$\omega(\zeta, t) = \sum_{i=1}^{t/\tau_G} w_G \exp\left[-\sum_{\alpha=1}^{N_{CV}} \frac{(\zeta_\alpha - \zeta_\alpha(i\tau_G))^2}{2\sigma_\alpha^2}\right] \quad (24.45)$$

In the approach of metadynamics, the reaction coordinate ζ_α is usually denoted as collective variables (CV). $\zeta_\alpha(i\tau_G)$ is the average of reaction coordinates in the time ith window $(i-1)\tau_G \rightarrow i\tau_G$, where τ_G is the time interval for adding a Gaussian function. w_G and σ are two parameters that determine the shape of the Gaussian.

Figure 24.4 demonstrates how the adaptive biased potential evolves during metadynamics. The simulation initiates from an unbiased MD simulation in the first time window of $0 \rightarrow \tau_G$, as shown in Figure 24.4a. The average position for the reaction coordinate is $\zeta_\alpha(\tau_G)$. In the next time window of $\tau_G \rightarrow 2\tau_G$, a repulsive Gaussian potential is added at $\zeta_\alpha(\tau_G)$, which pushes the system to a new equilibrium position $\zeta_\alpha(2\tau_G)$, as shown in Figure 24.4b. Subsequently, an additional Gaussian potential is added in the equilibrium position $\zeta_\alpha(2\tau_G)$ in the time window $2\tau_G \rightarrow 3\tau_G$, as shown in Figure 24.4c. After a number of time windows τ_G, the potential well of reactant states is flattened, which pushes the sampling point to product states, as shown in Figure 24.4d. The potential well of product states can also be flattened with the same procedure

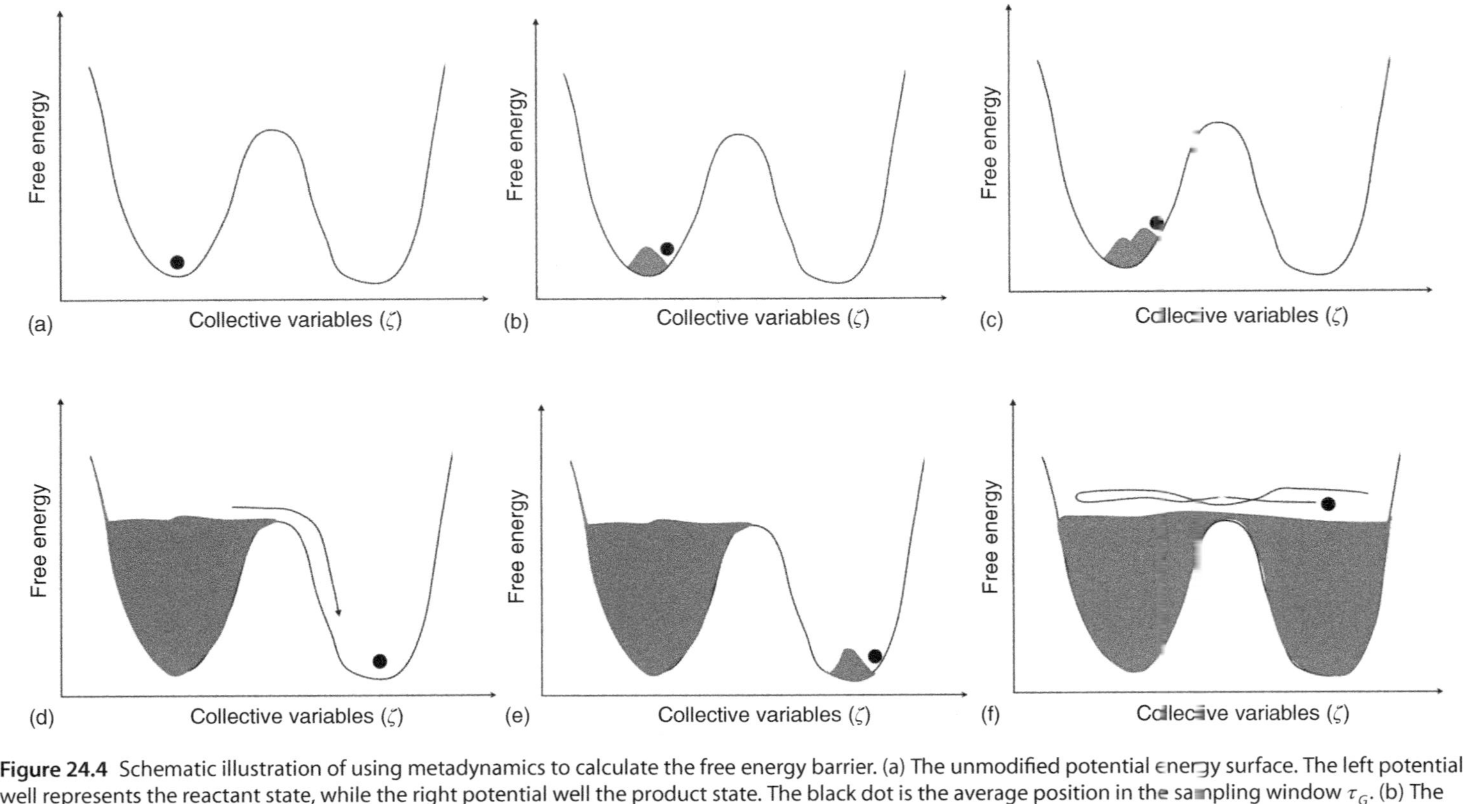

Figure 24.4 Schematic illustration of using metadynamics to calculate the free energy barrier. (a) The unmodified potential energy surface. The left potential well represents the reactant state, while the right potential well the product state. The black dot is the average position in the sampling window τ_G. (b) The potential energy surface after adding one repulsive Gaussian potential (the blue area) in the first time window. (c) The potential energy surface after adding another repulsive Gaussian potential in the second time window. (d) The potential energy surface when the biased potentials push the sampling point from the reactant state to the product state. (e) The potential energy surface after adding the first repulsive Gaussian potential in the product state. (f) The converged potential energy surface where a uniform sampling is achieved. (See online version for color figure).

(see Figure 24.4e–f), and uniform sampling is achieved. The free energy profile is then calculated by Eq. (24.39) as in umbrella sampling, where $\boldsymbol{F}_i$ need not to be calculated explicitly.

24.6 Accelerating MD Simulations by Neural Network

No matter which MD technique used, ab initio calculations are very time consuming. Most of the computational power is to calculate the energy and force by solving the Schrödinger equation, which hinders ab initio MD simulation to be applied on a large scale. One promising approach to speed up MD simulations is to use NN which recently has become a powerful technique in various fields. In quantum chemistry, one important application of NN is to fit ab initio PES. The data from biased ab initio MD simulations, such as metadynamics, can be used to train the NN. Then, the trained NN can then be used to perform long-time MD simulations to predict the physical and chemical properties. Combining this technique of NN with MD simulations, the cost of MD simulations can be reduced by several orders of magnitude. To understand the underlying architecture for NN, one may refer to the online manual written by Nielsen (http://neuralnetworksanddeeplearning.com). NN can be considered a particular type of force field, which gives the energy and force based on the Cartesian coordinates of the system with mathematical formula containing empirical parameters. On the other hand, NN can have the same precision of ab initio methods, which is higher than that of the empirical force field.

The most popular NN architecture in quantum chemistry is the one proposed by Behler and Parrinello [13]. In this scheme, it is assumed that the total energy of a system can be represented by the sum of energies on atoms as

$$E_{\text{tot}} = \sum_{i=1}^{N_{\text{atom}}} E_i \tag{24.46}$$

With this assumption, they use NN to predict the energy on each atom and then sum them together to get the total energy E_{tot}.

Figure 24.5 shows the basic framework of NN proposed by Behler and Parrinello to calculate the energy of atoms. To calculate the energy of the atom, we need first to produce a set of symmetry functions G_m for each atom using the Cartesian coordinates $\mathbf{r}_N$. The symmetry functions have the properties that their values do not change by the operations of rotation and permutation. Then these symmetry functions enter the fully connected layers to generate the energy of atoms. For atoms belonging to the same element, they enter the same NN, where the parameters in the fully connected layers are the same. The symmetry functions distinguish the difference between different chemical environments. But for atoms with different elements, they enter different NN. The advantage of this NN architecture is that the complexity of NN is scaled with the number of elements and can be easily applied to the system that contains a large number of atoms.

The challenges for NN to fit PES of ab initio methods come from the following three aspects: (i) generate NN training database that contains representative

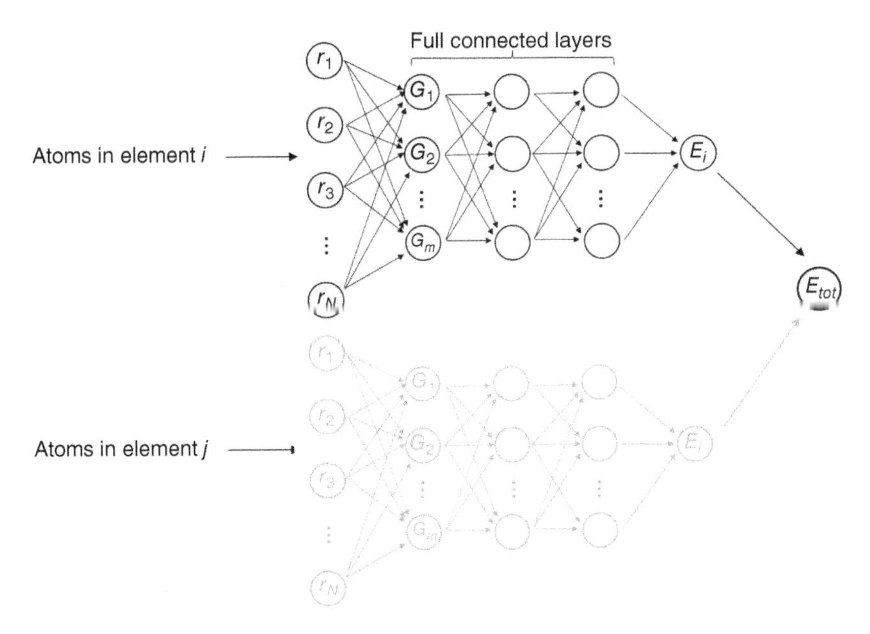

Figure 24.5 NN architecture proposed by Behler and Parrinello. $\mathbf{r}_N$ is the Cartesian coordinates, while G_m is the symmetry functions.

configurations, (ii) design symmetry functions that can capture the characteristics of structures, and (ii) find the optimal parameters with optimization method.

24.7 Examples for MD Simulations

In this section, we discuss two examples to show how MD simulations are applied to study complex catalytic systems. First, we use TiO_2/H_2O interface as a model system. The atomic structure is shown in Figure 24.6, which is anatase (101) and H_2O. All calculations are performed by the NN module as implemented in the LASP package [14] (http://www.lasphub.com). Figure 24.6 depicts the MD trajectories of energy in the NVE ensemble with two selected time steps Δt, namely, 0.5 femtoseconds (fs: 1 fs $= 10^{-15}$ seconds) and 1.2 fs. When $\Delta t = 0.5$ fs, the trajectory of total energy is well conserved in the 100 ps MD simulation (1 ps $= 10^{-12}$ seconds), and meanwhile, the potential energy fluctuates within a 5 eV energy window. The conserved total energy indicates that the time integration in this MD simulation is reliable.

On the other hand, when Δt increases to 1.2 fs, the total energy drifts within the 10 ps MD simulation. This phenomenon indicates that the time step is too large for this system. The time step required for an MD simulation depends on the highest frequency of vibration in the system. In our TiO_2/H_2O model system, a small Δt, e.g. 0.5 fs, is required in the time integration algorithm, due to the high frequency of O–H stretching vibration ($\sim$3000 cm^{-1}). For the systems not involving H, $\Delta t = 1$ fs is usually a safe parameter.

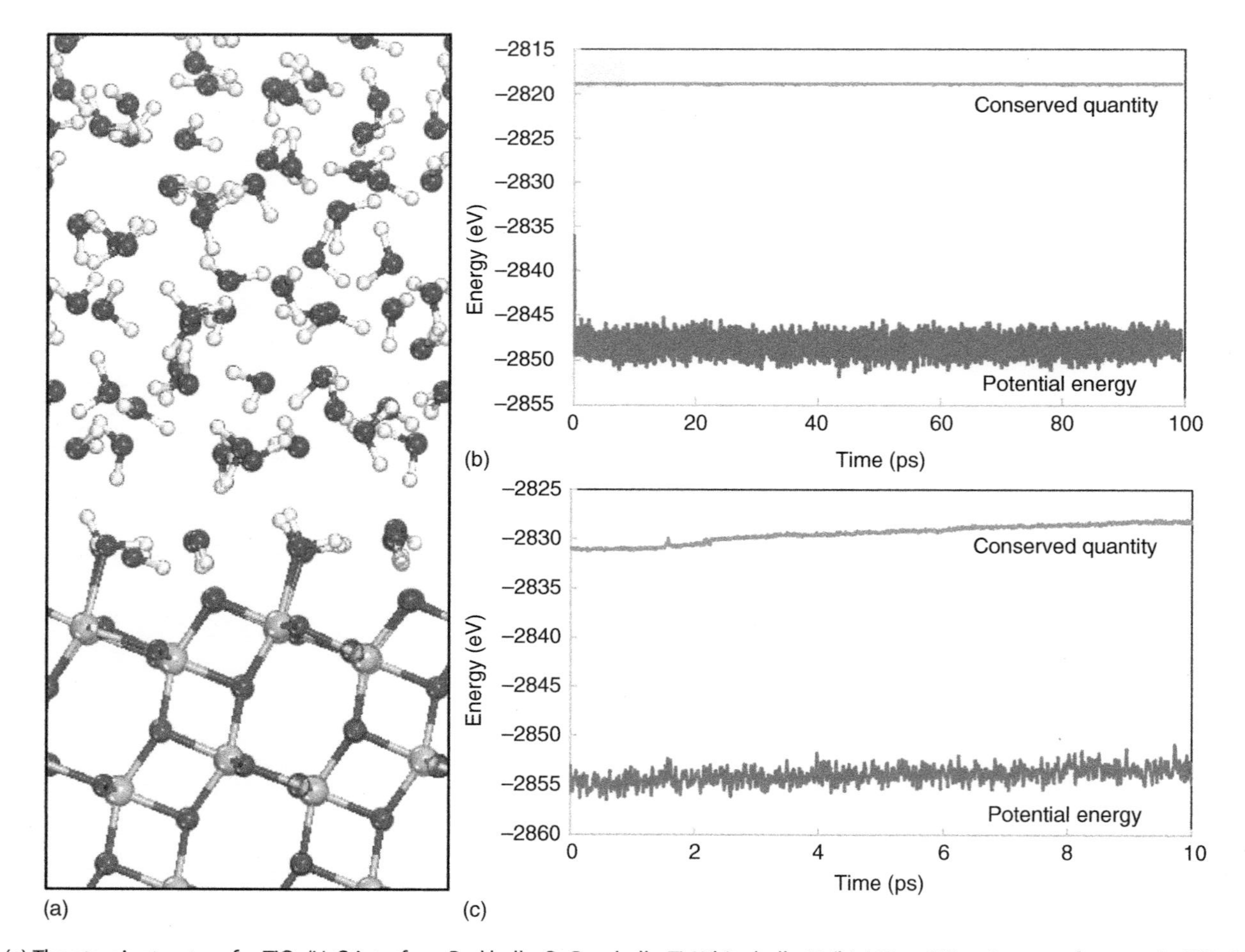

Figure 24.6 (a) The atomic structure for TiO_2/H_2O interface. Red balls: O. Gray balls: Ti. White balls: H. (b) 100 ps MD trajectory of energy for TiO_2/H_2O interface in NVE ensemble using time step $\Delta t = 0.5$ fs. The orange line denotes the conserved quantity, and the blue line the potential energy. (c) 10 ps MD trajectory of energy for TiO_2/H_2O interface in NVE ensemble using time step $\Delta t = 1.2$ fs. The orange line represents the conserved quantity, and the blue line the potential energy. (See online version for color figure).

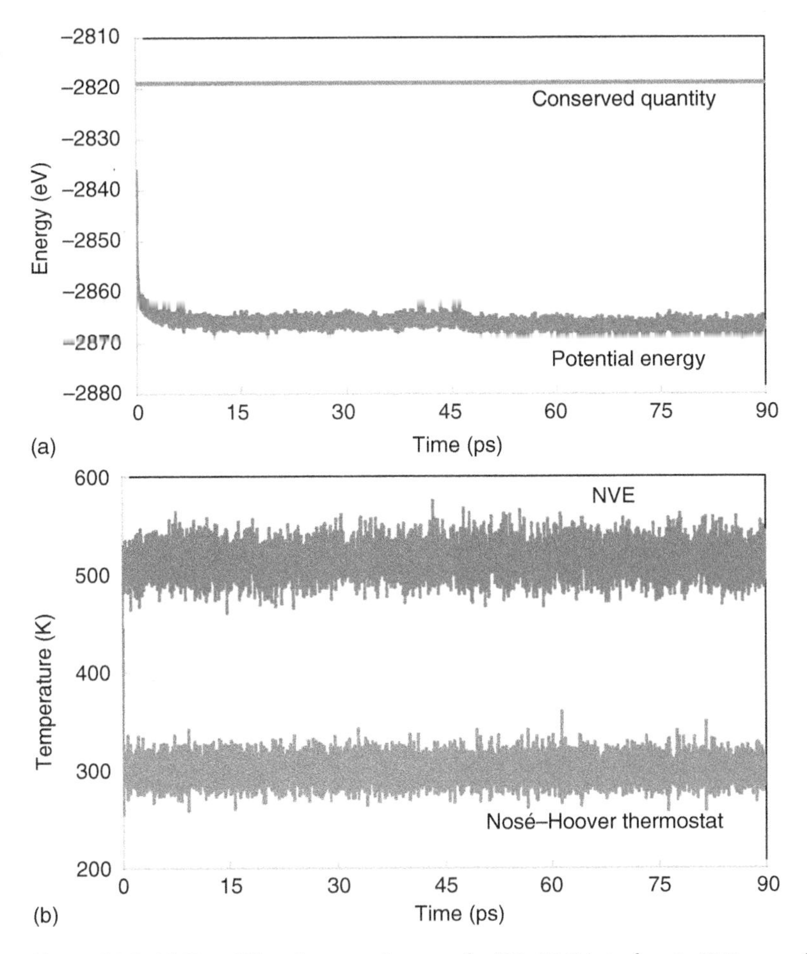

Figure 24.7 (a) 90 ps MD trajectory of energy for TiO_2/H_2O interface in NVT ensemble using time step $\Delta t = 0.5$ fs. The orange line denotes the conserved quantity; the blue line the potential energy. (b) 90 ps MD trajectory of temperature for TiO_2/H_2O interface in NVE (blue) and NVT (orange) ensembles. (See online version for color figure).

Next, we apply the Nosé–Hoover thermostat to this system to sample the NVT ensemble, and the MD trajectories of energies are shown in Figure 24.7a. The pattern of the MD trajectory is similar to that in the NVE ensemble: the potential energy fluctuates within a small energy window, while the total energy is conserved during the simulation. It should be noted that the conserved quantity in the Nosé–Hoover thermostat involves the kinetic and potential energies of the explicit TiO_2/H_2O interface plus the virtual energy of the thermostat.

The evident difference between the NVE and Nosé–Hoover thermostat is the MD trajectories of the temperature. Figure 24.7b illustrates the trajectories of the temperature of the NVE and NVT MD starting from the same configuration and the same initial temperature (300 K). The results show that the temperature will quickly diverge from the target temperature in the NVE MD; see the blue line in Figure 24.7b. But with the Nosé–Hoover thermostat, the temperature of

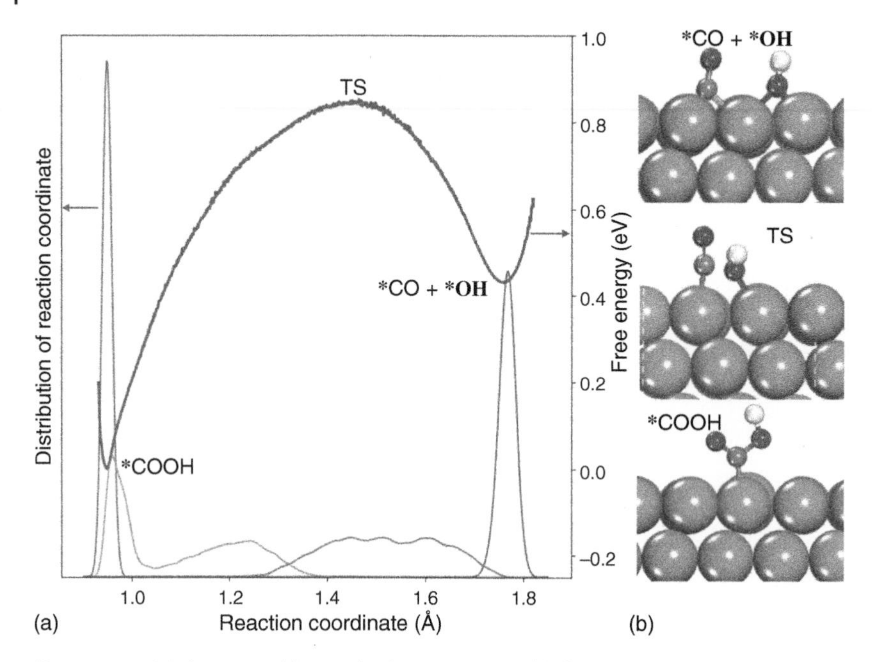

Figure 24.8 (a) The top red line is the free energy profile for *CO + *OH → *COOH. The reaction coordinate is defined as the distance between C in *CO and O in *OH. The right basin is the initial state of *CO+*OH, the left basin the product *COOH, and the top of the free energy profile the transition state (TS). The bottom peaks are the probability distributions of the reaction coordinate in four selected windows. Blue peak: *COOH. Orange peak: MS. Green peak: TS. Pink peak: *CO + *OH. (b) The snapshots for MD simulations in *CO + *OH, TS, and *COOH. Red balls: O. Gray balls: C. White balls: H. Orange balls: Cu. (See online version for color figure).

the system is well converged to the target temperature of 300 K, as shown in the orange line in Figure 24.7b. Therefore, with the Nosé–Hoover thermostat, the system can reach the target temperature quickly.

Next, we give another example to demonstrate how MD simulations can be used to resolve the free energy landscape of heterocatalysis. Here, we use (24.47) as an example, which is a key elemental step in the water-gas shift reaction ($CO + H_2O \rightarrow CO_2 + H_2$):

$$^*CO + {}^*OH \rightarrow {}^*COOH \qquad (24.47)$$

The single asterisk indicates a reactive site on the surface of Cu. We use umbrella sampling to establish the free energy profile of this reaction. We first define the distance between C in CO and O in OH as the reaction coordinate ζ, which is the geometrical parameter closely correlated with formation of product *COOH and ranges from 1.85 to 0.92 Å. Figure 24.8a illustrates the calculated free energy profile on Cu(111) surface at 300 K. The free energy of product *COOH is set as reference zero, and the free energies of the initial state (IS) (*CO + *OH) and the transition state (TS) are 0.42 and 0.84 eV, respectively. The free energy barrier of (24.47) is then the difference between TS and IS, that is, 0.42 eV.

The bottom peaks are the probability distributions $P(\zeta)$ of the reaction coordinate in four selected windows, namely, IS, transition state (TS), final state (FS), and an intermediate state (MS) at ζ=1.28 Å. Each probability distribution is obtained from a single MD simulation. The set of $P(\zeta)$ is then converted into the overall free energy profile (see the red line in Figure 24.8a) by the WHAM [11]. Figure 24.8b displays the snapshots for IS, TS, and FS.

24.8 Conclusions

This chapter gave a short overview of the MD-based methods to calculate the free energy profile in heterogeneous catalysis. The fundamental idea of the methods we introduced in this chapter is to add constraints or restraints to make a system away from its equilibrium state. With the calculated free energy profile, the free energy barrier and reaction rate constant can be obtained. This information provides us with significant details about the activity and selectivity of reaction of interest at the atomic level, which makes MD simulations a powerful tool to understand heterogeneous catalysis. At the end of the chapter, we introduced the recent advances for the application of NN in MD simulations, which we believe may be an important technique to accelerate MD simulations. MD simulation is a vast subject, and our discussion is, of course, incomplete. Nevertheless, the introduction in this chapter will be a good starting point for learning other MD techniques.

References

1 Goldstein, H., Charles, P.P., and John, L.S. *Classical Mechanics*. Addison Wesley.
2 McQuarrie, D.A. *Statistical Mechanics*. University Science Books.
3 Nosé, S. (1984). *J. Chem. Phys.* 81 (1): 511–519.
4 Hoover, W.G. (1986). *Phys. Rev. A* 34 (3): 2499–2500.
5 Frenkel, D. and Smit, B. *Understanding Molecular Simulation From Algorithms to Applications*. Elsevier Pte Ltd.
6 Eyring, H. (1935). *J. Chem. Phys.* 3 (2): 107–115.
7 Ryckaert, J.-P., Ciccotti, G., and Berendsen, H.J.C. (1977). *J. Comput. Phys.* 23 (3): 327–341.
8 Kräutler, V., van Gunsteren Wilfred, F., and Hünenberger Philippe, H. (2001). *J. Comput. Chem.* 22 (5): 501–508.
9 Tao, P., Wu, X., and Brooks, B.R. (2012). *J. Chem. Phys.* 137 (13): 134110.
10 Torrie, G.M. and Valleau, J.P. (1977). *J. Comput. Phys.* 23 (2): 187–199.
11 Kumar, S., Rosenberg John, M., Bouzida, D. et al. (1992). *J. Comput. Chem.* 13 (8): 1011–1021.
12 Laio, A. and Parrinello, M. (2002). *Proc. Natl. Acad. Sci. U.S.A.* 99 (20): 12562.
13 Behler, J. and Parrinello, M. (2007). *Phys. Rev. Lett.* 98 (14): 146401.
14 Huang, S.-D., Shang, C., Kang, P.-L., and Liu, Z.-P. (2018). *Chem. Sci.* 9: 8644.

25

First Principles Simulations of Electrified Interfaces in Electrochemistry

Stephen E. Weitzner[1] *and Ismaila Dabo*[2]

[1] *Lawrence Livermore National Laboratory, Materials Science Division, Physical and Life Sciences Directorate, Livermore, CA 94551, USA*
[2] *Materials Research Institute, Department of Materials Science and Engineering, Millennium Science Complex, University Park, PA 16802, USA*

25.1 Toward Stable and High-Performance Electrocatalysts

Electrochemical catalysts or *electrocatalysts* are used to interconvert electrical and chemical energy in electrochemical cells. This class of catalysts serves as the enabling technology for electrolysis cells, fuel cells, and a variety of electrochemical reactors that can be used to either store electrical energy in the chemical bonds of molecules or alternatively produce commodity chemicals via electrochemical reactions. Historically, platinum group metals have been used as the active material in electrochemical cells in light of their intrinsically high catalytic activity for a multitude of chemical and electrochemical reactions. However, the widespread adoption and commercialization of a number of electrochemical cells has been hindered due to the high cost of platinum group metals. A key focus of modern research has been to design new active materials that are either free of expensive platinum group metals or contain negligible amounts of them while simultaneously developing an understanding of how certain reactions proceed over catalyst surfaces so as to improve the mass activity and selectivity for particular reaction pathways [1]. A number of approaches have been pursued to lower the cost of electrocatalysts, such as the preparation of shape-controlled metal nanoparticles and nanoporous metal foams, both of which boast high specific surface areas, effectively minimizing the amount of active material needed to achieve a threshold mass activity [2]. Through these efforts, it was also discovered that nanostructuring electrocatalysts can lead to further performance improvements through finite size effects that occur as a result of the unique electronic properties of nanoscale metals [3, 4]. Further cost reductions may be realized by alloying the expensive catalytically active components with cheaper coinage metals such as silver and copper or base metals such as nickel or aluminum. The incorporation of multiple metals can lead to additional performance enhancement by way of ligand, strain, and ensemble effects that arise as a result of the

Heterogeneous Catalysts: Advanced Design, Characterization and Applications, First Edition.
Edited by Wey Yang Teoh, Atsushi Urakawa, Yun Hau Ng, and Patrick Sit.

difference in electronic structure of the alloyed metals, atomic size differences, and preferential organization of components within the surface layer [5–7]. Similarly, performance enhancements may arise through bifunctional effects where different alloy components present along the surface may work cooperatively to enhance the rate of an electrochemical step in a reaction pathway [8]. Each of these effects is generally interrelated, and the extent to which they may be present can be assessed through the combined use of electroanalytical methods such as cyclic voltammetry, scanning probe techniques such as scanning tunneling microscopy, and first principles approaches such as density functional theory (DFT). While it has been widely observed that the activity and selectivity of electrocatalysts can be improved through nanostructuring and alloying, an ongoing challenge has been the ability to rationally design high-performance electrocatalysts that are additionally stable in corrosive environments and capable of withstanding large applied voltages [9–11]. As a motivating example, we briefly consider core–shell catalyst architectures, where the active component is present as a thin shell on a cheaper core metal. As shown schematically in Figure 25.1, core–shell catalysts may be subject to *metal migration* effects where the catalytically active shell components diffuse into the core of the electrocatalyst, thereby reducing the concentration of active sites along the surface, or may similarly experience *dissolution* effects where catalytically active components dissolve into the surrounding electrolyte, again effectively diluting the number of sites available for electrocatalysis.

Obtaining a detailed understanding of the conditions under which particular electrocatalysts degrade is an important preliminary step to designing scalable and commercially viable electrochemical cells with durable and cost-effective electrocatalysts.

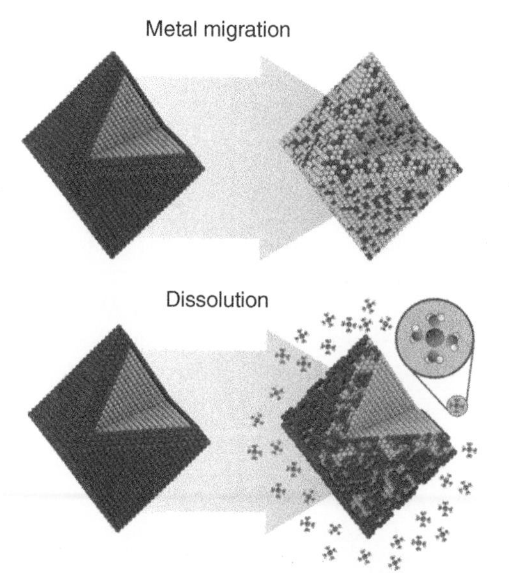

Figure 25.1 Modern electrocatalysts based on transition metal alloy chemistries can degrade through metal migration and dissolution effects. Mitigation of these and other related degradation mechanisms is a prominent focus of modern research.

In this chapter, we discuss some of the recent advances made in the first principles modeling of electrochemical catalysts. The key development, as will be discussed, is the ability to explicitly treat the effects of surface electrification due to electrochemical processes and applied voltages in a computationally efficient manner. The latter is achieved using a multiscale quantum-continuum approach that leverages an embedded polarizable continuum model to efficiently treat the electrolyte. We begin the discussion by first introducing the thermodynamics and statistical mechanics of electrified metal–solution interfaces. We subsequently discuss the structure of the electrode–electrolyte interface and the effects of applied voltages. And finally, we conclude with an overview of the first principles model and a motivating example.

25.2 A Brief Thermodynamic Detour

Before we proceed to discuss some of the modern developments in the modeling of electrochemical catalysts, it is worthwhile to touch on a few important concepts in chemical thermodynamics and electrochemistry. In Sections 25.2.1 and 25.2.2, we provide a brief summary of classical thermodynamics. In Section 25.3, we build on these concepts and discuss the statistical mechanics of electrified electrode-electrolyte interfaces. The following is by no means intended to provide a full overview of these subjects, but rather to merely provide a basis through which the subsequently developed models can be discussed. We furthermore assume that the reader has had little exposure to thermodynamics prior to reading this text and therefore present the material at a level suitable for first or second year undergraduates. For a more thorough overview of chemical thermodynamics and statistical mechanics, we highly recommend Refs. [12, 13].

25.2.1 The Fundamental Relation

Throughout the study of thermodynamics, we are fundamentally concerned with the flow of energy and matter between a *system* and its *surroundings*. As shown in Figure 25.2, we can place certain restrictions on the boundary of our defined system, such that it can be an *open* system so that heat, work, and mass can be freely exchanged with its surroundings, a completely *isolated* system that prevents any type of energy or mass transfer, and several other types of boundaries that selectively allow heat, work, and mass transfer.

This notion of boundary permeability to energy transfer is critically important to understand and to define since energy and matter are conserved quantities and can therefore only be transferred or transformed. This is stipulated by the *first law of thermodynamics*, which states that the change in internal energy dU of a system is equal to the amount of heat and work that is exchanged with its surroundings

$$dU = \delta Q + \delta W \tag{25.1}$$

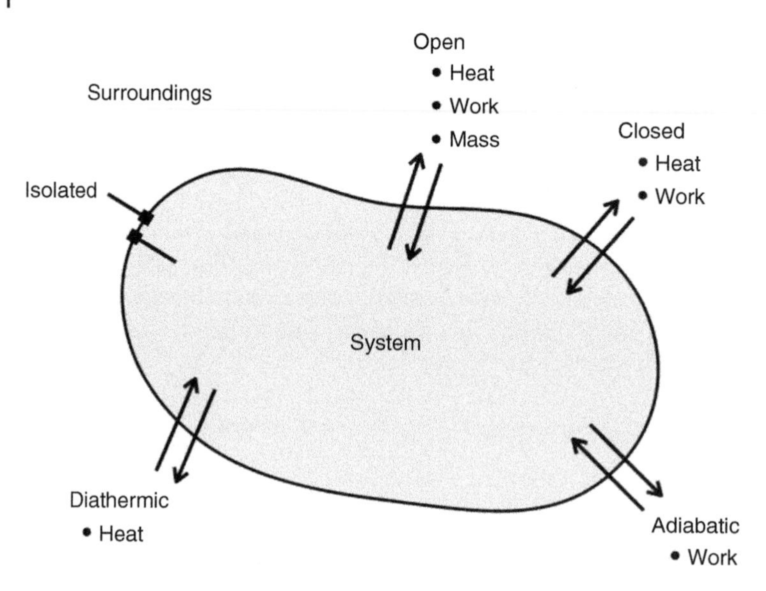

Figure 25.2 Depending on the type of thermodynamic wall being considered, we can alter how a system interacts with its surroundings. For example, a system with open boundaries allows energy to be exchanged between the system and surroundings in the form of heat, work, and matter, while diathermic boundaries that are rigid and impermeable allow only heat to be exchanged.

where δQ and δW are infinitesimal amounts of heat and work, respectively.[1] While the first law ensures the conservation of energy, it does not actually place any limits on the exchange of heat or work between a system and its surroundings. Understanding how energy is transferred and defining the rules by which it can transfer is useful since there are a multitude of ways in which the internal energy change dU may be realized.

It is straightforward to first think about what kind of work can be performed on or by a system. We can express the total amount of work as $\delta W = \sum_i \delta W_i$, whose value depends on the types of work δW_i that are possible. One can, for example, consider the *mechanical* work $-PdV$ that is done to compress or expand a volume by an amount dV at a pressure P, the *chemical* work μdN that is done by adding or removing dN atoms or molecules to a system at a chemical potential of μ, and the *electrical* work ΦdQ that is done by adding or removing an electronic charge dQ to the system at a voltage of Φ. In the case that each of these processes is allowed to occur, we can rewrite Eq. (25.1) as

$$dU = \delta Q - PdV + \sum_i \mu_i dN_i + \Phi dQ \tag{25.2}$$

1 Here, we differentiate between an exact differential quantity dX and an inexact differential quantity δY such that X is defined to be a state function whose cyclic integral vanishes $\oint dX = 0$, while no such condition is guaranteed for an inexact differential $\oint \delta Y \neq 0$. This is to say, the extent of heat transfer and work done by or to a system is path dependent or depends on the manner in which the processes are carried out, while the change in a thermodynamic potential such as the internal energy U or entropy S is path independent and thus depends only upon the initial and final states of the system.

While the work contributions to the change in internal energy are straightforward to specify, the extent of heat transfer is less obvious and requires a somewhat more in-depth discussion. In order to define δQ, it is useful to first recall that real processes that are observed to occur in nature such as the mixing of two different gases or heat transfer from hot regions to colder regions occur spontaneously or *irreversibly*. That is, for a given system in an initially nonequilibrium state, the system will tend to evolve toward an equilibrium state in which it will remain until acted upon by some externally applied force. Thus, once a system is at equilibrium with its surroundings, the driving force to move to a different state vanishes. In the context of the examples provided above, gases will continue to mix until their constituent particles are homogeneously distributed, creating a spatially uniform composition, and heat will continue to transfer until a spatially uniform temperature in the system is achieved. The extent to which a process is considered to be spontaneous or irreversible is characterized by the change in the *entropy* S of the system. Formally, entropy is defined to be a state function and a property of a system similar to the way in which internal energy is a property of a system. The change in entropy can be expressed as

$$dS = \frac{\delta Q}{T} \tag{25.3}$$

where T is the temperature at which the heat transfer is conducted. Physically speaking, Eq. (25.3) states that the gain in entropy due to a transfer of heat is greater for a system at low temperature than for the same system at an elevated temperature. Furthermore, a system is said to be at equilibrium with its surroundings when the total entropy of the system and surroundings is maximized upon which $dS = 0$. A simple example demonstrating this fact is shown in Figure 25.3.

Here, we consider an isolated system that is partitioned into two subsystems via an adiabatic wall, and we notice initially that they have two different temperatures T_1 and T_2, where $T_1 < T_2$. We thus have a "hot" region and a "cold" region inside of our isolated system. The total energy of the isolated composite system is simply the sum of the energy of the two subsystems, $U_0 = U_1 + U_2$. If we now replace the adiabatic wall with a diathermal wall, an amount of heat δQ will flow between the two subsystems due to the initial temperature gradient. The effect of this heat transfer is that the energy of subsystem 1 has increased by δQ and the energy of subsystem 2 has decreased by δQ. The heat transfer stops only when the temperature of the composite system is uniform everywhere at a

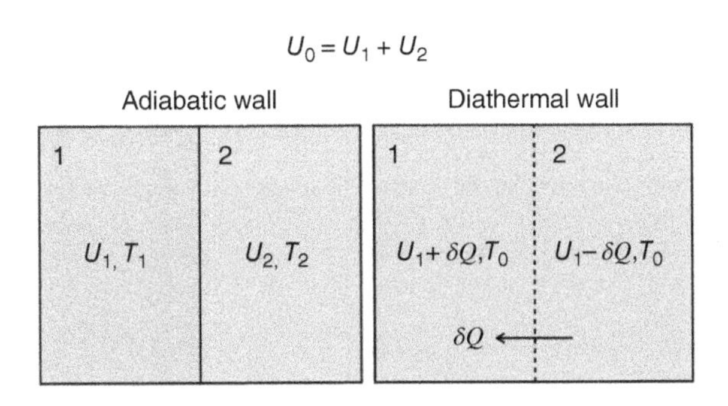

Figure 25.3 Two subsystems contained in adiabatic enclosures with different temperatures T_1 and T_2 with $T_1 < T_2$ will experience a spontaneous heat transfer when the adiabatic wall is replaced with a diathermal wall.

value of T_0, which we speculate to lie somewhere in the interval $[T_1, T_2]$. After this process is complete, the total energy of the composite system is unchanged: $U_0 = (U_1 + \delta Q) + (U_2 - \delta Q) = U_1 + U_2$. However, if we consider the change in entropy of the system due to this spontaneous process $\Delta S = \delta Q/T_1 - \delta Q/T_2$, we find that the overall entropy of the system has increased since $T_1 < T_2$. The heat transfer stops only when $T_1 = T_2 = T_0$ since the driving force for heat transfer (i.e. the temperature gradient) vanishes; in other words, once the system is at equilibrium, any additional heat transfer would fail to increase the entropy any further since $\Delta S = \delta Q/T_0 - \delta Q/T_0 = 0$ J/K.[2] This leads us to the *second law of thermodynamics*, which states that for an isolated system, the total entropy of the system can never decrease; the total entropy can only increase or remain constant. Using the definition of the second law, we can write the combined first and second law of thermodynamics as

$$dU = TdS - PdV + \sum_i \mu_i dN_i + \Phi dQ \tag{25.4}$$

which indicates that the internal energy of a system is a function of a set of *natural variables* $U = U(S, V, N_i, Q)$. This particular set of variables is said to be *extensive variables*, since their values scale proportionally with system size, while their conjugate variables T, P, μ, and Φ are said to be *intensive variables* that are independent of system size and have the additional property that they are equal everywhere within a system at equilibrium. Furthermore, Eq. (25.4) is said to be a *fundamental relation* that one can use to completely define the state of a thermodynamic system. The caveat to this is that it is also necessary to identify the *equations of state*[3] that relate the intensive variables to the extensive variables in the system:

$$\begin{aligned}
T &= T(S, V, N_i, Q) = \left(\frac{\partial U}{\partial S}\right)_{V,N_i,Q} \\
P &= P(S, V, N_i, Q) = -\left(\frac{\partial U}{\partial V}\right)_{S,N_i,Q} \\
\mu_i &= \mu_i(S, V, N_i, Q) = \left(\frac{\partial U}{\partial N_i}\right)_{S,V,N_j \neq N_i,Q} \\
\Phi &= \Phi(S, V, N_i, Q) = \left(\frac{\partial U}{\partial Q}\right)_{S,V,N_i}
\end{aligned} \tag{25.5}$$

Thus, given a fundamental relation of a system and its associated equations of state, one can fully understand a thermodynamic system.

---

2 Here we consider the thermal energy to be in units of joules (1 J = 1 N m = 1 kg/m^2/s^2) and temperature to be in units of Kelvin (K).

3 Here we see that the intensive variables can be equivalently expressed as partial derivatives of the internal energy with respect to their extensive conjugate variables. This follows from the fact that the internal energy is a state function so we consider an infinitesimal change in internal energy to be an exact differential:

$$dU = \left(\frac{\partial U}{\partial S}\right)_{V,N_i,Q} dS + \left(\frac{\partial U}{\partial V}\right)_{S,N_i,Q} dV + \sum_i \left(\frac{\partial U}{\partial N}\right)_{S,V,N_j \neq N_i,Q} dN_i + \left(\frac{\partial U}{\partial Q}\right)_{S,V,N_i} dQ$$

25.2.2 Alternative Forms of the Fundamental Relation

While Eq. (25.4) is an important result embodying the first two laws of thermodynamics, it is not always the most convenient form of the fundamental relation to use when discussing chemical or electrochemical thermodynamics. This stems primarily from the fact that the internal energy depends on extensive quantities such as entropy, volume, particle number, and charge, which are infeasible to rigorously control in laboratory settings. Instead, it is more useful to consider forms of the fundamental relation that depend on temperature, pressure, chemical potential, or voltage since these quantities can be more readily manipulated in experiments. These alternative forms can be derived using a mathematical technique called *Legendre transformation*, which provides an alternative but equivalent representation of a function $f(x)$ in terms of its derivative $m(x) = f'(x)$ and its associated tangent line's y-intercept b:

$$b(m) = f(x) - m(x)x \tag{25.6}$$

In Figure 25.4, we demonstrate the procedure for a one-dimensional parabolic function [12]. For higher-dimensional cases, such as the thermodynamic potentials we are considering here, one can perform an analogous procedure using partial derivatives. Because the intensive variables we aim to replace the extensive variables with are defined as partial derivatives of the internal energy (Eq. (25.5)), we are able to employ this technique to derive a variety of fundamental relations. To demonstrate this, we must first consider the internal energy in its Euler form[4]

$$U = TS - PV + \sum_i \mu_i N_i + \Phi Q \tag{25.7}$$

As an example, if we wish to obtain a fundamental relation with the set of natural variables (S, P, N_i, Q), we may compute the Legendre transform of the internal energy with respect to the volume:

$$H = U - \left(\frac{\partial U}{\partial V}\right) V = U + PV \tag{25.8}$$

where the new thermodynamic potential H introduced above is known as the *enthalpy* of a system. Using Eq. (25.4), we can obtain the desired fundamental relation by computing the differential of Eq. (25.8):

$$dH = dU + PdV + VdP = TdS + VdP + \sum_i \mu_i dN_i + \Phi dQ \tag{25.9}$$

We can perform single Legendre transforms as we have done above, or we can perform multiple Legendre transforms to obtain a fundamental relation that has more than one intensive variable in its set of natural variables. For example, the

4 Extensive quantities such as the internal energy are said to be first-order homogeneous functions, which obey the property $f(\lambda x_1, \lambda x_2, \dots, \lambda x_n) = \lambda f(x_1, x_2, \dots, x_n)$, where λ is an arbitrary constant. If we compute the derivative of both sides with respect to λ, we obtain $\sum_i \left(\frac{\partial f}{\partial \lambda x_i}\right) x_i = f(x_1, x_2, \dots, x_n)$. Because λ is just a constant, we can consider the case of $\lambda = 1$, which yields $\sum_i \left(\frac{\partial f}{\partial x_i}\right) x_i = f(x_1, x_2, \dots, x_n)$. The latter is referred to as an Euler relation as it obeys Euler's theorem for homogeneous functions [12].

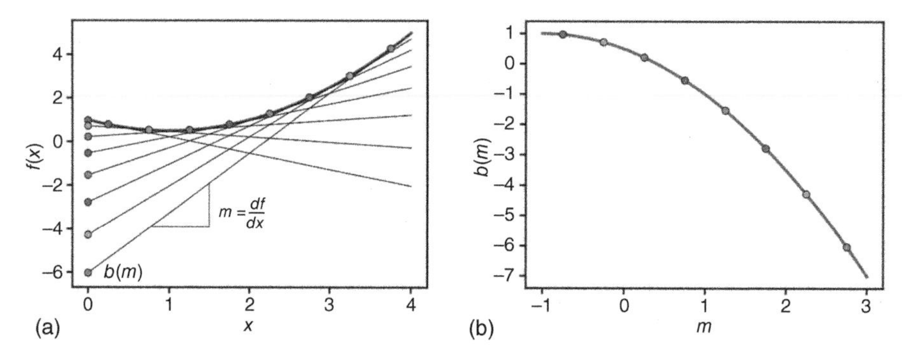

Figure 25.4 The procedure for computing the Legendre transform of a function $f(x)$ involves (a) obtaining the set of tangent line slopes $m = df/dx$ of the underlying function $f(x)$ and their associated set of y-intercepts $b = b(m)$ for the entire domain of $f(x)$. The original function $f(x)$ is transformed (b) to $b(m)$, which can be viewed as an alternative yet equivalent representation of the underlying function $f(x)$. (See online version for color figure).

Gibbs free energy G has the set of natural variables (T, P, N_i, Q), and its fundamental relation can be obtained via the double Legendre transform:

$$G = U - TS + pV \rightarrow dG = -SdT + VdP + \sum_i \mu_i dN_i + \Phi dQ \tag{25.10}$$

Using this approach, a variety of fundamental relations can be derived from the combined first and second law, several of which have been summarized in Table 25.1.

Table 25.1 Summary of important thermodynamic potentials, their Legendre transforms, and their associated fundamental relations

Thermodynamic potential	Legendre transform	Fundamental relation
Enthalpy	$H = U + PV$	$dH = TdS + VdP + \sum_i \mu_i dN_i + \Phi dQ$
Helmholtz free energy	$A = U - TS$	$dA = -SdT - PdV + \sum_i \mu_i dN_i + \Phi dQ$
Gibbs free energy	$G = U - TS + PV$	$dG = -SdT + VdP + \sum_i \mu_i dN_i + \Phi dQ$
Electrochemical enthalpy	$F = U - \Phi Q$	$dF = TdS - PdV + \sum_i \mu_i dN_i - Qd\Phi$
Electrochemical free energy	$\mathcal{F} = U - TS - \Phi Q$	$d\mathcal{F} = -SdT - PdV + \sum_i \mu_i dN_i - Qd\Phi$
Grand potential	$\phi = U - TS - \mu_j N_j$	$d\phi = -SdT - PdV - N_j d\mu_j + \sum_{i \neq j} \mu_i dN_i + \Phi dQ$
Electrochemical grand potential	$\varphi = U - TS - \mu_j N_j - \Phi Q$	$d\varphi = -SdT - PdV - N_j d\mu_j + \sum_{i \neq j} \mu_i dN_i - Qd\Phi$

25.3 Statistical Mechanics

Up until this point, we have mainly considered the thermodynamics of macroscopic systems and introduced several useful fundamental relations. While these equations enable us to at least in principle identify equilibrium states, they do not provide any information about the microscopic details that are needed to effectively understand the properties of functional materials. Such a detailed view can be obtained by considering the *statistical mechanics* of a system, which provides a link between the configuration and motion of a set of particles and the observed macroscopic properties of a thermodynamic system. In what follows, a brief overview of the statistical mechanics of electrochemical interfaces will be provided.

25.3.1 Preliminaries

In statistical mechanics, we consider a system to exist in a certain *macrostate*, which is a particular thermodynamic state specified by a set of fixed properties such as constant particle number N, constant volume V, and constant temperature T. Each macrostate is associated with a vast set or *ensemble* of *microstates*, where a microstate can be viewed as a copy of the system whose constituent particles adopt different configurations and momenta. The set of microstates corresponding to a given macrostate spans all of the possible configurations and momenta that conform to the particular macrostate. We can furthermore associate each microstate with a set of $6N$ coordinates $\{\boldsymbol{r}^N, \boldsymbol{p}^N\}$ describing the positions $\boldsymbol{r}$ and the momenta $\boldsymbol{p}$ of the N particles in the system, as well as an energy $E = \mathcal{H}(\boldsymbol{r}^N, \boldsymbol{p}^N)$ that is determined by a suitable *Hamiltonian* $\mathcal{H} = \mathcal{T} + \mathcal{V}$ that accounts for the kinetic energy $\mathcal{T}(\boldsymbol{p}^N) = \sum_i^N |\boldsymbol{p}_i|^2/2m_i$ and the potential energy $\mathcal{V}(\boldsymbol{r}^N)$ of the particles. Individual microstates can additionally be viewed as points in a $6N$-dimensional space that is referred to as the *phase space* of the system. We can therefore quantify the "size" of a macrostate by the volume it occupies in phase space, or equivalently the number of microstates it contains. Thus within any given macrostate, there are an enormous number of possible microstates that the system could be in at any given time; however, as we will discuss shortly, certain microstates may be more probable than others. It follows then that the subset of microstates with the highest probabilities can be identified to be the equilibrium states of a system.

The probability of the system being in a particular microstate is determined by the Boltzmann distribution, which, for the example (N, V, T) ensemble discussed above, would be expressed as

$$P(\boldsymbol{r}^N, \boldsymbol{p}^N) = \frac{\exp[-\beta\mathcal{H}(\boldsymbol{r}^N, \boldsymbol{p}^N)]}{\frac{1}{h^{3N}N!}\int_\Omega d\boldsymbol{p}^N d\boldsymbol{r}^N \exp[-\beta\mathcal{H}(\boldsymbol{r}^N, \boldsymbol{p}^N)]} = \frac{\exp[-\beta\mathcal{H}(\boldsymbol{r}^N, \boldsymbol{p}^N)]}{\mathcal{Q}} \tag{25.11}$$

where $\mathcal{Q}$ is the partition function of the system that involves $6N$ integrals over the region of phase space Ω occupied by the macrostate and $\beta = 1/k_B T$ is known as the thermodynamic beta, where k_B is the Boltzmann constant. Because each volume element of phase space $\boldsymbol{dpdr}$ has dimensions of action, it is common by convention to introduce a factor of h^{-3} for each element to obtain a dimensionless probability, where h is Planck's constant. Furthermore, in the case that we consider particles to be indistinguishable from one another, a permutation factor $(N!)^{-1}$ is introduced to account for the identical configurations that arise from interchanging the positions of indistinguishable particles. Macroscopic properties of the system such as the internal energy, volume, and charge can be computed as an expectation value or an *ensemble average* of the system. For example, the ensemble average of an arbitrary property A of a classically interacting system is computed by integrating a weighted probability distribution over the region of phase space occupied by the macrostate

$$\langle A \rangle = \frac{1}{h^{3N}N!}\int_{\Omega} d\boldsymbol{p}^N d\boldsymbol{r}^N A(\boldsymbol{r}^N,\boldsymbol{p}^N)P(\boldsymbol{r}^N,\boldsymbol{p}^N) \tag{25.12}$$

where here we see that the pre-factor $(h^{3N}N!)^{-1}$ cancels when the expectation value is computed. In practice, this expression can be simplified by analytically solving the momentum integrals over the phase space of the system.[5] This enables us to focus solely on the contribution of the configuration-dependent potential energy $E_i = \mathcal{V}(\boldsymbol{r}_i^N)$ when computing expectation values, which means that the probability distribution that must be considered depends only on the configurational space of the system and is thus proportional to the overall probability $\mathcal{P}(\boldsymbol{r}^N) \propto P(\boldsymbol{r}^N,\boldsymbol{p}^N)$. Similarly, we also consider a configuration-dependent partition function $Z \propto Q$. And finally, the integrals over the configurational space of the system may be replaced by a summation over the individual microstates that are accessible within the given macrostate. This allows us to rewrite the probability distribution as

$$\mathcal{P}_i = \frac{\exp[-\beta E_i]}{\sum_j \exp[-\beta E_j]} = \frac{1}{Z}\exp[-\beta E_i] \tag{25.13}$$

Accordingly, ensemble averages are analogously obtained by computing a weighted sum over the phase space of the system

$$\langle A \rangle = \sum_i A_i \mathcal{P}_i \tag{25.14}$$

25.3.2 The Electrochemical Canonical (N, V, T, Φ) Ensemble

In the electrochemical canonical ensemble, we consider a closed system to be in contact with a large reservoir that is held at a fixed temperature T and a fixed electrode potential Φ [14]. The system is able to freely exchange heat and electrons

5 For a single-component system, it can be shown that the momentum integrals $I = \int_{-\infty}^{\infty} d\boldsymbol{p}^N \exp\left[-\beta\sum_i|\boldsymbol{p}_i|^2/2m\right]$ evaluates to $I = (\sqrt{2m\pi k_B T})^{3N}$. This allows the partition function to be expressed as $Q = (\Lambda^{3N}N!)^{-1}\int d\boldsymbol{r}^N \exp[-\beta\mathcal{V}(\boldsymbol{r}^N)] = Z/\Lambda^{3N}N!$. Here $\Lambda = \sqrt{h^2/2\pi m k_B T}$ is the thermal de Broglie wavelength, and Z is referred to as the configurational integral. It is straightforward to show that this result may be generalized to treat classically interacting multicomponent systems.

with the reservoir, and at equilibrium, the energy and the charge of the system will fluctuate around their equilibrium values. Furthermore, as in classical thermodynamics, the temperature and the electrode potential of the system will be equal to that of the reservoir at equilibrium. It can additionally be shown[6] that the Boltzmann distribution for the (N, V, T, Φ) ensemble can be expressed as

$$\mathcal{P}_i = \frac{1}{Z}\exp[-\beta(E_i - \Phi Q_i)] \tag{25.15}$$

The validity of this definition for the Boltzmann distribution can be demonstrated by calculating the entropy of the system via the Gibbs entropy formula:

$$S = -k_B \sum_i \mathcal{P}_i \ln \mathcal{P}_i \tag{25.16}$$

Inserting Eq. (25.15) into Eq. (25.16), we obtain

$$S = \frac{\langle E\rangle}{T} - \frac{\Phi\langle Q\rangle}{T} + k_B \ln Z \tag{25.17}$$

where we have made use of Eq. (25.14). After some minor algebra, we find

$$-k_B T \ln Z = \langle E\rangle - TS - \Phi\langle Q\rangle \tag{25.18}$$

Comparing this equation to the Euler relation for the electrochemical free energy, $\mathcal{F} = U - TS - \Phi Q$, we can readily identify $\mathcal{F} = -k_B T \ln Z$, $U = \langle E\rangle$, and $Q = \langle Q\rangle$. Thus, the electrochemical free energy $\mathcal{F}$ and the entropy S depend explicitly on the partition function Z, while the internal energy and the charge of the system are obtained as expectation values of fluctuating quantities. It is also straightforward to show that the average charge and the internal energy can be expressed in terms of logarithmic derivatives of the partition function:

$$\langle Q\rangle = \sum_i Q_i \mathcal{P}_i = \frac{1}{\beta}\frac{\partial \ln Z}{\partial \Phi} = -\left(\frac{\partial \mathcal{F}}{\partial \Phi}\right)_{N,V,T} \tag{25.19}$$

$$\langle E\rangle = \sum_i E_i \mathcal{P}_i = \Phi\langle Q\rangle - \frac{\partial \ln Z}{\partial \beta} \tag{25.20}$$

At equilibrium, heat and charge fluctuate randomly between the system and the reservoir, indicating that the fluctuations $\Delta E = E_i - \langle E\rangle$ and $\Delta Q = Q_i - \langle Q\rangle$ may also be considered to be random variables. For a general fluctuation $\Delta A = A_i - \langle A\rangle$, it can be shown that the mean of the fluctuation vanishes $\langle \Delta A\rangle = 0$, verifying that the energy and the charge distributions are centered on their means. The spread of the distributions can be quantified as the mean square of the fluctuations $\langle(\Delta A)^2\rangle = \langle A_i^2 - 2A_i\langle A\rangle + \langle A\rangle^2\rangle = \langle A^2\rangle - \langle A\rangle^2$, which is nothing other than the *variance* of the distribution of fluctuations. It can be shown that the variance of a fluctuation is related to the second-order logarithmic derivative of the

6 The Boltzmann distribution can be obtained in a straightforward manner using Lagrange multipliers to maximize the Gibbs entropy of the system, subject to the constraints that the probability remains normalized $\sum_i \mathcal{P}_i = 1$ and that the probability distribution recovers the expectation values of the energy $\langle E\rangle = \sum_i \mathcal{P}_i E_i$ and charge $\langle Q\rangle = \sum_i \mathcal{P}_i Q_i$.

partition function with respect to the intensive conjugate variable of the fluctuating quantity. For example, when considering the fluctuation of the charge in the system, we observe the following relation to hold:

$$\frac{\partial \langle Q \rangle}{\partial \Phi} = \frac{1}{\beta}\frac{\partial^2 \ln Z}{\partial \Phi^2} = \frac{1}{\beta}\frac{\partial}{\partial \Phi}\left[\frac{1}{Z}\frac{\partial Z}{\partial \Phi}\right] = \frac{1}{\beta}\left[\frac{1}{Z}\frac{\partial^2 Z}{\partial \Phi^2} - \frac{1}{Z^2}\left(\frac{\partial Z}{\partial \Phi}\right)^2\right] = \beta[\langle Q^2 \rangle - \langle Q \rangle^2] \tag{25.21}$$

The latter is quite interesting as it provides a connection between the macroscopic charge–voltage response of an electrochemical interface and the fluctuating charge on the electrode. In fact, this is the definition of the differential capacitance of the electrode–solution interface, which allows us to write

$$C_0 = \frac{d\sigma}{d\Phi} = \frac{1}{A}\frac{d\langle Q \rangle}{d\Phi} = \frac{\beta}{A}[\langle Q^2 \rangle - \langle Q \rangle^2] \tag{25.22}$$

where $\sigma = Q/A$ is the surface charge density on the electrode and C_0 is the *areal differential capacitance*. The exactness of this definition is guaranteed in the thermodynamic limit where the total number of particles $N_{tot} = \sum_i N_i$ in the system approaches Avogadro's number. To show this, we consider the fact that the charge and the capacitance of a system are extensive quantities and can be expressed in terms of a scalar λ as $\langle Q \rangle = \lambda \langle Q \rangle_{ref}$ and $AC_0 = \lambda A_{ref} C_0$. Here, we define λ in terms of a reference system that has a fixed stoichiometry and a total number of particles $N_{ref} = \sum_i N_i^{ref}$ so that the total number of particles in the scaled system is $N_{tot} = \sum_i N_i = \lambda N_{ref}$. Next, we define the amplitude of the charge fluctuation to be the root mean square fluctuation $\Delta Q_{rms} = (\langle Q^2 \rangle - \langle Q \rangle^2)^{1/2}$, which is nothing other than the standard deviation of the charge fluctuations. If we now consider the ratio of the fluctuation amplitude to the average charge in the system

$$\frac{\Delta Q_{rms}}{\langle Q \rangle} = \sqrt{\frac{AC_0}{\beta \langle Q \rangle^2}} = \sqrt{\frac{\lambda A_{ref} C_0}{\beta \lambda^2 \langle Q \rangle_{ref}^2}} = \sqrt{\frac{A_{ref} C_0}{\beta \langle Q \rangle_{ref}^2}}\frac{1}{\sqrt{\lambda}} \tag{25.23}$$

we find that the charge fluctuations become vanishingly small in comparison with the average charge of the system in the thermodynamic limit where $\lambda \rightarrow \infty$. As a consequence of this, the charge distributions become very sharply peaked about their means.

The differential capacitance is one example of what is known as a *response function* or a *susceptibility* in statistical mechanics. Several other response functions can be defined in the (N, V, T, Φ) ensemble that consist of the response of an extensive quantity to variations in either to their conjugate intensive variable or to another intensive variable that is controlled by the reservoir. It can be shown that given two sets of conjugate variables (X, F_X) and (Y, F_Y), the following relation generally holds:

$$\langle \Delta X \Delta Y \rangle = \langle XY \rangle - \langle X \rangle \langle Y \rangle = \frac{1}{\beta}\frac{\partial \langle X \rangle}{\partial F_Y} = \frac{1}{\beta}\frac{\partial \langle Y \rangle}{\partial F_X} \tag{25.24}$$

which states that the *covariance* of fluctuations in X and Y is related to the response of $\langle X \rangle$ to variations in the externally controlled potential F_Y and vice versa. A number of response functions can be derived for the (N, V, T, Φ) ensemble, several of which have been summarized in Table 25.2.

Table 25.2 Summary of several useful response functions in the electrochemical canonical (N, V, T, Φ) ensemble

Response function	
Internal energy – voltage	$\dfrac{\partial\langle E\rangle}{\partial\Phi} = \beta[\langle EQ\rangle - \langle E\rangle\langle Q\rangle]$
Differential capacitance,AC_0	$\dfrac{\partial\langle Q\rangle}{\partial\Phi} = \beta[\langle Q^2\rangle - \langle Q\rangle^2]$
Constant volume heat capacity,C_V	$\dfrac{\partial\langle E\rangle}{\partial T} = k_B\beta^2\left[\langle E^2\rangle - \langle E\rangle^2 - \dfrac{\Phi}{\beta}\dfrac{\partial\langle E\rangle}{\partial\Phi}\right]$
Charge – temperature	$\dfrac{\partial\langle Q\rangle}{\partial T} = k_B\beta^2\left[\dfrac{1}{\beta}\dfrac{\partial\langle E\rangle}{\partial\Phi} - \dfrac{\Phi}{\beta}\dfrac{\partial\langle Q\rangle}{\partial\Phi}\right]$

25.3.3 The Electrochemical Grand Canonical (μ, V, T, Φ) Ensemble

In the electrochemical grand canonical ensemble, we consider an open system that is in contact with a large reservoir that is held at a fixed temperature T, a fixed electrode potential Φ, and a fixed chemical potential μ_j for a species j [15]. The system is able to exchange heat, electrons, and j particles with the reservoir, and at equilibrium, the energy, charge, and particle number of the system will fluctuate around their equilibrium values. The Boltzmann distribution in the (μ, V, T, Φ) ensemble can be expressed as

$$\mathcal{P}_k = \frac{1}{\mathcal{Z}}\exp[-\beta(E_k - \mu_j(N_j)_k - \Phi Q_k)] = \frac{1}{\mathcal{Z}}\exp[-\beta(F_k - \mu_j(N_j)_k)] \quad (25.25)$$

where we show explicitly that the Boltzmann factor can be rewritten in terms of the electrochemical enthalpy $F_k = E_k - \Phi Q_k$. Following the same procedure we employed in Section 25.3.2, we compute the Gibbs entropy of the system as

$$S = \frac{\langle E\rangle}{T} - \frac{\mu_j\langle N_j\rangle}{T} - \frac{\Phi\langle Q\rangle}{T} + k_B \ln Z \quad (25.26)$$

which upon rearrangement leads to the definition of the electrochemical grand potential:

$$\varphi = -k_B T \ln \mathcal{Z} = \langle E\rangle - TS - \mu_j\langle N_j\rangle - \Phi\langle Q\rangle \quad (25.27)$$

Similar to the energy and charge in the (N, V, T, Φ) ensemble, the mean particle number can be expressed in terms of a logarithmic derivative of the partition function:

$$\langle N_j\rangle = \frac{1}{\beta}\frac{\partial \ln \mathcal{Z}}{\partial \mu_j} = -\left(\frac{\partial\varphi}{\partial\mu_j}\right)_{T,V,\Phi,\mu_k\neq\mu_j} \quad (25.28)$$

Finally, a number of response functions can be defined in the (μ, V, T, Φ) ensemble that describe the response of unconstrained extensive quantities to variations in the electrode potential, chemical potential, and temperature. Several of these response functions are summarized in Table 25.3.

Table 25.3 Summary of several useful response functions in the electrochemical grand canonical (μ, V, T, Φ) ensemble

Response function	
Particle number – voltage	$\dfrac{\partial\langle N_j\rangle}{\partial\Phi} = \beta[\langle N_jQ\rangle - \langle N_j\rangle\langle Q\rangle]$
Internal energy – chemical potential	$\dfrac{\partial\langle E\rangle}{\partial\mu_j} = \beta[\langle EN_j\rangle - \langle E\rangle\langle N_j\rangle]$
Particle number – chemical potential	$\dfrac{\partial\langle N_j\rangle}{\partial\mu_j} = \beta[\langle N_j^2\rangle - \langle N_j\rangle^2]$
Charge – chemical potential	$\dfrac{\partial\langle Q\rangle}{\partial\mu_j} = \beta[\langle QN_j\rangle - \langle Q\rangle\langle N_j\rangle]$
Constant volume heat capacity,C_V	$\dfrac{\partial\langle E\rangle}{\partial T} = k_B\beta^2\left[\langle E^2\rangle - \langle E\rangle^2 - \dfrac{\mu_j}{\beta}\dfrac{\partial\langle E\rangle}{\partial\mu_j} - \dfrac{\Phi}{\beta}\dfrac{\partial\langle E\rangle}{\partial\Phi}\right]$
Particle number – temperature	$\dfrac{\partial\langle N_j\rangle}{\partial T} = k_B\beta^2\left[\dfrac{1}{\beta}\dfrac{\partial\langle E\rangle}{\partial\mu_j} - \dfrac{\mu_j}{\beta}\dfrac{\partial\langle N_j\rangle}{\partial\mu_j} - \dfrac{\Phi}{\beta}\dfrac{\partial\langle N_j\rangle}{\partial\Phi}\right]$
Charge – temperature	$\dfrac{\partial\langle Q\rangle}{\partial T} = k_B\beta^2\left[\dfrac{1}{\beta}\dfrac{\partial\langle E\rangle}{\partial\Phi} - \dfrac{\mu_j}{\beta}\dfrac{\partial\langle N_j\rangle}{\partial\Phi} - \dfrac{\Phi}{\beta}\dfrac{\partial\langle Q\rangle}{\partial\Phi}\right]$

a) The response functions listed in Table 25.2 (excluding the (N, V, T, Φ) heat capacity) also exist in the (μ, V, T, Φ) ensemble, but are omitted for brevity.

25.3.4 Computational Methods

Leading up to this point, we have recapitulated thermodynamics and statistical mechanics, and we have additionally demonstrated how the effects of an applied electrode potential may be included within the developed theoretical framework. We note that while statistical mechanics provides a formally exact answer for how a collection of particles behaves under a set of externally applied potentials, it stops short of providing an approach for performing practical calculations to obtain useful numerical results. Such approaches are necessary to guide the design and understanding of physical material systems. That being said, a significant effort has been made since the inception of statistical mechanics to develop the computational tools and methods needed to obtain reliable estimates of material properties from atomistic models. In general, two major obstacles must be overcome in order to achieve satisfactory results. The first challenge to address is how to define the many-body potential $\mathcal{V}(\mathbf{r}^N)$ that describes the interactions between the particles in the system. In condensed matter systems such as the electrode–electrolyte interfaces we consider herein, quantum effects generally become important, and computationally demanding quantum mechanical calculations must be performed to explicitly treat the electrons in the system. As will be discussed in Section 25.4, modern approaches rely on *DFT* calculations, which offer a balance between computational efficiency and accuracy for the ground-state properties of most material systems. The second challenge to be addressed is that the thermodynamic averages we aim to compute require integrals to be carried out over the $6N$-dimensional phase space of the system. Because of this, attempts to directly compute the partition function in the thermodynamic limit where statistical mechanics becomes precise require the solution of an astronomically large number of integrals on the order of 10^{23}. We therefore regard the partition function to be an essentially unknowable quantity, and as will be discussed shortly, several methods have been developed to circumvent the need for its direct evaluation.[7]

One popular approach to obtaining the equilibrium properties of an atomistic system is to perform a *molecular dynamics* simulation, which evolves a collection of atoms through space and time by integrating Newton's equations of motion:

$$\boldsymbol{F}_i(t) = m_i \boldsymbol{a}_i(t) = -\nabla_{\boldsymbol{r}_i} \mathcal{V}(\boldsymbol{r}^N, t) \tag{25.29}$$

$$\boldsymbol{v}_i(t) = \int_{t_0}^{t} \boldsymbol{a}_i(s)\, ds + \boldsymbol{v}_i(t_0) \tag{25.30}$$

$$\boldsymbol{r}_i(t) = \int_{t_0}^{t} \boldsymbol{v}_i(s)\, ds + \boldsymbol{r}_i(t_0), \tag{25.31}$$

where $\boldsymbol{F}_i(t)$, $\boldsymbol{a}_i(t)$, $\boldsymbol{v}_i(t)$, and $\boldsymbol{r}_i(t)$ are the time-dependent force, acceleration, velocity, and position of atom i, respectively. Ensemble averages are then approximated as a time average across the trajectory that the system takes through phase space after the system has been adequately equilibrated.[8] The time-dependent potential energy of the system $\mathcal{V}(\boldsymbol{r}^N, t)$ that appears in Eq. (25.29) may be defined *ab initio* using DFT as we have discussed above or may be approximated using a suitably parametrized model *force field*. Because of their analytic form, the use of force fields considerably extends the length and time scales that may be considered in a molecular dynamics simulation, thereby enabling the simulation of a broader range of phenomena. However, the use of force fields becomes challenging when modeling electrochemical interfaces since the electronic degrees of freedom in the system must be accounted for to accurately model the effects of an applied electrode potential. A number of efforts have been made recently to effectively describe the behavior of electrons within classical force fields, resulting in the development of polarizable force fields, charge-optimized many-body force fields, and reactive force fields that can capture the effects of instantaneous molecular dipole formation, the variation in atomic oxidation states, and the formation and breaking of chemical bonds, respectively [16–18]. Unfortunately, these approaches tend to be *ad hoc* by design and therefore

7 There are, however, some exceptional cases where the partition function can be written as a closed-form expression, such as the 1D and 2D Ising models that describe the magnetization of a lattice of binary spins. The latter are often regarded to be "toy models" in view of their simplicity.
8 The equivalence of time and ensemble averages is posited by the *ergodic hypothesis*, which states that over a long enough simulation time, the microstates associated with a certain energy become equally likely. Thus, over a long enough trajectory of an equilibrated system, the simulation samples the most probable microstates of a system that constitute a large fraction of the phase space volume, thereby approximating the ensemble average.

require extensive parametrization and validation work each time a new material system is considered. However, once parametrized, large simulations may be conducted with an accuracy that approaches that of ab initio molecular dynamics at a fraction of the computational cost. Nevertheless, in cases where force fields do not yet exist or are challenging to parametrize, ab initio molecular dynamics may still be performed to provide a rigorous description of the atomic and electronic degrees of freedom in the system. Of course, the downside to this gain in accuracy and rigor is a highly demanding calculation that may presently require hundreds of processors operating for weeks or even months at a time in order to obtain sufficiently accurate statistics.

While molecular dynamics simulations provide a perfectly valid basis for modeling materials, certain types of phenomena or properties may be challenging to study due to the limited time and length scale of a simulation. Fortunately, we are not restricted to using purely dynamical simulations to sample the microstates of a thermodynamic system. Alternatively, one may employ a *Monte Carlo* method, which facilitates the efficient computation of ensemble averages via a random sampling of the configurational space of a system. Monte Carlo approaches are deemed to be superior to dynamical simulations when studying infrequent events such as the binding of a substrate to an active site in enzyme catalysis or the penetration of radiation through dense media such as the shielding materials used in nuclear reactors. Similarly, Monte Carlo methods are useful for predicting properties that require a large degree of conformational or configurational sampling such as when determining the most probable structure of a protein based on how its underlying peptide chain may fold in on itself or the composition and structure of an alloy electrode surface at a given temperature, pressure, or electrode potential. In general, Monte Carlo methods excel in any application that requires the evaluation of large multidimensional integrals such as the integrals frequently encountered in statistical physics [19]. The utility of Monte Carlo methods stems from the fact that the accuracy of an estimator $\overline{A}_N$ for an expected value $\langle A \rangle$ improves systematically as the number of independent random samples N drawn from a probability distribution P approaches infinity as a result of the *law of large numbers* and similarly that the statistical error $\Delta_{\overline{A}_N} = (\langle \overline{A}_n^2 \rangle - \langle A \rangle^2)^{1/2}$ that describes the standard deviation of the estimator for the expected value decreases as $N^{-1/2}$ as a consequence of the *central limit theorem.* In other words, we can always improve the accuracy and precision of our estimate for an expected value by drawing more random samples so long as the samples are independent and identically distributed. This is at variance with conventional numerical quadrature techniques, which require the integration domain to be discretized into a mesh of uniformly spaced points at which a function is evaluated. The total number of points in this mesh grows exponentially as N^d, where N is the number of points along each dimension and d is the number of dimensions to be considered. Because the accuracy and precision of quadrature techniques improve with increasingly dense meshes, high-dimensional systems become prohibitively expensive to treat due to the exponential growth in computational effort. Compared to analogous Monte Carlo integration schemes, achieving systematic improvements in the accuracy and precision of integral estimates via numerical quadrature methods can be challenging. While a number of Monte Carlo algorithms exist in the literature, we will concern ourselves with just one approach in this work that is known widely as the Metropolis algorithm. We will provide a more detailed discussion of this approach in Section 25.4.3.

25.4 The Quantum-Continuum Approach

25.4.1 Overview

In Section 25.3, a thermodynamic framework that accounts for the effects of an electrode potential on the state of a thermodynamic system was introduced and discussed at a high level. The connection between the net electronic charge and the microscopic properties of the system was underscored, and notable quantities such as the differential capacitance of an electrode–electrolyte interface were defined in terms of statistical fluctuations of unconstrained extensive quantities. The latter quantity is of particular interest in interfacial electrochemical modeling since it describes the response of the electrode surface charge to variations in the applied electrode potential. Physically, these charge variations may occur as the result of either *faradaic* or *non-faradaic* processes that may be measured experimentally via a suitable electroanalytical technique. In a faradaic process, electronic charge is transferred across the electrode–electrolyte interface to participate in electrochemical reactions, whereas in a non-faradaic process, electronic charge accumulates along the electrode surface, and the structure of the electrolyte in the vicinity of the interface adjusts in response to the excess charge on the electrode surface. In the limit where non-faradaic processes prevail, the electrode is referred to as an *ideally polarizable electrode*, and the electrode–electrolyte interface behaves as a capacitor where electronic charge stored on the electrode surface is compensated by a buildup of ionic charge within the electrolyte near the surface. In the opposite limit where faradaic current dominates, the electrode is referred to as an *ideally non-polarizable electrode*, and no capacitive charging along the interface takes place. Generally, real electrodes exhibit properties somewhere in between these two extremes, and we can take these to be idealized limits; however in certain cases, near-ideal polarizability can be observed in certain voltage windows that are referred to as *double layer ranges*. It is evident, however, that in order to understand the charge–voltage response of electrodes, we must first understand the structure and properties of what is referred to as the *electric double layer* (EDL), which is composed of the charged electrode surface and the structured electrolyte near the surface.

As depicted schematically in Figure 25.5, the EDL generally consists of a *compact layer* and a *diffuse layer*. Within the compact layer, water and specifically adsorbed ions with broken hydration shells may be present proximal to the surface forming what is referred to as an *inner Helmholtz plane*, and hydrated ions and water may be present above this forming what is referred to as an *outer Helmholtz plane*, or *Stern layer*. The ions present within the compact layer are generally regarded to be immobile and form planes of charge that are situated

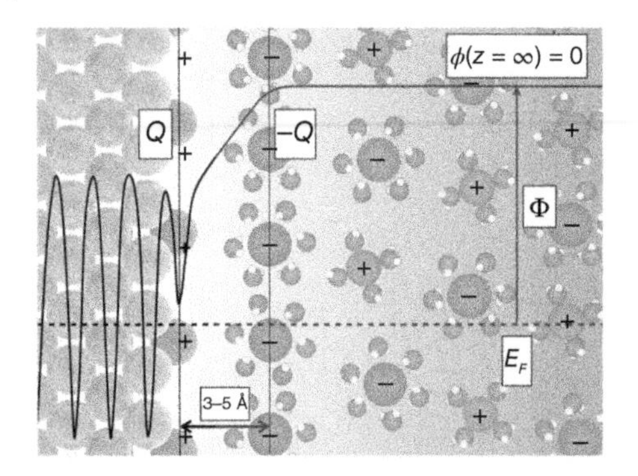

Figure 25.5 The electric double layer consists of a charged electrode surface and often a combination of a compact layer of ions adjacent to the surface and a diffuse layer of ions that extend into the bulk of the electrolyte. The electric double layer screens the interfacial electric field generated by the charged surface, as depicted by the electrostatic potential of the interface shown as the solid black curve. The converged potential $\phi(z = \infty)$ is aligned to zero by convention in the bulk of the electrolyte to simplify the computation of the voltage to $\Phi = -E_F$, where E_F is the Fermi level of the electrode denoted by the dashed line.

at a distance that is on the order of the ionic radius or hydrated ionic radius from the surface (or roughly 3–5 Å). Unlike the ions in the compact layer, the ions present in the diffuse layer are fairly mobile and form a space charge region that can extend approximately 10–100 nm into the bulk of the electrolyte [20]. This length scale is problematic for first principles calculations, however, since it would require an immense number of water molecules and ions to be considered introducing an astronomical number of degrees of freedom into the calculation. As will be discussed shortly, it is a common practice to employ a multiscale modeling approach in which only a small portion of the calculation domain is treated quantum mechanically to retain sufficient accuracy, while the remainder of the calculation domain is treated in a simplified manner to greatly improve the speed of the calculation. In the examples that will be discussed below, we will see how an embedded polarizable continuum model can be employed to simplify the treatment of the electrode–electrolyte interface. The latter forms the basis of the *quantum-continuum approach* to first principles electrochemical modeling.

25.4.2 Electric Double Layer (EDL) Models

To date, several models with varying complexity have been proposed to describe the distribution of ionic charge within an aqueous electrolyte. In order to preface our discussion of the main EDL models and how they describe the charge–voltage response of model electrochemical interfaces, it is necessary to first discuss how the voltage on an electrode is defined in this context. As shown in Figure 25.5, the voltage Φ on an electrode is defined as the work required to transfer an electron from the electrode to the bulk of the electrolyte $\Phi = -e_0\phi(z = \infty) - E_F$, where $-e_0$ is the charge of an electron, $\phi(z = \infty)$ is the electrostatic potential of the interfacial system far from the electrode surface in the bulk of the electrolyte, and

E_F is the Fermi level (or chemical potential) of the electrons in the electrode. In practice, we align the potential to be zero in the bulk of the electrolyte so that the voltage can be simply computed as $\Phi = -E_F$, where the Fermi level can be readily obtained for a model electrode via quantum-continuum calculations. Here we note that the latter is the definition of the voltage on an absolute scale and that it is necessary for practical reasons to report this voltage on a relative scale to a well-established reference electrode, such as the standard hydrogen electrode (SHE). This can be achieved by subtracting 4.44 V, which is an estimated value for the absolute SHE potential, or alternatively by aligning the *potential of zero charge* (PZC) of the model electrode to an experimentally measured PZC [14, 15, 21, 22]. Here, the PZC is the voltage at which zero net charge exists on the electrode surface. In general, the PZC of an electrode is sensitive to the composition of the electrolyte it is in contact with. However, if the PZC is observed to have a concentration dependence in a given electrolyte, then specific adsorption effects may be prominent, and greater care must be taken in order to properly describe the electrode surface [20].

The earliest model introduced for the EDL, which is referred to as the *Helmholtz model,* treats the ionic countercharge as a plane of charge sitting a distance λ_H from the electrode surface. The electrode itself is modeled as having an infinite permittivity $\epsilon_M = \infty$, so the potential within the electrode is a constant ϕ_M. In effect, this model treats the electrode–electrolyte interface as a parallel plate capacitor where the two plates are separated by a dielectric medium. As shown in Figure 25.6, the electrostatic potential varies linearly between the two plates, and the interfacial electric field ($E = -d\phi/dz$) is completely screened ($E = 0$) by the ionic countercharge at $z = \lambda_H$, leading to the following piecewise dependence of the electrostatic potential:

$$\phi(z) = \begin{cases} \phi_M = 4\pi\sigma\lambda_H/\epsilon_S, & z < 0 \\ 4\pi\sigma(\lambda_H - z)/\epsilon_S, & 0 \leq z < \lambda_H \\ 0, & \lambda_H \leq z \end{cases} \tag{25.32}$$

where $z = 0$ defines the top of the metal surface, σ is the surface charge density of the metal, and ϵ_S is the permittivity of the solvent layer separating the electronic and ionic charge.

The electrode potential is therefore modeled to vary linearly with the surface charge density $\Phi = -4\pi\sigma\lambda_H/\epsilon_S$, and as depicted in Figure 25.7, the differential capacitance is a constant parametrized by the Helmholtz layer thickness and the dielectric permittivity:

$$C_H = \frac{d\sigma}{d\Phi} = \frac{\epsilon_S}{4\pi\lambda_H} \tag{25.33}$$

Shortly after the introduction of the Helmholtz model, it was quickly realized that it provided an incomplete picture of the EDL as real electrodes exhibit a more complex charge–voltage response that leads to a variable capacitance that passes through a minimum at the PZC of the electrode. The Gouy–Chapman model solves some of these deficiencies by considering the ionic countercharge to be composed of a diffuse layer of ions. In this framework, the ions are assumed to be

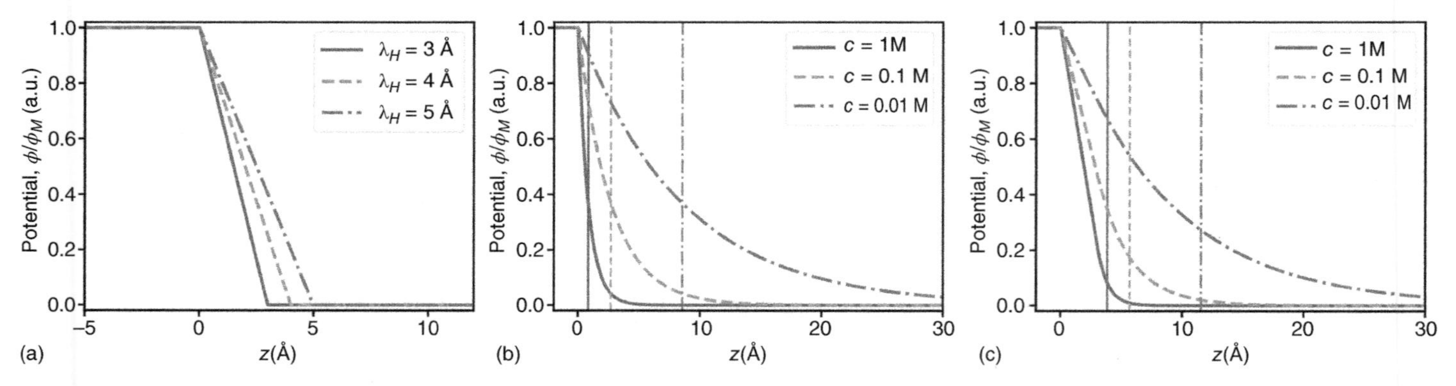

Figure 25.6 Electrostatic potential profiles for the (a) Helmholtz, (b) Gouy–Chapman, and (c) Gouy–Chapman–Stern models plotted on a relative potential scale. The vertical lines indicate the Debye lengths for 1 M, 0.1 M, and 0.01 M symmetric electrolytes with monovalent ions.

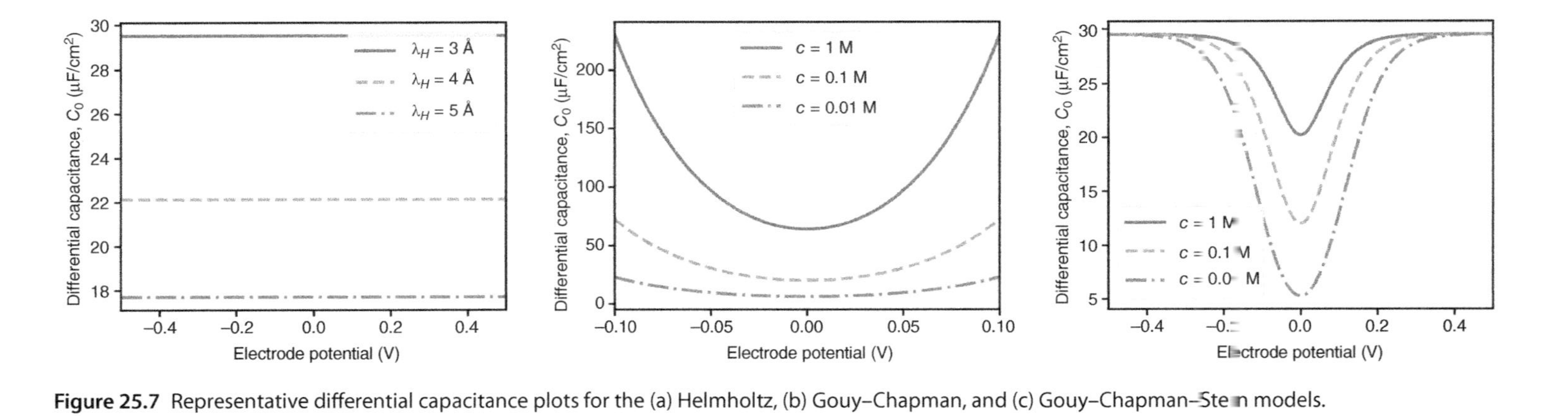

Figure 25.7 Representative differential capacitance plots for the (a) Helmholtz, (b) Gouy–Chapman, and (c) Gouy–Chapman–Stern models.

point particles that are distributed within the electrolyte according to Boltzmann statistics, leading to a diffuse layer charge density:

$$\rho_D(z) = z_D c_D[\exp(-\beta z_D \phi(z)) - \exp(\beta z_D \phi(z))] \tag{25.34}$$

where z_D is the valence of the ions (for now we assume symmetric electrolytes with ion charges of $\pm z_D$) and c_D is the concentration of the ions. Here we see that the ionic charge distribution depends upon the interfacial potential, which is generally not known *a priori*. In order to obtain the potential (and therefore the ionic charge distribution), we can insert the Boltzmann distribution density $\rho_D(z)$ into a Poisson equation and solve the resulting *Poisson–Boltzmann equation*:

$$\frac{d^2\phi}{dz^2}(z) = -4\pi\frac{\rho_D(z)}{\epsilon_S} = \frac{8\pi z_D c_D}{\epsilon_S}\sinh(\beta z_D \phi(z)) \tag{25.35}$$

By integrating this equation with the boundary conditions $\phi(z = 0) = \phi_M$ and $d\phi/dz \rightarrow 0$ as $z \rightarrow \infty$, we obtain the derivative of the potential as

$$\frac{d\phi}{dz}(z) = -\left(\frac{32\pi c_D}{\beta\epsilon_S}\right)^{\frac{1}{2}}\sinh\left(\frac{\beta z_D \phi(z)}{2}\right) \tag{25.36}$$

where we note that the potential and its derivative have opposite signs, indicating that the potential is correctly being screened by the diffuse ionic charge density. By employing Gauss' law ($\oiint \vec{E}\cdot d\vec{S} = 4\pi Q/\epsilon$), we can show that the magnitude of the electric field is proportional to the surface charge density as $d\phi/dz = 4\pi\sigma/\epsilon_S$, allowing us to express the surface charge density at $z = 0$ as

$$\sigma = -\left(\frac{2\epsilon_S c_D}{\pi\beta}\right)^{\frac{1}{2}}\sinh\left(\frac{\beta z_D \phi_M}{2}\right) \tag{25.37}$$

From here, we can obtain the differential capacitance (Figure 25.7) of the diffuse ionic countercharge as

$$C_D = \frac{d\sigma}{d\Phi} = C_{D,PZC}\cosh\left(\frac{\beta z_D \Phi}{2}\right) \tag{25.38}$$

where $C_{D,PZC}$ is the differential capacitance at the PZC of the electrode and the absolute voltage in this model is taken to be $\Phi = -\phi_M$. The former quantity can be expressed as

$$C_{D,PZC} = \left(\frac{\epsilon_S \beta c_D z_D^2}{2\pi}\right)^{\frac{1}{2}} = \frac{\epsilon_S}{4\pi\lambda_D} \tag{25.39}$$

where we have introduced the *Debye length* $\lambda_D = (\epsilon_S/8\pi\beta c_D z_D^2)^{1/2}$, which describes the characteristic screening length of the diffuse countercharge that is sensitive to both the composition and concentration of the bulk electrolyte. Furthermore, by integrating Eq. (25.36), it can be shown that the potential (Figure 25.6) in the system decays exponentially into the electrolyte,

$$\phi(z) = \phi_M \exp\left(-\frac{z}{\lambda_D}\right) \tag{25.40}$$

and that the Debye length is the distance at which the potential has decayed by a factor of $1/e$. Therefore, as the concentration or valence of the ions is increased, the differential capacitance of the interface increases, while the Debye length shrinks. Similarly, the capacitance increases at potentials both above and below the PZC as shown in Figure 25.7, providing an enhanced qualitative agreement with experimental measurements.

The Gouy–Chapman model is successful to some extent in that it predicts a minimum in the differential capacitance at the PZC; however the model becomes less applicable at modest electrode potentials due to the predicted exponential growth in the capacitance, a feature that is clearly at variance with experimental results. The model also fails in situations where ions become specifically adsorbed to the surface since the point-like ions are free to approach infinitely close to the electrode surface. These issues were rectified by Stern who considered the EDL to contain both Helmholtz and Gouy–Chapman layers. In the Gouy–Chapman–Stern model, the interfacial capacitance is modeled as two capacitors placed in series, leading to the overall capacitance:

$$C_S = \left(\frac{1}{C_H} + \frac{1}{C_D}\right)^{-1} = \frac{\epsilon_S}{4\pi}\left(\lambda_H + \frac{\lambda_D}{\cosh(z_d \beta \Phi/2)}\right)^{-1} \tag{25.41}$$

where C_H and C_D are the Helmholtz and Gouy–Chapman capacitance, respectively. As shown in Figure 25.7, the capacitance exhibits a minimum about the PZC of the electrode and attains a finite value at electrode potentials higher and lower than the PZC in closer agreement with measurements. The electrostatic potential of the Gouy–Chapman–Stern model is depicted in Figure 25.6, where we see that the model behaves similarly to the Helmholtz model close to the surface where the potential decays linearly but then begins to decay exponentially at the interface of the compact and diffuse layers adopting a Gouy–Chapman-like response.

25.4.3 Example: Silver Monolayer Stripping on Au(100)

In this section, we demonstrate how the quantum-continuum approach may be applied to model interfacial electrochemical phenomena using plane-wave DFT and a polarizable continuum model [23]. As an example, we consider the electrochemical stability of an atomically thin layer of silver on the gold (100) surface [15]. The calculations that will be described in this section are performed in a periodic cell and consist of a model gold (100) surface that is composed of seven atomic layers, where the adsorbed silver atoms are included symmetrically on top of the exterior layers of gold. This type of model is commonly referred to as a *symmetric slab*, and we would say that the gold surface is modeled within the *slab-supercell approximation*. The geometry of a typical cell is shown in Figure 25.8, along with the electrostatic potential profile of a silver-covered slab with several different surface charges.

In these calculations, the electrode surface is embedded in a polarizable dielectric cavity that enables the description of solvation effects in a computationally

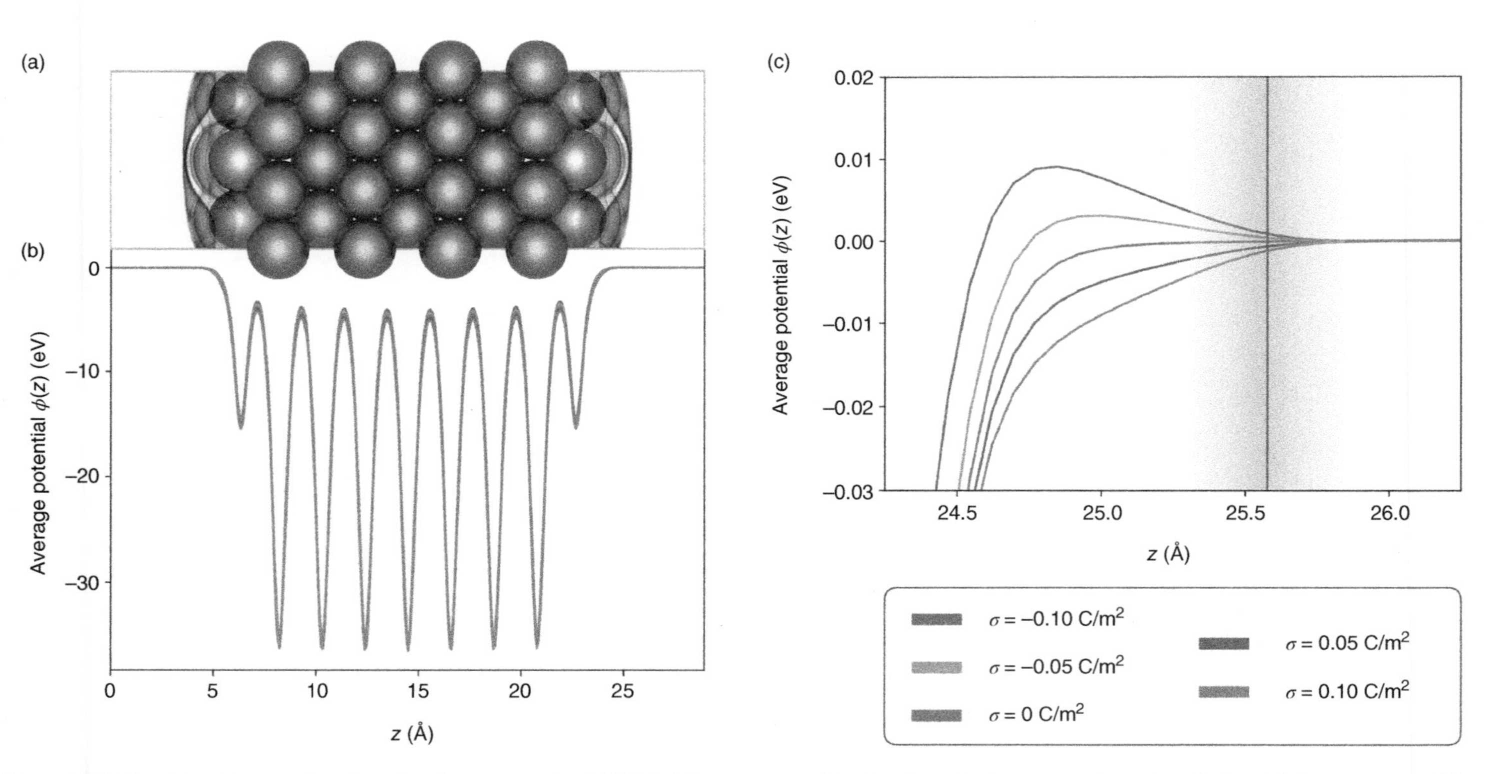

Figure 25.8 The finite charges placed on the silver-covered gold (100) slab are screened by the planar ionic countercharge in solution. (a) The response of the continuum dielectric at the cavity interface is visualized. Positive/negative polarization charges are shown in red/blue. (b) The electrostatic potential of the electrified slabs is aligned to zero at the edges of the supercell. (c) The Helmholtz layer placed 3 Å from the silver monolayer fully screens the surface charge. Source: Weitzner and Dabo 2017 [15]. Reproduced with permission of American Physical Society. (See online version for color figure).

efficient manner.[9] Specifically, the mutual interaction of the slab and the continuum dielectric induces the dielectric medium and the electron density to polarize along the interface of the cavity (as shown in Figure 25.8a), which mimics the interaction of the metal surface and the dipoles of interfacial water [23]. We additionally consider the electrolyte to be sufficiently concentrated so that the ionic countercharge can be effectively modeled as a Helmholtz layer. The planar ionic countercharge is positioned 3 Å from the electrode surface and fully screens the surface charge density as shown in Figure 25.8c.

The electrochemical stability of silver monolayers on single-crystal gold surfaces has been thoroughly studied experimentally. For example, cyclic voltammetry measurements of silver monolayer stripping on the gold (100) surface indicate that the silver monolayer desorbs in three distinct stages, with desorption peaks at 1.25 V, 0.93 V, and 0.72 V vs. the SHE in a solution of 5×10^{-3} M Ag_2SO_4 + 0.5 M H_2SO_4 at $T = 298$ K [25]. In order to gain insight into how the presence of surface charge stabilizes the silver monolayer on the surface, we can simulate the adsorption process under different environmental conditions. To begin, we must consider the following electrochemical reaction:

$$Ag^+ + e^- + * \leftrightharpoons Ag^* \tag{25.42}$$

where $*$ is an available surface site and Ag^* indicates an adsorbed silver atom on the gold surface. At equilibrium, we have

$$\mu_{Ag^+} - e_0\Phi = \mu_{Ag^*}(\Phi, \theta) \tag{25.43}$$

where we note that the equilibrium chemical potential of the adsorbed silver atoms $\mu_{Ag^*}(\Phi, \theta)$ has an explicit dependence on the applied electrode potential Φ and silver surface coverage θ. Because the silver atoms may adopt an inordinate number of configurations for a given coverage, it is necessary to sample a number of these configurations to gain a sense of how key thermodynamic properties of the interface may vary. For example, by sampling a series of configurations with neutral surface charge, we may directly probe the variation in the PZC of the electrode as a function of surface coverage, as shown in Figure 25.9a.

Here we see the general trend that as the surface coverage increases, the PZC initially begins to shift toward negative potentials, but then begins to increase again after the silver surface coverage exceeds 50%. We also observe that surfaces with compact island-like silver configurations tend to have more positive PZCs as compared with surfaces with dispersed configurations. This indicates that dispersed configurations are accompanied with a larger change in the surface dipole upon monolayer formation as compared with compact configurations. The effect of this is most readily observed in the differential capacitance, which is shown in Figure 25.9b for Helmholtz layer thicknesses of 3 and 5 Å. As anticipated, we observe that increasing the Helmholtz layer thickness decreases the capacitance in agreement with Eq. (25.33). Here we also observe the trend that configurations with lower PZCs tend to have higher capacitance. The origin of this relationship

9 Specifically, these calculations were performed with the plane-wave DFT code PWscf that is implemented in Quantum ESPRESSO and the *self-consistent continuum solvation model* that is implemented in the Environ module, which serves as a plugin to the PWscf code [23, 24].

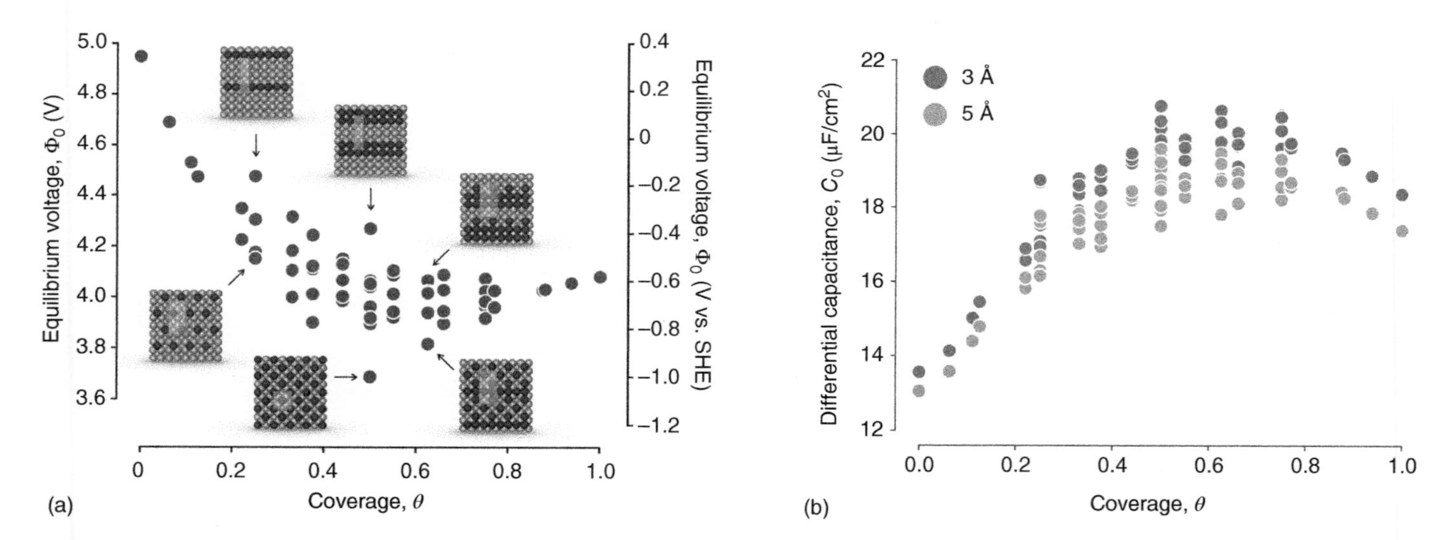

Figure 25.9 (a) Variation in PZCs of silver-covered Au(100). (b) Variation in the Helmholtz differential capacitance of a variety of silver monolayer configurations on an Au(100) surface. Source: Weitzner and Dabo 2017 [15]. Reproduced with permission of American Physical Society. (See online version for color figure).

between PZC shift and capacitance enhancement is due to the charge transfer that takes place between the silver adatoms and the gold surface. In general, larger extents of charge transfer occur for dispersed silver adlayers as compared with condensed adlayers, which is additionally associated with an enhancement in the density of states near the Fermi level of the electrode. The latter enhancement is important as it leads to a concomitant increase in the *quantum capacitance* of the electrode, which is typically on the order of 100–1000 μF/cm^2 (as compared with 30 μF/cm^2 for electrolytes) for transition metals. Because the overall capacitance of the interface depends on the capacitance of the electrode and the capacitance of the electrolyte $C_{tot} = (1/C_Q + 1/C_H)^{-1}$, enhancements in the quantum capacitance increase the overall capacitance of the interface.

Another key thermodynamic property to be assessed is the binding energy of the monolayer per surface site as a function of the silver adlayer coverage. Here we would like to point out that binding energies are enthalpies and not free energies as they do not include entropic contributions, which, as we saw previously, requires knowledge of the partition function. Furthermore, understanding how the electrode potential affects binding energies is a prerequisite for determining the equilibrium coverage at a given electrode potential. With this said, we can calculate the binding energy E_B per site of a neutral symmetric slab as

$$E_B = \frac{1}{2N_{sites}}[E(N_{Ag^*}) - E_{clean} - N_{Ag^*}\mu^\circ_{Ag}] \tag{25.44}$$

where N_{sites} is the total number of sites on one face of the symmetric slab, $E(N_{Ag^*})$ is the total energy of the neutral silver-covered slab with N_{Ag^*} adsorbed silver atoms, E_{clean} is the total energy of the pristine gold (100) slab, and μ°_{Ag} is the chemical potential of silver in its standard reference state as a bulk metal. A charge-dependent binding energy can then be obtained by Taylor expanding the neutral binding energy to second order:

$$E_B(Q) = E_B + \left(\frac{\partial E_B}{\partial Q}\right)_{Q=0} Q + \frac{1}{2}\left(\frac{\partial^2 E_B}{\partial Q^2}\right)_{Q=0} Q^2 = E_B + \Phi_0 Q + \frac{Q^2}{2AC_0} \tag{25.45}$$

where, by definition, the first- and second-order coefficients are the PZC and the inverse of the differential capacitance. We can subsequently convert the charge-dependent binding energy of each adlayer configuration to an electrochemical enthalpy by the Legendre transform $F = E_B(Q) - \Phi Q$, where $Q = AC_0(\Phi - \Phi_0)$ and the PZC Φ_0 in this expression is configuration dependent. In principle, the differential capacitance could also be configuration dependent; however we will just consider the case where it is a constant parameter for all configurations to understand how its value influences the overall thermodynamics of the system. As shown in Figure 25.10a, if we set the capacitance to 0 μF/cm^2, which is equivalent to considering each surface with a neutral charge, we find that the binding energies are insensitive to the applied electrode potential.

Furthermore, because DFT treats electronic systems at 0 K, only the lowest energy states or *ground states* would be available to the surface at this temperature. The lower convex hull or lower envelope formed by these ground

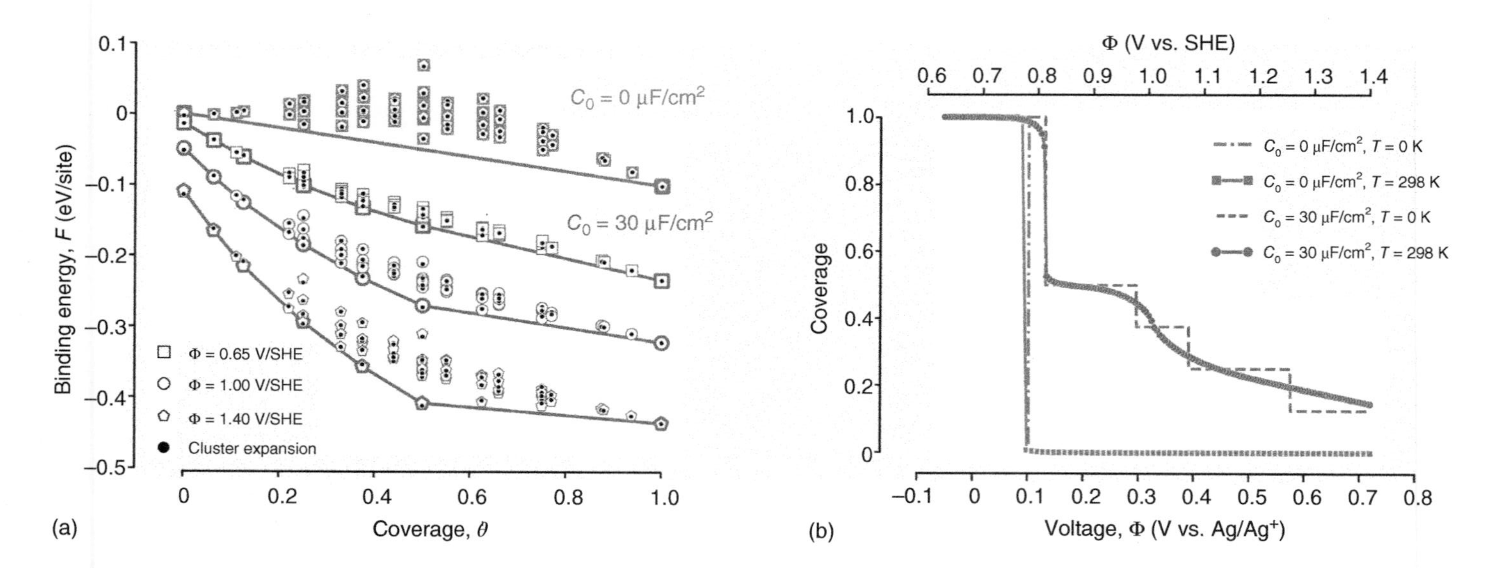

Figure 25.10 (a) Variation in the silver monolayer binding energies on Au(100) as a function of silver coverage and applied electrode potential for $\Phi = 0.65$ V vs. SHE (squares), $\Phi = 1.00$ V vs. SHE (circles), and $\Phi = 1.40$ V vs. SHE (pentagons). (b) Silver adsorption isotherms computed for $C_0 = 0\ \mu F/cm^2$ and $C_0 = 30\ \mu F/cm^2$ for $T = 0$ K (dashed lines) and $T = 298$ K (solid lines with markers). The 298 K results were obtained via grand canonical Monte Carlo simulations. Source: Weitzner and Dabo 2017 [15]. Reproduced with permission of American Physical Society.

states shown as the solid line in Figure 25.10a indicates that only two stable states exist in this case: the clean gold (100) surface and the same surface with a full silver monolayer. All of the configurations with intermediate coverages are higher in energy and are therefore inaccessible to the system at 0 K. If we alternatively consider the capacitance to be 30 μF/cm^2, we see distinctly different behavior. In this case, the binding energies become potential dependent, and we see that several new structures with intermediate coverages migrate to the convex hull, indicating that more configurations with sub-monolayer coverage become stable in the presence of finite surface charge than compared with the neutral case.

In the interest of determining the equilibrium coverage as a function of applied potential, i.e. the *adsorption isotherm* of silver on the gold (100) surface, it is necessary to consider the chemical potential of the adsorbed silver. At 0 K, this is a straightforward task since the coverage-dependent chemical potential can be determined directly by computing the derivative of the convex hull of the binding energy data. The equilibrium surface coverage can then be obtained for a range of electrode potentials through the equilibrium condition defined by Eq. (25.43). As shown in Figure 25.10b, if we consider the 0 μF/cm^2 case, the silver monolayer is predicted to desorb in a single step at an electrode potential of 0.78 V vs. the SHE. This is in line with what was found with the binding energy data since the only ground states present were the clean and fully covered surface. Alternatively, for the 30 μF/cm^2 case, we find that the silver monolayer is predicted to desorb over the course of several steps as suggested by the higher number of ground states present on the voltage-sensitive convex hulls. While it is useful to consider 0 K results to gain a general understanding of adsorption trends, it is considerably more interesting to study adsorption behavior at room temperature since experiments are usually carried out under ambient conditions. This can be achieved by employing a force field or an analogous model that can be trained to predict the binding energies determined from quantum-continuum calculations. In this example, we have employed a *cluster expansion* approach that models configuration-dependent quantities as a series expansion of interacting sites or *clusters* on a lattice [15, 26]. Using the cluster expansion approximation for the electrochemical enthalpies, grand canonical Monte Carlo simulations can be performed to rapidly sample the configurational space and to subsequently estimate the average surface coverage at a fixed ion chemical potential, temperature, and electrode potential. In these simulations, a lattice of adsorption sites is modeled to be in contact with a reservoir of silver ions, along with a potentiostat and a thermostat that the system can exchange silver atoms, electronic charge, and thermal energy with. A trajectory is generated by randomly proposing new configurations through a series of trial moves and accepting or rejecting them according to the criteria set forth by the *Metropolis algorithm* [19, 27]. A trial move would consist of selecting a lattice site at random and adding (removing) a silver atom if the site is empty (occupied). The electrochemical enthalpy of the newly proposed configuration is computed, which is then used to compute the relative Boltzmann probability of the proposed configuration (P_p) to the current configuration (P_c):

$$\frac{P_p}{P_c} = \frac{\exp[-\beta(F_p - N_p\mu_{\mathrm{Ag^+ + e^-}})]}{\exp[-\beta(F_c - N_c\mu_{\mathrm{Ag^+ + e^-}})]} = \exp[-\beta(\Delta F - \Delta N\mu_{\mathrm{Ag^+ + e^-}})] \quad (25.46)$$

where $\mu_{\mathrm{Ag^+ + e^-}} = \mu_{\mathrm{Ag^+}} - e_0\Phi$ is the coupled chemical potential of the silver ions in solution and electrons in the electrode. The trajectory is updated at each step of the simulation by drawing a random number from a uniform probability distribution over the range [0, 1]. If the random number is less than or equal to the probability ratio, the newly proposed configuration is accepted and added to the trajectory as the next configuration; otherwise, the proposed configuration is rejected and the current configuration is added in its place.[10] The average coverage can then be computed as

$$\langle\theta\rangle = \frac{1}{N}\sum_{i=1}^{N}\theta_i \quad (25.47)$$

where N is the number of steps over which the average is being computed and θ_i is the coverage of configuration i in the trajectory. The adsorption isotherms for the 0 and 30 μF/cm^2 cases computed at 298 K are plotted on top of the 0 K isotherms in Figure 25.10b. Here we observe that at finite temperature, the 0 μF/cm^2 isotherm still predicts the monolayer to desorb in a single step, while several of the predicted steps in the 30 μF/cm^2 case have been smoothed to reveal an overall two-step desorption process. The finite capacitance case is more closely aligned with the experimental voltammetry that revealed a three-step desorption process. While the first two peak positions of the voltammetry are predicted fairly well with this model, we find that the presence of the third peak cannot be attributed to surface charge effects alone. A likely explanation for this may be found in Figure 25.9, where we observe that the PZC of the gold (100) surface with low silver surface coverage is predicted to be less than 0.4 V vs. the SHE. Therefore, at electrode potentials positive of this, the silver-covered surface is predicted to have a net positive charge. The presence of a positive surface charge would provide a driving force for the co-adsorption of anions such as the bisulfate or sulfate anions present in the electrolyte. Although this is a feature that has not been presently considered in this example, it is highly anticipated to improve the agreement of the predictions with voltammetric results.

Acknowledgments

Ismaila Dabo acknowledges primary financial support from the US Department of Energy, Office of Science, Basic Energy Sciences, CPIMS Program, under Award No. DE-Sc2518646. A portion of this work was performed under the auspices of the US Department of Energy by Lawrence Livermore National Laboratory under Contract DE-AC52-07NA27344.

10 It is crucial in the event of a rejection to recycle the previous configuration to ensure the configurations are sampled according to the Boltzmann distribution [19, 27].

References

1 Greeley, J., Stephens, I., Bondarenko, A. et al. (2009). Alloys of platinum and early transition metals as oxygen reduction electrocatalysts. *Nat. Chem.* 1 (7): 552.

2 Seh, Z.W., Kibsgaard, J., Dickens, C.F. et al. (2017). Combining theory and experiment in electrocatalysis: insights into materials design. *Science* 355 (6321): eaad4998.

3 Hayden, B.E. (2013). Particle size and support effects in electrocatalysis. *Acc. Chem. Res.* 46 (8): 1858–1866.

4 Li, L., Larsen, A.H., Romero, N.A. et al. (2012). Investigation of catalytic finite-size-effects of platinum metal clusters. *J. Phys. Chem. Lett.* 4 (1): 222–226.

5 Liu, P. and N& oslash;rskov, J.K. (2001). Ligand and ensemble effects in adsorption on alloy surfaces. *Phys. Chem. Chem. Phys.* 3 (17): 3814–3818.

6 Mavrikakis, M., Hammer, B., and N& oslash;rskov, J.K. (1998). Effect of strain on the reactivity of metal surfaces. *Phys. Rev. Lett.* 81 (13): 2819.

7 Slanac, D.A., Hardin, W.G., Johnston, K.P., and Stevenson, K.J. (2012). Atomic ensemble and electronic effects in Ag-rich AgPd nanoalloy catalysts for oxygen reduction in alkaline media. *J. Am. Chem. Soc.* 134 (23): 9812–9819.

8 Watanabe, M. and Motoo, S. (1975). Electrocatalysis by ad-atoms: Part III. enhancement of the oxidation of carbon monoxide on platinum by ruthenium ad-atoms. *J. Electroanal. Chem. Interfacial Electrochem.* 60 (3): 275–283.

9 Greeley, J. and Mavrikakis, M. (2004). Alloy catalysts designed from first principles. *Nat. Mater.* 3 (11): 810.

10 Greeley, J., Jaramillo, T.F., Bonde, J. et al. (2011). Computational high-throughput screening of electrocatalytic materials for hydrogen evolution. In: *Materials for Sustainable Energy: A Collection of Peer-Reviewed Research and Review Articles from Nature Publishing Group* (ed. Vincent Dusastre), 280–284. World Scientific.

11 Kim, Y.-T., Lopes, P.P., Park, S.-A. et al. (2017). Balancing activity, stability and conductivity of nanoporous core–shell iridium/iridium oxide oxygen evolution catalysts. *Nat. Commun.* 8 (1): 1449.

12 Callen, H.B. (1985). *Thermodynamics and an introduction to thermostatistics* 2e. John Wiley & Sons. ISBN-13: 978-0471862567.

13 Chandler, D. (1987). *Introduction to modern statistical mechanics* 1e. Oxford University Press. ISBN-13: 978-0195042771.

14 Weitzner, S.E. and Dabo, I. (2019). Voltage effects on the stability of Pd ensembles in Pd–Au/Au (111) surface alloys. *J. Chem. Phys.* 150 (4): 041715.

15 Weitzner, S.E. and Dabo, I. (2017). Voltage-dependent cluster expansion for electrified solid–liquid interfaces: application to the electrochemical deposition of transition metals. *Phys. Rev. B* 96 (20): 205134.

16 Warshel, A., Kato, M., and Pisliakov, A.V. (2007). Polarizable force fields: history, test cases, and prospects. *J. Chem. Theory Comput.* 3 (6): 2034–2045.

17 Yu, J., Sinnott, S.B., and Phillpot, S.R. (2007). Charge optimized many-body potential for the Si/SiO_2 system. *Phys. Rev. B* 75 (8): 085311.

18 Van Duin, A.C., Dasgupta, S., Lorant, F., and Goddard, W.A. (2001). ReaxFF: a reactive force field for hydrocarbons. *J. Phys. Chem. A* 105 (41): 9396–9409.

19 Landau, D.P. and Binder, K. (2014). *A Guide to Monte Carlo Simulations in Statistical Physics*. Cambridge University Press.

20 Bard, A.J., Faulkner, L.R., Leddy, J., and Zoski, C.G. (1980). *Electrochemical Methods: Fundamentals and Applications*, vol. 2. New York: Wiley.

21 Trasatti, S. (1986). The absolute electrode potential: an explanatory note (recommendations 1986). *Pure Appl. Chem.* 58 (7): 955–966.

22 Weitzner, S.E. and Dabo, I. (2017). Quantum–continuum simulation of underpotential deposition at electrified metal–solution interfaces. *NPJ Comput. Mater.* 3 (1): 1.

23 Andreussi, O., Dabo, I., and Marzari, N. (2012). Revised self-consistent continuum solvation in electronic-structure calculations. *J. Chem. Phys.* 136 (6): 064102.

24 Giannozzi, P., Baroni, S., Bonini, N. et al. (2009). Quantum espresso: a modular and open-source software project for quantum simulations of materials. *J. Phys. Condens. Matter* 21 (39): 395502.

25 Garcia, S., Salinas, D., Mayer, C. et al. (1998). Ag UPD on Au (100) and Au (111). *Electrochim. Acta* 43 (19–20): 3007–3019.

26 Sanchez, J.M., Ducastelle, F., and Gratias, D. (1984). Generalized cluster description of multicomponent systems. *Physica A* 128 (1–2): 334–350.

27 Metropolis, N., Rosenbluth, A.W., Rosenbluth, M.N. et al. (1953). Equation of state calculations by fast computing machines. *J. Chem. Phys.* 21 (6): 1087–1092.

26

Time-Dependent Density Functional Theory for Excited-State Calculations

Chi Yung Yam

Beijing Computational Science Research Center, ZPark II, No.10 East Xibeiwang Road, Haidian District, Beijing 100193, PR, China

26.1 Introduction

Density functional theory (DFT) [1, 2] is the popular method in chemistry, physics, materials science, and many other areas to determine the ground state of quantum systems and to evaluate related properties. The theory replaces an interacting many-body problem with a much simpler single-particle problem. An extension of ground-state DFT, time-dependent density functional theory (TDDFT), on the other hand, is a theoretical approach to dynamical quantum many-body systems. Similarly, the complicated many-body time-dependent Schrödinger equation is replaced by a set of time-dependent single-particle Kohn–Sham equations, which is done by mapping the real system of interacting electrons to a fictitious system of noninteracting particles whose orbitals yield the same time-dependent density $n(\mathbf{r}, t)$. The formal basis of TDDFT is the Runge–Gross theorem [3], which shows that for a system initially in its ground state subjected to a time-dependent perturbation, determines the potential up to an additive time-dependent function. While all the equations are formally exact, practically, TDDFT calculation requires an approximation for the exchange–correlation (XC) potential. One of the major ongoing challenges is the development of XC functionals. In principle, the exact XC potential is nonlocal both in space and time. The adiabatic approximation, which approximates the XC potential at point $\mathbf{r}$ and time t by that of the ground state with density $n(\mathbf{r}, t)$, however, works surprisingly well in many cases and is the most widely used approximation for the XC potential.

In the past decades, there is growing interest in the use of TDDFT in areas where the interaction of systems with time-dependent external fields is important. TDDFT is an efficient way to study excited-state properties as well as the real-time dynamics of many-body systems. It has become a routine tool for description of photoabsorption [4, 5], electron-ion dynamics [6–8], transport phenomena [9–11], and other excited-state properties of molecules, nanostructures, and solids. For photocatalysis, semiconducting materials have been generally used as catalysts, which perform as light absorbers and stimulate

Heterogeneous Catalysts: Advanced Design, Characterization and Applications, First Edition.
Edited by Wey Yang Teoh, Atsushi Urakawa, Yun Hau Ng, and Patrick Sit.

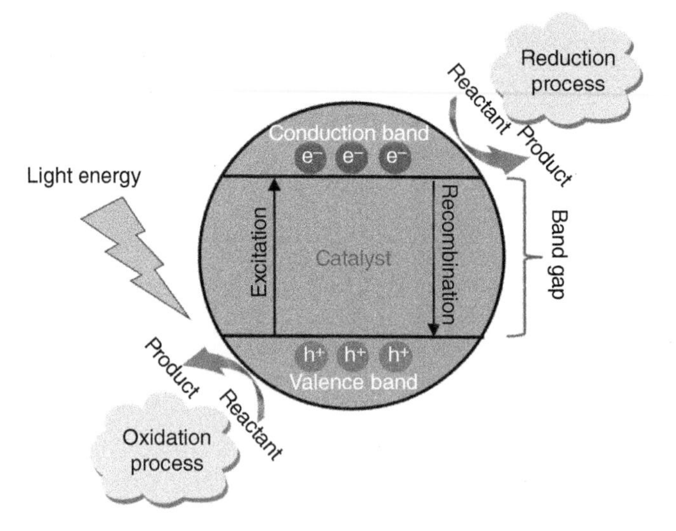

Figure 26.1 Schematic diagram of typical semiconductor photocatalytic mechanism.

redox processes. Figure 26.1 is a schematic representation of semiconductor photocatalytic mechanism. The fundamental steps in photocatalysis are as follows: (i) light is absorbed when its energy is equivalent to or larger than the bandgap of the semiconductor; (ii) excited electrons and holes are generated in the conduction band and valence band, respectively; (iii) valence band holes and conduction band electrons induce the corresponding redox reaction. From the theoretical point of view, absorption spectra of semiconductor catalysts can be simulated by solving the time-dependent Kohn–Sham (TDKS) equations. This chapter gives an overview of the time-dependent density functional theory and focuses largely on the fundamentals of the theory. The chapter is organized as follows. Section 26.2 presents the theoretical foundation of TDDFT, a Hohenberg–Kohn-like theorem derived by Runge and Gross in 1984 for the time-dependent Schrödinger equation. The TDKS equations can be solved in different ways. In the linear regime where the time-dependent potential is weak, excited-state properties can be obtained via a linear response treatment in frequency domain. Section 26.3 presents the linear response theory for the calculation of photoabsorption spectra. Section 26.4 discusses an approach where the TDKS equations are directly propagated in time domain, which is valid beyond the linear response regime. Section 26.5 discusses a class of computational approaches tailored to investigate photochemical and photophysical processes – nonadiabatic mixed quantum/classical dynamics methods. In these methods, a system is partitioned into two subsystems, where one is treated quantum mechanically and the other is treated classically. Combined with TDDFT for excited states, nonadiabatic mixed TDDFT/classical dynamics methods address a wider range of applications, including photochemical reactions [12, 13], strong-field ionization [14], high harmonic generation [15], quantum control [16], and the simulation of optical breakdown in semiconductors [17].

26.2 Theoretical Foundation of TDDFT

The central statement of TDDFT, the Runge–Gross theorem, proves that there exists a one-to-one correspondence between the time-dependent density and the external potential, $v_{ext}(\boldsymbol{r}, t)$ for many-body systems evolving from an initial ground state [3]. By solving the time-dependent Schrödinger equation with various potentials, one can obtain many-body wave functions and electron densities. The Runge–Gross theorem implies that one can uniquely determine the external potential for a given density. With the external potential, the time-dependent Schrödinger equation can be solved, and, thus, all observables of a quantum system can then be written as functionals of density only. Precisely, it states that for any two potentials $v(\boldsymbol{r}, t)$ and $v'(\boldsymbol{r}, t)$ that differ by more than a constant at any time $t \geq t_0$, the corresponding time-dependent densities are different. The proof can be divided into two parts. First, using the equation of motion for the current density, $\boldsymbol{j}(\boldsymbol{r}, t)$, one can show

$$\frac{\partial^{k+1}}{\partial t^{k+1}}[\boldsymbol{j}(\boldsymbol{r}, t) - \boldsymbol{j}'(\boldsymbol{r}, t)]\bigg|_{t=t_0} = n(\boldsymbol{r}, t_0)\,\nabla\,\frac{\partial^k}{\partial t^k}[v(\boldsymbol{r}, t) - v'(\boldsymbol{r}, t)]\bigg|_{t=t_0} \neq 0 \quad (26.1)$$

This demonstrates that $\boldsymbol{j}(\boldsymbol{r}, t)$ and $\boldsymbol{j}'(\boldsymbol{r}, t)$ will become different infinitesimally after t_0 since

$$\frac{\partial^k}{\partial t^k}[v(\boldsymbol{r}, t) - v'(\boldsymbol{r}, t)]\bigg|_{t=t_0} \neq \text{const} \quad (26.2)$$

Next, by using the current continuity equation, one can demonstrate that if two systems have different current densities, they must also possess different time-dependent densities:

$$\frac{\partial^{k+2}}{\partial t^{k+2}}[n(\boldsymbol{r}, t) - n'(\boldsymbol{r}, t)]\bigg|_{t=t_0} = -\nabla\cdot\left[n(\boldsymbol{r}, t_0)\,\nabla\,\frac{\partial^k}{\partial t^k}[v(\boldsymbol{r}, t) - v'(\boldsymbol{r}, t)]\bigg|_{t=t_0}\right] \quad (26.3)$$

It has been proven in Ref. [3] that it is impossible to have the right-hand side of Eq. (26.3) equals to zero on the entire $\boldsymbol{r}$ space. Thus, this proves that the time-dependent densities $n(\boldsymbol{r}, t)$ and $n'(\boldsymbol{r}, t)$ are different if the potentials $v(\boldsymbol{r}, t)$ and $v'(\boldsymbol{r}, t)$ are different.

Analogous to ground-state DFT, a TDKS scheme is constructed by introducing an auxiliary noninteracting system subjected to an effective local Kohn–Sham potential, v_{KS}. The effective potential is chosen such that the electron density of the Kohn–Sham system is the same as the density of a real interacting system. The system is governed by one-particle TDKS equations:

$$i\frac{\partial}{\partial t}\varphi_i(\boldsymbol{r}, t) = \left[-\frac{\nabla^2}{2} + v_{KS}(\boldsymbol{r}, t)\right]\varphi_i(\boldsymbol{r}, t) \quad (26.4)$$

where $\varphi(\boldsymbol{r}, t)$ are Kohn–Sham orbitals, and the density of the interacting system can be calculated from the occupied Kohn–Sham orbitals:

$$n(\boldsymbol{r}, t) = \sum_i^{\text{occ}} |\varphi_i(\boldsymbol{r}, t)|^2 \quad (26.5)$$

Similar to static DFT, the TDKS potential can be written as

$$v_{\rm KS}[n](\boldsymbol{r},t) = v_{\rm ext}(\boldsymbol{r},t) + \int d\boldsymbol{r}' \frac{n(\boldsymbol{r}',t)}{|\boldsymbol{r}-\boldsymbol{r}'|} + v_{\rm xc}[n](\boldsymbol{r},t) \tag{26.6}$$

where $v_{\rm ext}(\boldsymbol{r},t)$ includes both the attractive potential of the nuclei and the time-dependent contribution of an external electromagnetic field. The second term accounts for the electrostatic Coulomb interaction between electrons. The XC potential, $v_{\rm xc}(\boldsymbol{r},t)$, is a functional of electron density and includes all nontrivial many-body effects. TDDFT is formally exact given the exact $v_{\rm xc}(\boldsymbol{r},t)$. However, the functional dependence of the exact TDDFT XC potential on electron density is extremely complex. It is nonlocal in space and has memory that depends on the history of electron density. As a local approximation in time, the adiabatic approximation of TDDFT assumes the XC potential responses instantaneously and without memory to any temporal change in the density:

$$v_{\rm xc}^{\rm adiabatic}[n](\boldsymbol{r},t) = v_{\rm xc}[n](\boldsymbol{r})|_{n=n(\boldsymbol{r},t)} \tag{26.7}$$

where $v_{\rm xc}[n]$ is ground-state XC functional. The accuracy of TDDFT depends crucially on the quality of the functional approximations involved. By using the local density approximation (LDA) potential in Eq. (26.7), one obtains the adiabatic local density approximation (ALDA), which is the most widely used and simplest functional in TDDFT. Naturally, the adiabatic XC potential inherits all the problems present in the ground-state XC potential. There have been attempts to derive a functional that is nonlocal in time; see, for example, Refs. [18–20]. Interested readers are referred to the review by Casida and Huix-Rotllant [21] for more details.

26.3 Linear Response Theory

In a true many-body treatment, electronic excited states can be obtained by diagonalizing the many-body Hamiltonian. Naturally, one would expect that the differences between the ground-state Kohn–Sham eigenvalues are good approximations to the excitation energies. However, it is well known that the Kohn–Sham states do not have any strict physical meaning, except the highest occupied one [22]. In general, differences in virtual and occupied Kohn–Sham orbital energies underestimate optical transitions. Instead, excited-states information can be obtained by determining the system's response to a time-dependent perturbation and can, therefore, be readily extracted from a TDDFT calculation. Working in frequency domain, the density response $n(\boldsymbol{r},\omega)$ is given by

$$\delta n(\boldsymbol{r},\omega) = \int d\boldsymbol{r}' \chi(\boldsymbol{r},\boldsymbol{r}',\omega)\delta v_{\rm ext}(\boldsymbol{r},\omega) \tag{26.8}$$

$\chi(\boldsymbol{r},\boldsymbol{r}',\omega)$ is defined as a linear density response function of the interacting electrons, which tells us how to calculate the change in density, δn, from a change in the external potential, δv_{ext}. The exact density response can be similarly obtained in the Kohn–Sham system:

$$\delta n(\boldsymbol{r},\omega) = \int d\boldsymbol{r}'\,\chi_{\text{KS}}(\boldsymbol{r},\boldsymbol{r}',\omega)\delta v_{\text{KS}}(\boldsymbol{r},\omega) \tag{26.9}$$

where $\chi_{\text{KS}}(\boldsymbol{r},\boldsymbol{r}',\omega)$ is the response function of the noninteracting Kohn–Sham system. The full density response function has poles at frequencies that correspond to the excitation energies of the interacting system, while the Kohn–Sham density response function has poles at frequencies that correspond to the Kohn–Sham orbital energy differences. Using Eq. (26.6), one can obtain the linear change of the Kohn–Sham potential:

$$\delta v_{\text{KS}}(\boldsymbol{r},\omega) = \delta v_{\text{ext}}(\boldsymbol{r},\omega) + \int d\boldsymbol{r}'\left(\frac{1}{|\boldsymbol{r}-\boldsymbol{r}'|} + f_{\text{xc}}(\boldsymbol{r},\boldsymbol{r}',\omega)\right)\delta n(\boldsymbol{r}',\omega) \tag{26.10}$$

where $f_{\text{xc}}(\boldsymbol{r},\boldsymbol{r}',\omega)$ is defined as the Fourier transform of the XC kernel:

$$f_{\text{xc}}(\boldsymbol{r},\boldsymbol{r}',t-t') = \frac{\delta v_{\text{xc}}(\boldsymbol{r},t)}{\delta n(\boldsymbol{r}',t')} \tag{26.11}$$

As mentioned in Section 26.2, most applications of TDDFT invoke the adiabatic approximation where memory, and thus the frequency dependence in XC kernel, is neglected. With the definitions of response functions and XC kernel, the linear density response in Eq. (26.9) can be rewritten as an integral equation:

$$\int d\boldsymbol{r}''\left[\delta(\boldsymbol{r}-\boldsymbol{r}'') - \int d\boldsymbol{r}'\,\chi_{\text{KS}}(\boldsymbol{r},\boldsymbol{r}',\omega)\left(\frac{1}{|\boldsymbol{r}'-\boldsymbol{r}'|} + f_{\text{xc}}(\boldsymbol{r}',\boldsymbol{r}',\omega)\right)\right]\delta n(\boldsymbol{r}'',\omega)$$
$$= \int d\boldsymbol{r}'\,\chi_{\text{KS}}(\boldsymbol{r},\boldsymbol{r}',\omega)\delta v_{\text{ext}}(\boldsymbol{r},\omega). \tag{26.12}$$

Considering that the linear density response on the left-hand side of Eq. (26.12) has poles at the true excitation energies Ω, this leads to an eigenvalue problem, which determines the true excitation spectrum of the interacting system [23].

$$\sum_{jb}[K_{ia,jb}(\Omega) + \omega_{ia}\delta_{ij}\delta_{ab}]F_{jb} = \Omega F_{ia}. \tag{26.13}$$

On the diagonal of the response matrix in Eq. (26.13), the Kohn–Sham energy differences provide a rough estimate of the true excited states and are corrected by the coupling matrix, $K_{ia,jb}$.

$$K_{ia,jb}(\omega) = \int d\boldsymbol{r}\int d\boldsymbol{r}'\varphi_i(\boldsymbol{r}')\varphi_b(\boldsymbol{r}')\left(\frac{1}{|\boldsymbol{r}-\boldsymbol{r}'|} + f_{\text{xc}}(\boldsymbol{r},\boldsymbol{r}',\omega)\right)\varphi_j(\boldsymbol{r}')\varphi_b(\boldsymbol{r}') \tag{26.14}$$

φ corresponds to the Kohn–Sham states, where i, j, k stand for occupied orbitals and a, b, c for virtual orbitals. The eigenvalue equation (26.13) is solved to yield the excited-state energies Ω and eigenvectors F_{ia}. The latter contains important information on the electronic structure and can be used to evaluate excited-state dipole moments, nonadiabatic couplings, and the oscillator strengths of the transition. The dimension of coupling matrix K is $O(N^2)$, where N is proportional to the system size. Thus, the computational effort in solving Eq. (26.13) scales formally as $O(N^6)$. In practice, iteration methods [24] such as Lanczos method or Davidson's algorithm may be used to solve the eigenvalue problem efficiently since in most cases only the few lowest transitions are of interest. However, for large systems that exhibit a high density of states, individual excited states may carry only a small fraction of the total oscillator strength of a given band and the iterative solution of the eigenvalue problem becomes inefficient. In that case, solving TDKS equations in time domain offers practical advantages over the frequency-domain approach.

26.4 Real-Time TDDFT

Direct propagation of TDKS equations (26.4) in time domain [4, 25, 26] addresses a wider range of physical problems. It can simulate fluctuations of charge density on the attosecond to femtosecond scale. In addition, the effects of an ultrafast, intense laser field beyond the linear response regime can be modeled. The real-time TDDFT method can also provide the frequency-dependent properties via a Fourier transform from time domain to frequency domain, and full optical spectra with meV resolution can be extracted from a single run of few tens of fs propagation time.

Instead of Kohn–Sham orbitals, the dynamics of the Kohn–Sham systems can also be written in terms of the reduced single-electron density matrix, which is given by

$$\rho(\boldsymbol{r}, \boldsymbol{r}', t) = \sum_{i}^{\text{occ}} \varphi_i(\boldsymbol{r}, t)\varphi_i^*(\boldsymbol{r}', t) \tag{26.15}$$

From the single-electron density matrix, one can obtain the interacting density and current density. The density is given by

$$n(\boldsymbol{r}, t) = \rho(\boldsymbol{r}, \boldsymbol{r}, t) \tag{26.16}$$

and the current density can be obtained from

$$\boldsymbol{j}(\boldsymbol{r}, t) = \frac{1}{2i}(\nabla_r - \nabla_{r'})\rho(\boldsymbol{r}, \boldsymbol{r}', t)\bigg|_{r=r'} \tag{26.17}$$

Using the TDKS equations (26.4), one can readily find that the time evolution of $\rho(\boldsymbol{r}, \boldsymbol{r}', t)$ is governed by the Liouvillevon Neumann equation:

$$i\frac{\partial}{\partial t}\rho(t) = [h(t), \rho(t)] \tag{26.18}$$

where $h(\boldsymbol{r}, t) = -\frac{\nabla^2}{2} + v_{\text{KS}}(\boldsymbol{r}, t)$ is the single-particle TDKS Hamiltonian. Since the Kohn–Sham Hamiltonian depends on the density, Eq. (26.18) is a set of nonlinear equations. As discussed in Section 26.2, the Kohn–Sham potential at time t, in principle, depends on the full history of density at all times $t' \leq t$, and in most cases, the adiabatic approximation removes the memory effects. Practically, Eq. (26.18) is written in terms of matrix–matrix products and integrated in real time domain. The computational effort in solving Eq. (26.18) scales as

$O(N^3)$, and linear-scaling TDDFT has been achieved by exploiting the locality of single-electron density matrix [27]. Numerically, the most commonly used propagator for Eq. (26.18) is arguably the family of Runge–Kutta methods [28]. Chebyshev expansion method is more efficient and stable compared to the fourth-order Runge–Kutta integrator [29]. Interested readers are referred to a recent article that examines the performance of different integration schemes [30].

26.5 Nonadiabatic Mixed Quantum/Classical Dynamics

Another class of problems that are amenable to TDDFT involves electron-ion dynamics, which allows one to investigate important processes in physics, chemistry, and materials science. The Born–Oppenheimer and extended Lagrangian molecular dynamics are extensively used to describe processes when the systems are constrained to the ground state or a certain excited-state potential energy surface [31]. A major limitation of these adiabatic approaches is that they are not applicable to reactions involving multiple potential energy surfaces. Two widely used approaches to account for nonadiabatic effects are the Ehrenfest method [32, 33] and the surface hopping method [34].

In Ehrenfest dynamics, the total wave functions Φ are factorized based on the mean-field approximation:

$$\Phi(\boldsymbol{r}, \boldsymbol{R}, t) = \Psi(\boldsymbol{R}, t)\psi(\boldsymbol{r}, t) \exp\left(\frac{i}{\hbar}\int^{t} E_r(t')\, dt'\right) \tag{26.19}$$

where Ψ and ψ are the wave functions for the nuclei and electrons, respectively. $\boldsymbol{R}$ and $\boldsymbol{r}$ are the coordinates of the nuclei and electrons, respectively. $E_r(t)$ is a phase factor. Substituting this wave function Ansatz into time-dependent Schrödinger equation results in two coupled time-dependent equations for Ψ and ψ, and the classical limit of the equation for Ψ gives rise to the Newton's equation of motion for the nuclei [35]:

$$\frac{d^2\overline{\boldsymbol{R}}_\alpha}{dt^2} = -\frac{1}{M_\alpha}\boldsymbol{F}(\overline{\boldsymbol{R}}) \tag{26.20}$$

where $\overline{\boldsymbol{R}}$ denotes the classical coordinate. M_α is the mass of nuclei α and

$$\boldsymbol{F}(\overline{\boldsymbol{R}}) = \nabla_\alpha \langle \psi(\boldsymbol{r}, t)|\hat{H}_e(\boldsymbol{r}; \overline{\boldsymbol{R}})|\psi(\boldsymbol{r}, t)\rangle_r \tag{26.21}$$

$\hat{H}_e$ is the electronic Hamiltonian and governs the evolution of the electrons:

$$i\hbar\frac{\partial\psi(\boldsymbol{r}, t; \overline{\boldsymbol{R}})}{\partial t} = \hat{H}_e(\boldsymbol{r}; \overline{\boldsymbol{R}})\psi(\boldsymbol{r}, t; \overline{\boldsymbol{R}}) \tag{26.22}$$

Equation (26.22) is solved by expanding

$$\psi(\boldsymbol{r}, t; \overline{\boldsymbol{R}}) = \sum_i c_i(t)\psi_i(\boldsymbol{r}; \overline{\boldsymbol{R}}) \tag{26.23}$$

where ψ_i is the electronic state i. Substituting Eq. (26.23) into Eq. (26.22) gives

$$i\hbar\frac{dc_i}{dt} = \sum_j c_j(\langle\psi_i|\hat{H}_e|\psi_j\rangle - i\hbar d_{ij}) \tag{26.24}$$

where d is the nonadiabatic coupling,

$$d_ij = \left\langle \psi_i \middle| \frac{\partial \psi_j}{\partial t} \right\rangle \tag{26.25}$$

In the surface hopping approach, the nuclear dynamics is propagated on a single Born–Oppenheimer potential energy surface, and nonadiabatic transitions between electronic states are determined based on a stochastic algorithm. In this approach, the electronic dynamics is governed by the same equation of motion given in Eq. (26.24), while the nuclear equation of motion is reduced to

$$\frac{d^2\overline{\boldsymbol{R}}_\alpha}{dt^2} = -\frac{1}{M_\alpha}\nabla_\alpha\langle\psi_i|\hat{H}_e|\psi_i\rangle \tag{26.26}$$

During the time evolution, the instantaneous probability that the trajectory hops from state i to state j is given by

$$P_{i,j} = -\frac{2\Delta t Re(d_{ij}c_j c_i^*)}{c_i c_i^*} \tag{26.27}$$

A hopping event is determined by a Monte Carlo procedure in which a random number $r \in [0, 1]$ is generated. The following condition has to be satisfied for a hop from state i to state j [34]:

$$\sum_{k\leq j-1} P_{i,k} < r < \sum_{k<j} P_{i,k} \tag{26.28}$$

Combined with TDDFT, the dynamics of electrons are governed by the TDKS equations, while nuclei follow classical Newtonian trajectories. The equations of motion for electrons and nuclei are solved self-consistently, which allows for free energy exchange between the electronic and ionic degrees of freedom as long as the total energy is conserved. Over the years, these methods have been successfully applied to investigate different problems. Yan et al. simulated the real-time dynamics of water dissociation under laser field using TDDFT with Ehrenfest dynamics [13]. In their work, the dynamic response of a water molecule adsorbed on a gold nanoparticle was followed. They found that the water molecule dissociates into a hydroxyl group (OH) and a hydrogen atom within 30 fs upon the plasmonic excitation of the gold nanoparticle. The hot electron transfer from the gold nanoparticle to the antibonding state of water molecule is attributed to the mechanism of water dissociation. However, the simulation results show that there is no direct correlation between the dissociation rate and the photoabsorption of the nanoparticle, and this suggests a selectivity of plasmonic mode on the photoinduced reaction. Figure 26.2a,b shows, respectively, the time-dependent change of the occupation of the Kohn–Sham states upon the excitation of an even and odd plasmonic mode of the nanoparticle. The occupations are obtained by projecting the TDKS state onto the ground state. It is obvious that the odd plasmonic mode decays to form hot electrons with energy well overlapped with the antibonding state of the water molecule, which greatly facilitates the charge transfer from the nanoparticle to the water molecule. Thus, the aforementioned example shows that with mixed TDDFT/classical dynamics simulations, a real-time picture of reaction mechanism at atomic scale can be obtained.

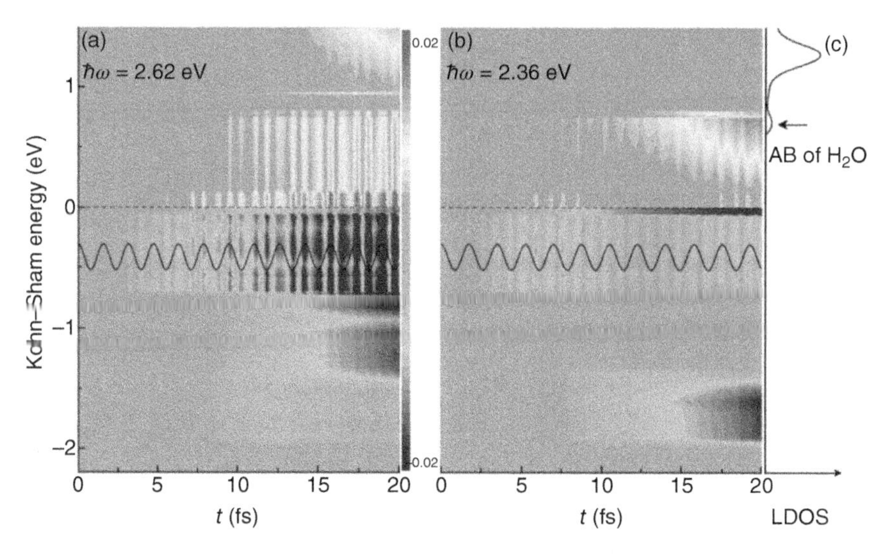

Figure 26.2 Time-dependent changes of the occupation of the Kohn–Sham states of the gold nanoparticle under continuous laser illumination with frequencies of (a) 2.62 eV and (b) 2.36 eV. (c) Calculated local density of states (LDOS) of the water molecule. Yan et al. 2016 [13]. Obtained with permission of American Chemical Society. (See online version for color figure).

By applying linear-response TDDFT to surface hopping dynamics, Tapavicza et al. studied the reaction mechanism of the photochemical ring opening in oxirane molecules and confirmed the main experimentally derived Gomer–Noyes mechanism [12]. In recent years, TDDFT has become a standard tool in computational photochemistry due to its computational competitiveness. However, being a single-reference method, TDDFT fails to describe the regions in the potential energy surface where a significant multireference trait is involved, for instance, at conical intersections where the ground and excited states become energetically degenerate. In these situations, nonadiabatic dynamics based on single-reference methods will not be adequate, and multireference methods such as complete active space self-consistent field (CASSCF) and complete active space second-order perturbation theory (CASPT2) should be used. Interested readers are referred to a recent review on the nonadiabatic mixed quantum/classical dynamics for more details [36].

References

1 Hohenberg, P. and Kohn, W. (1964). Inhomogeneous electron gas. *Phys. Rev.* 136: B864.
2 Kohn, W. and Sham, L.J. (1965). Self-consistent equations including exchange and correlation effects. *Phys. Rev.* 140: A1133.
3 Runge, E. and Gross, E.K.U. (1984). Density-functional theory for time-dependent systems. *Phys. Rev. Lett.* 52: 997–1000.
4 Yabana, K. and Bertsch, G.F. (1999). Time-dependent local-density approximation in real time: application to conjugated molecules. *Int. J. Quantum Chem.* 75 (1): 55–66.

5 Marques, M.A.L., Castro, A., and Rubio, A. (2001). Assessment of exchange–correlation functionals for the calculation of dynamical properties of small clusters in time-dependent density functional theory. *J. Chem. Phys.* 115 (7): 3006–3014.
6 Send, R. and Furche, F. (2010). First-order nonadiabatic couplings from time-dependent hybrid density functional response theory: consistent formalism, implementation, and performance. *J. Chem. Phys.* 132 (4): 044107.
7 Tapavicza, E., Tavernelli, I., and Rothlisberger, U. (2007). Trajectory surface hopping within linear response time-dependent density-functional theory. *Phys. Rev. Lett.* 98 (2): 023001.
8 Craig, C.F., Duncan, W.R., and Prezhdo, O.V. (2005). Trajectory surface hopping in the time-dependent Kohn–Sham approach for electron-nuclear dynamics. *Phys. Rev. Lett.* 95 (16): 163001.
9 Kurth, S., Stefanucci, G., Almbladh, C.O. et al. (2005). Time-dependent quantum transport: a practical scheme using density functional theory. *Phys. Rev. B* 72 (3): 035308.
10 Zheng, X., Wang, F., Yam, C.Y. et al. (2007). Time-dependent density-functional theory for open systems. *Phys. Rev. B* 75 (19): 195127.
11 Zheng, X., Chen, G.H., Mo, Y. et al. (2010). Time-dependent density functional theory for quantum transport. *J. Chem. Phys.* 133 (11): 114101.
12 Tapavicza, E., Tavernelli, I., Rothlisberger, U. et al. (2008). Mixed time-dependent density-functional theory/classical trajectory surface hopping study of oxirane photochemistry. *J. Chem. Phys.* 129 (12). https://doi.org/10.1063/1.2978380.
13 Yan, L., Wang, F., and Meng, S. (2016). Quantum mode selectivity of plasmon-induced water splitting on gold nanoparticles. *ACS Nano* 10 (5): 5452–5458.
14 Bauer, D. and Ceccherini, F. (2001). Time-dependent density functional theory applied to nonsequential multiple ionization of Ne at 800 nm. *Opt. Express* 8 (7): 377–382.
15 Chu, X. and Chu, S.I. (2001). Time-dependent density-functional theory for molecular processes in strong fields: study of multiphoton processes and dynamical response of individual valence electrons of N_2 in intense laser fields. *Phys. Rev. A* 64 (6): 063404.
16 Castro, A., Werschnik, J., and Gross, E.K.U. (2012). Controlling the dynamics of many-electron systems from first principles: a combination of optimal control and time-dependent density-functional theory. *Phys. Rev. Lett.* 109 (15): 153603.
17 Otobe, T., Yamagiwa, M., Iwata, J.-I. et al. (2008). First-principles electron dynamics simulation for optical breakdown of dielectrics under an intense laser field. *Phys. Rev. B* 77 (16): 165104.
18 vanLeeuwen, R. (1996). The Sham–Schluter equation in time-dependent density-functional theory. *Phys. Rev. Lett.* 76 (19): 3610–3613.
19 Dobson, J.F., Bunner, M.J., and Gross, E.K.U. (1997). Time-dependent density functional theory beyond linear response: an exchange–correlation potential with memory. *Phys. Rev. Lett.* 79 (10): 1905–1908.

20 Hirata, S., Ivanov, S., Grabowski, I., and Bartlett, R.J. (2002). Time-dependent density functional theory employing optimized effective potentials. *J. Chem. Phys.* 116 (15): 6468–6481.

21 Casida, M.E. and Huix-Rotllant, M. (2012). Progress in time-dependent density-functional theory. In: *Annual Review of Physical Chemistry*, vol. 63 (ed. M.A. Johnson and T.J. Martinez, 287–323. Annual Reviews.

22 Almbladh, C.O. and Vonbarth, U. (1985). Exact results for the charge and spin-densities, exchange–correlation potentials, and density-functional eigenvalues. *Phys. Rev. B* 31 (6): 3231–3244.

23 Grabo, T., Petersilka, M., and Gross, E.K.U. (2000). Molecular excitation energies from time-dependent density functional theory. *J. Mol. Struct. THEOCHEM* 501: 353–367.

24 Saad, Y. (1992). *Numerical Methods for Large Eigenvalue Problems*. United Kingdom: Manchester University Press.

25 Baer, R. (2000). Accurate and efficient evolution of nonlinear Schrodinger equations. *Phys. Rev. A* 62 (6): 063810.

26 Castro, A., Marques, M.A.L., and Rubio, A. (2004). Propagators for the time-dependent Kohn–Sham equations. *J. Chem. Phys.* 121 (8): 3425–3433.

27 Yam, C.Y., Zhang, Q., Wang, F., and Chen, G.H. (2012). Linear-scaling quantum mechanical methods for excited states. *Chem. Soc. Rev.* 41 (10): 3821–3838.

28 Press, W.H., Teukolsky, S.A., Vetterling, W.T., and Flannery, B.P. (1992). *Numerical Recipes in FORTRAN*, 2e. New York: Cambridge University Press.

29 Wang, F., Yam, C.Y., Chen, G.H., and Fan, K. (2007). Density matrix based time-dependent density functional theory and the solution of its linear response in real time domain. *J. Chem. Phys.* 126 (13): 134104.

30 Gomez, P.A., Marques, M.A.L., Rubio, A., and Castro, A. (2018). Propagators for the time-dependent Kohn–Sham equations: multistep, Runge–Kutta, exponential Runge–Kutta, and Commutator free magnus methods. *J. Chem. Theory Comput.* 14 (6): 3040–3052.

31 Deumens, E., Diz, A., Longo, R., and Ohrn, Y. (1994). Time-dependent theoretical treatments of the dynamics of electrons and nuclei in molecular-systems. *Rev. Mod. Phys.* 66 (3): 917–983.

32 Li, X.S., Tully, J.C., Schlegel, H.B., and Frisch, M.J. (2005). Ab initio Ehrenfest dynamics. *J. Chem. Phys.* 123 (8): 084106.

33 Wang, F., Yam, C.Y., Hu, L.H., and Chen, G.H. (2011). Time-dependent density functional theory based Ehrenfest dynamics. *J. Chem. Phys.* 135 (4): 044126.

34 Tully, J.C. (1990). Molecular-dynamics with electronic-transitions. *J. Chem. Phys.* 93 (2): 1061–1071.

35 Tully, J.C. (1998). Mixed quantum-classical dynamics. *Faraday Discuss.* 110: 407–419.

36 Crespo-Otero, R. and Barbatti, M. (2018). Recent advances and perspectives on nonadiabatic mixed quantum-classical dynamics. *Chem. Rev.* 118 (15, SI): 7026–7068. https://doi.org/10.1021/acs.chemrev.7b00577.

27

The *GW* Method for Excited States Calculations

Paolo Umari

Department of Physics and Astronomy, University of Padova, via Marzolo 8, 35122 Padova, PD, Italy

27.1 Introduction

The origin of modern quantum mechanics traces back to the pioneering work done in the early twentieth century aiming at giving an explanation to measured light emission and absorption spectra in rarefied atomic gases. Indeed, light is absorbed and emitted only at certain particular frequencies yielding to the well-known spectral lines. Such behavior finds no possible explanation within classical physics. The genius of Niels Bohr could assign these lines to the photons that are either absorbed or emitted when an atom "jumps" from one of its "quantum" states to another. As each state has its own energy, commonly referred to as the energy level, the photon one equates the energy difference of the two states. The ground state is referred to as the one with the lowest energy, while all the others are excited states. Hence, the appearance of quantum states different from the fundamental one is a key concept of quantum mechanics.

For a newcomer into the field of first principles simulations, it could seem strange that excited states seem not to attract so much interest in textbooks. Indeed, density functional theory (DFT) is a ground-state theory that is in principle exact only for the ground state. From one side its large degree of applicability to materials engineering stems from the use of the Born–Oppenheimer approximation, which permits to treat atomic positions as classical coordinates. From the other it is a consequence of the so-called $2n + 1$ theorem. This states that the knowledge of the nth derivatives of the wavefunctions with respect to a perturbation up to the nth order is enough for expressing the derivatives of the system energy up to the $2n + 1$ order. For $n = 0$ this principle is also known as the Hellman–Feynamnn theorem, which is routinely used for obtaining atomic forces from the ground-state wavefunctions. Its $n = 1$ formulation is used in density functional perturbation theory for achieving quantities as the dielectric tensor or the phononic dispersion.

However, there are other fundamental material properties that for being determined require the explicit evaluation of excited states. These include,

Heterogeneous Catalysts: Advanced Design, Characterization and Applications, First Edition.
Edited by Wey Yang Teoh, Atsushi Urakawa, Yun Hau Ng, and Patrick Sit.

among others, electronic band gaps and band dispersion, optical responses such as the frequency-dependent dielectric function, and electron and hole lifetimes and mobilities. Moreover, some chemical reactions involve one or more chemical components in one of their excited states.

Heterogeneous photocatalysis is a paradigmatic case where the correct alignment of the main excited-state levels determines the global functioning [1, 2]. In Figure 27.1 the main physical processes involved in overall photocatalytic water splitting are depicted. First, a photon of energy $\hbar\omega$ impinges on a semiconducting nanoparticle. Only if $\hbar\omega$ is large enough the photon is adsorbed by the nanoparticle promoting this into its excited state whose energy is $\hbar\omega$ above that of the ground state. This involves the creation of an in-principles bound electron–hole pair. Thermal excitations or internal electric fields can furnish enough energy for separating electron–hole pairs into noninteracting electrons and holes. Only if the hole energy level is lower than the oxidation O_2/H_2O potential water can be oxidized to oxygen, and only if the electron level is higher than the reduction H^+/H_2 potential protons can be reduced to molecular hydrogen.

In this chapter, we will introduce a class of robust theoretical approaches based on many-body perturbation theory (MBPT) for calculating from first principles quantum excited states and their properties.

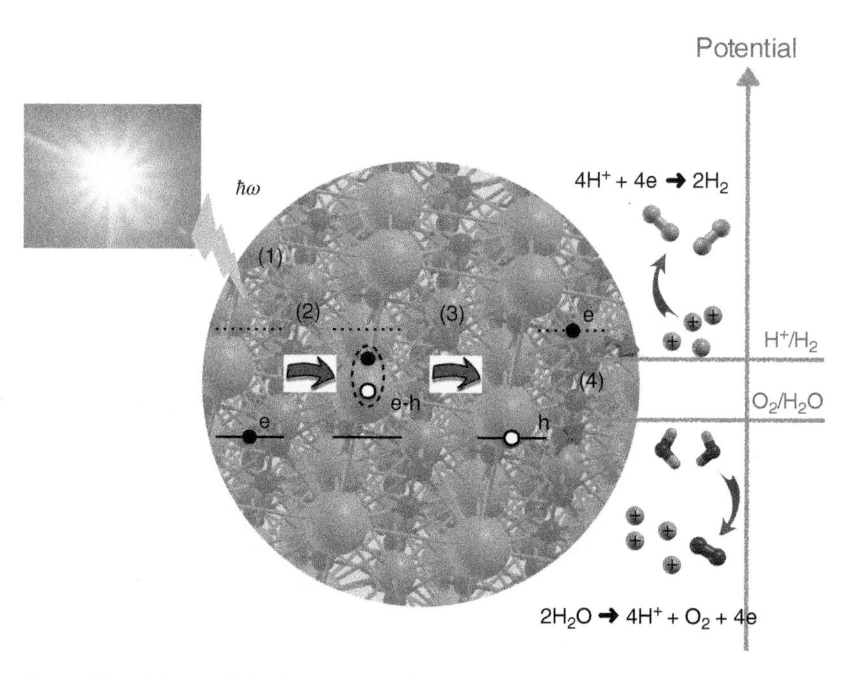

Figure 27.1 Scheme of the functioning of one-step photocatalytic water splitting: (1) one photon of energy $\hbar\omega$ is adsorbed by the photocatalyst creating (2) a bound electron–hole pair (2) that is then split (3) into separate noninteracting electron and hole. Holes can then oxidize water to oxygen, while electrons can reduce atomic to molecular hydrogen (4) provided that the hole energy level is lower than the oxidation O_2/H_2O potential and the hole energy level is higher than the reduction H^+/H_2 potential.

27.2 Excitations in Many-Electron Systems

When dealing with many-electron systems, ranging from isolated molecules to bulk extended materials, it is very advantageous to discriminate between two classes of excitations. In the following we will use the word excitation as a synonymous of excited state. The simplest case happens when upon excitation the number of electrons does not change. One of the clearest examples is the absorption of a photon from a system that initially is in the ground state and at the end of the process will be found in an excited state. It is worthy of illustrating such process within a simple noninteracting electron picture. First, in the ground state, all the energy levels below the lowest unoccupied molecular orbital (LUMO) one are doubly occupied (see Figure 27.2a). Then, the final electronic state will be determined by the incoming photon energy and by the relative oscillator strengths. For example, a possible final states is depicted in Figure 27.2b: one electron has been promoted from the highest occupied molecular orbital (HOMO) level to an empty one. If the photon energy is large enough, one electron can be promoted from a level deeper than the HOMO (Figure 27.2c). Final states involving more than one electronic transition can also be achieved.

At first sight, it could seem a bit surprising that neutral excitations require a more elaborate theoretical approach than the one that is used to describe another class of excitations, which takes the name of charged excitations. An excitation is said to be charged when the number of electrons before and after the perturbation differs. Probably the best example of a charge excitation is charge transport. If we focus on a very small volume of a metal or a semiconductor, initially it will be found in its ground state. We indicate with N its number of electrons. Then, one electron is injected in the volume bringing it into an excited state of the quantum

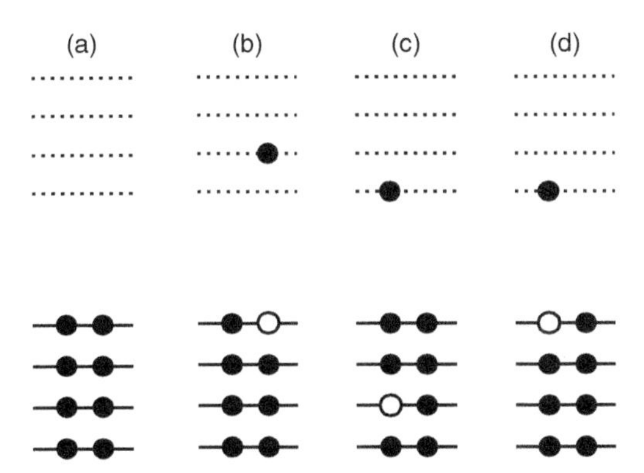

Figure 27.2 Representation of the occupancy of energy levels in case of noninteracting electrons. Full discs denote electrons. Void discs denote removed electrons. (a) The ground-state configuration is illustrated: all the valence energy levels (bold lines) are doubly occupied, while all the empty (conduction) levels (dotted lines) are unoccupied. (b) One electron has been promoted from the HOMO to one empty level. (c) One electron has been promoted from a deeper valence level to the LUMO one. (d) One electron has been promoted from the HOMO to the LUMO so that the system is in its first excited state.

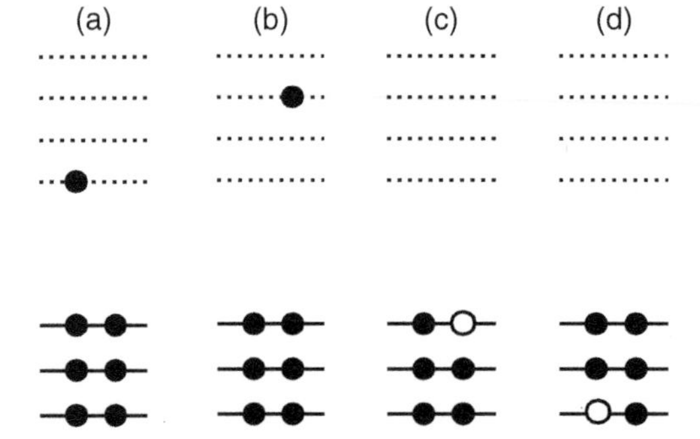

Figure 27.3 Representation of the occupancy of energy levels in case of noninteracting electrons. Full discs denote electrons. Void discs denote removed electrons. (a) An electron has been added to the LUMO. (b) An electron has been added to a higher empty level. (c) An electron has been removed from the HOMO. (d) An electron has been removed from a deeper valence level.

system containing $N + 1$ electrons. This is the case when electrons are the charge carriers. If holes are the charge carriers, an electron is removed from the volume, letting it in one of the quantum states of the system with $N - 1$ electrons.

In this chapter we will consider only charged transitions involving the gain or the loss of a single electron (namely, $N \rightarrow N + 1$ and $N \rightarrow N - 1$ excitations). It is didactically profitable to analyze these processes in a simple noninteracting electron picture. Initially, in the initial ground state, all the valence electronic levels are doubly occupied. Then, in the case of a $N \rightarrow N + 1$ excitation, an additional electron will occupy one of the empty levels. If it occupies the LUMO, the quantum $N + 1$ system will be in its ground state (Figure 27.3a). Conversely, if it occupies a higher empty level, the system will be in an excited state (Figure 27.3b). A similar situation happens for electron removal. If one electron is removed from the HOMO level, the final $N - 1$ system will be left in its ground state (Figure 27.3c). Instead, it will be in an exited state if the electron is taken from a lower valence level (Figure 27.3d).

Probably, the best experimental setup for probing charged ($N \rightarrow N - 1$) excitations is electron photoemission. This technique is based on the measurement of electrons emitted from a sample subsequent to photon absorption. From the measurement of electron energy and momentum, the quantum state of the $N - 1$ system can be derived from conservation laws. ($N \rightarrow N + 1$) excitations can be probed through inverse photoemission, which analyzes the photons emitted after electron absorption.

27.3 Green's Functions

Green's functions offer an extremely powerful albeit compact description of excitations [3, 4]. Let us start with charged $N \rightarrow N + 1$ and $N \rightarrow N - 1$ transitions. We indicate with $E_{N,i}$ the energy of the system with N electrons in its i-th quantum state. So that $E_{N-1,0}$ stays for the ground state energy of the $N - 1$ system. The frequency, or Lehmann's representation, of the one-particle Green's function gives access to the $E_{N-1,i}$ and $E_{N+1,j}$ levels. Indeed, it exhibits peaks when the frequency ω is in correspondence of the $E^{N+1,i} - E^{N,0}$ and $E^{N-1,j} - E^{N,0}$ energy transitions:

$$G^1(\mathbf{r}_1, \mathbf{r}_2;\ \omega) = \sum_i \frac{A^i(\mathbf{r}_1)A^{i*}(\mathbf{r}_2)}{(E^{N+1,i} - E^{N,0} - \omega - i\eta_i)} - \sum_j \frac{B^j(\mathbf{r}_2)B^{j*}(\mathbf{r}_1)}{(E^{N-1,j} - E^{N,0} + \omega - i\eta_j)} \tag{27.1}$$

where the functions A^i and B^j are called the quasiparticle amplitudes and depend on the electron positions $\mathbf{r}_1$ and $\mathbf{r}_2$. The terms η_i and η_j are the inverse lifetime associated to the i and j transitions. They are infinitesimal for long-living excitations in isolated (molecular) systems while they usually are finite in extended ones. The energy differences $\mathcal{E}_i = E^{N+1,i} - E^{N,0}$ and $\mathcal{E}_j = E^{N-1,j} - E^{N,0}$ take the name of quasiparticle energies.

Similarly, neutral excitations are described by the two-particle Green's function. However, as this contains also contributions from couples of charged $N \rightarrow N + 1$ and $N \rightarrow N - 1$ excitation, it is simpler to subtract them out obtaining the two-particle correlation functions:

$$\begin{aligned} &L^2(\mathbf{r}_1, \mathbf{r}_2, \mathbf{r}'_1, \mathbf{r}'_2,;\ \omega) \\ &= -\left(\frac{i}{\hbar}\right) \sum_s \left(\frac{\chi_s(\mathbf{r}_2, \mathbf{r}'_2)\chi_s^*(\mathbf{r}'_1, \mathbf{r}_1)}{E_{N,s} - E_{N,0} - \hbar\omega - i\eta_s} + \frac{\chi_s(\mathbf{r}_1, \mathbf{r}'_1)\chi_s^*(\mathbf{r}'_2, \mathbf{r}_2)}{E_{N,s} - E_{N,0} + \hbar\omega - i\eta_s} \right) \end{aligned} \tag{27.2}$$

Each excitation s is defined spatially by the two-particle amplitude χ_s and by the inverse lifetime η_s, which is infinitesimal for long-living excited states as in isolated molecular systems. Dealing with a two-particle processes, two couples of spatial coordinates $\mathbf{r}_1, \mathbf{r}_2$ and $\mathbf{r}'_1, \mathbf{r}'_2$ are needed.

Although Green's functions are the key for understanding basically every excitation process, they are many-body objects, and their exact computation would require the solution of the genuine N-electron problem being prohibitive but for the smallest systems. Luckily, approximations that can alleviate the computational burden are available.

27.4 Many-Body Perturbation Theory

As the direct computation of the exact G^1 is generally not feasible, theoretical approaches have been introduced for deriving approximations to it in terms of the one-particle Green's function G_0^1 of a noninteracting electron system. For most cases relevant for the reader, the starting unperturbed system is described by DFT:

$$G_0^1(\mathbf{r}_1, \mathbf{r}_2;\ \omega) = -\sum_v \frac{\phi_v(\mathbf{r}_1)\phi_v^*(\mathbf{r}_2)}{\omega - \epsilon_v - i\eta} - \sum_c \frac{\phi_c(\mathbf{r}_1)\phi_c^*(\mathbf{r}_2)}{\omega - \epsilon_c + i\eta} \tag{27.3}$$

where ϕ_v (ϕ_c) are the valence (conduction) KS orbitals of energy ϵ_v (ϵ_c). The ideal perturbation that is introduced for finding G^1 switches off the DFT exchange and correlation potential and at the same time switches on the (*exact*) electron–electron Coulomb interaction. A robust framework for dealing with such perturbation is provided by MBPT [5]. This is based on the second quantization formalism of quantum mechanics, and even a basic treatment is well beyond the scope of this chapter. Usually in MBPT, for alleviating the complexity of the formalism, integrals are represented in terms of diagrams (as the well-known Feynman's diagrams). The order of a diagram is determined by the number of times the *bare* Coulomb interaction appears in it. As the other component of diagrams (or integrals) is G_0^1, one would be tempted to sum up diagrams up to a given order for getting the many-body G^1. However, because of the long-range $1/r$ nature of the Coulomb potential, such sums are, in general, not absolutely convergent, meaning that summing terms up to a finite order does not give a good approximation to G^1. This difficulty can be overcome identifying an opportune class of diagrams and summing them up to infinite order. Such procedure is called renormalization and is used in the *GW approximation*. This furnishes a complete scheme for evaluating quasiparticle energies and amplitudes. We start from the Dyson's equation for G^1:

$$G^1(\mathbf{r}_1, \mathbf{r}_2;\ \omega) = G_0^1(\mathbf{r}_1, \mathbf{r}_2;\ \omega) + \int d\mathbf{r}' d\mathbf{r}'' G_0^1(\mathbf{r}_1, \mathbf{r};\ \omega)\Sigma(\mathbf{r}', \mathbf{r}'';\ \omega)G^1(\mathbf{r}'', \mathbf{r}_2;\ \omega) \tag{27.4}$$

The name GW stems from the approximation used for the self-energy operator Σ:

$$\Sigma(\mathbf{r}_1, \mathbf{r}_2;\ \omega) \cong \frac{i}{2\pi}\int d\omega' e^{i\omega'\eta} G^1(\mathbf{r}_1, \mathbf{r}_2;\ \omega + \omega')W(\mathbf{r}_1, \mathbf{r}_2;\ \omega') \tag{27.5}$$

where in the frequency convolution of G^1 with the screened Coulomb interaction W the term $e^{i\omega'\eta}$, with η positive infinitesimal, discriminates among the poles of G^1 while integrating in the complex plane. It is common to express W in terms of the bare Coulomb potential v and of the inverse dielectric matrix ϵ^{-1} as $W = \epsilon^{-1}v$.

Few steps allow to write W as a set of integrals dealing only with G_1^0, G_1, and v. These provide a self-consistent scheme, called the GW method, for calculating G^1 from G_0^1. However, its solution is too difficult to be achieved but for relatively small molecules because of its self-consistent nature and the need of working with large matrices as $W(\mathbf{r}_1, \mathbf{r}_2;\ \omega)$, which are not diagonal either in real or reciprocal space. The so-called G_0W_0 approximation is the simplest way for making such calculations doable. Self-consistency is totally avoided substituting G^1 with G_0^1. The resulting screened Coulomb interaction is usually referred to as W_0, hence the name of such approximation. As consequence, the final results will depend on the choice for the DFT exchange and correlation functional adopted in the starting DFT calculation. It is common, for solving Eq. (27.4), to represent the quasiparticle amplitudes on a basis of KS orbitals. Moreover, for lowering further the computational cost, the additional diagonal approximation is made: the quasiparticle amplitudes are directly approximated with KS orbitals. This leads to the following self-consistent one-variable equation for the quasiparticle energies:

$$\mathcal{E}_i = \epsilon_i - \langle\phi_i|\hat{V}_{\mathrm{xc}}|\phi_i\rangle + \langle\phi_i|\hat{\Sigma}(\mathcal{E}_i)|\phi_i\rangle \tag{27.6}$$

where $\hat{V}_{xc}$ is the DFT exchange and correlation potential operator. Although the diagonal G_0W_0 scheme is the *entry level* MBPT tool for calculating charged excitations, we will see that results improve significantly upon DFT.

27.5 *GW* in Practice

The *GW* method can be seen as a post-processing, albeit complex, elaboration that follows an initial DFT calculation. Hence, *GW* implementations, such as DFT, can be based on diverse choices for representing wavefunctions as plane-waves or localized basis sets. Several popular DFT distributions contains their own *GW* code. For citing just a few, comprising both open-source and proprietary software, *GW* packages are embedded in QUANTUM ESPRESSO [6, 7], ABINIT [8, 9], VASP [10], and FHI-aims [11]. Other *GW* softwares instead are distributed alone and can be interfaced with several DFT codes. A few examples are Yambo [12], Berkeley*GW* [13], and West [14].

As the *GW* scheme, even within the simplest G_0W_0 approximation, is highly complex, several strategies have been developed for making calculations feasible in systems large enough to be of interest. The first *GW* codes developed in the 1980s had to cope with limited computational power: they were limited to the G_0W_0 scheme and also inserted an additional approximation. The screened frequency- dependent Coulomb interaction W_0 instead of being calculated on a set of frequencies sampling the entire frequency axis is evaluated only at a couple of frequencies and then extrapolated using a plasmon-pole model [15]. This leads to an additional benefit of the plasmon-pole approximation: it permits to perform analytically the integration in Eq. (27.5). The plasmon-pole approximation, based on advanced models for the W, is still extensively in use nowadays.

Approaches, intended to be more accurate, that calculate exactly W over the frequencies' axis have to deal with the problem of the integrals as in Eq. (27.5). Indeed, the involved quantities have poles in proximity of the *real* frequency axis so that extremely fine sampling would be required. To ease this burden two different strategies are widely used. The simplest one requires the calculation of the screened Coulomb interaction W and the self-energy Σ on the imaginary frequency axis where such quantities are smooth. Finally, the expectation values of the self-energy entering Eq. (27.6) are analytically continued to the real frequency axis by fitting either with a multipole formula or with a Padé expression [16]. Hence, the final result will depend on the quality of the fit. The other popular strategy, called the contour deformation method [17], permits the evaluation of the self-energy on real frequency values performing the integration in Eq. (27.5) on a clever contour in the complex plane. Although more demanding than analytic continuation, the latter approach is in most cases more reliable.

With respect to DFT, *GW* requires more care for assuring the convergence of results. In particular, it is common in plane-wave implementations to evaluate the screened Coulomb interaction on a plane-wave basis sets defined by an energy cutoff well smaller than the one used in the foregoing DFT run. Although the computational cost is largely reduced, a too small cutoff would lead to

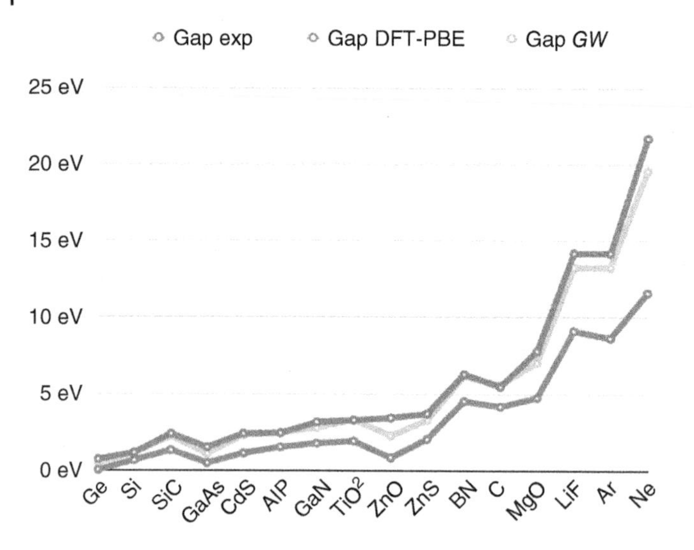

Figure 27.4 Experimental and theoretical electronic band gaps for series of semiconductors. Values are taken from Ref. [20]. DFT calculations are based on the PBE approximation for the exchange and correlation functional. *GW* calculations are based on the G_0W_0 approximation. Source: Adapted from Ren et al. 2012 [11].

unconverged *GW* energy levels [18]. The problem persists when the matrix *W* is represented on different basis sets. Setting the degree of completeness of the basis is crucial for controlling the accuracy to computational cost ratio. As the cost of *GW* is significantly higher than that of DFT, this is crucial for making a calculation doable.

The other main factor controlling convergence is the number of empty conduction orbitals included in the initial determination of G_0^1. In principles, all the empty states should be included. In practice, convergence can be achieved either including a limited number of empty states and extrapolating the resulting energy levels or adopting strategies, related to density functional perturbation theory, which avoid completely explicit sums over empty orbitals [19]. The issue with empty levels is less severe when adopting localized basis sets for developing wavefunctions.

Probably the rapid popularity of *GW* has been determined by the success in predicting energy band gaps of semiconductors where DFT notoriously fails. As illustrated in Figure 27.4 even the simple G_0W_0 scheme yields gap values in bulk semiconductors generally within few tenths of eV from their experimental counterparts. Reoccurring to more advanced *GW* methods, involving some degree of self-consistency [21], permits to further improve such agreement. Also the choice of the flavor of the starting DFT calculation can be important. It has been shown that hybrid functionals can provide a better starting point in particular for systems containing atoms whose d-orbitals take part to the valence bands as it is the case of transition metals [22]. *GW* not only provides band gaps, but also the entire band structure is generally well reproduced as can be shown comparing with

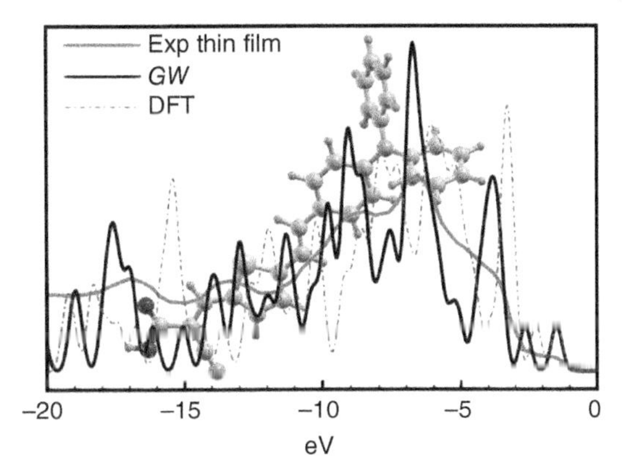

Figure 27.5 Electronic densities of states calculated with the DFT and the G_0W_0 methods for the triphenylamine (TPA)-based organic dye L_2 together with the experimental photoemission spectra for a thin film. All the data are from Ref. [24]. DFT calculations are based on the BLYP exchange and correlation functional. Source: Strinati 1988 [4]. Reproduced with permission of Springer Nature.

experimental band spectra measured through angle-resolved photoemission. It is worth noting that *GW* gives satisfactory results for the relative alignment of electronic bands at interfaces so that it can be used for predicting a variety of properties of electronic devices.

By definition the *GW* approach, in the case of a spatially isolated system, gives energy levels for the HOMO and the LUMO that corresponds to the vertical ionization energy and vertical electron affinity, respectively. Indeed, the latter are usually referred to as electron removal and electron addition energies. It has been shown that *GW* gives these values with average errors on the order of few tenths of eV [23]. The discrepancies with respect to experiment are smaller in the case the foregoing DFT calculation is based on hybrid exchange and correlation functionals. In the last two decades *GW* gained a strong reputation as a tool for simulating photoemission and inverse photoemission spectra.

As shown in Figure 27.5, *GW*, in contrast with DFT, provides electronic densities of states in general nice agreement with photoemission data. *GW* can then be used for assigning to specific orbitals the features appearing in the recorded spectra.

Being accurate both in the opposite limits of isolated molecular systems and of extended bulk ones, *GW* can give accurate energy level alignment at interfaces. In particular, *GW* fairly reproduces the HOMO–LUMO gap reduction due to the image charge effect that molecules exhibit in proximity of a metallic or semiconducting surface. This makes *GW* one of the best and frequently the only reliable theoretical method for assessing the energetics of complex systems.

27.6 The Bethe–Salpeter Equation

As we have seen *GW* handles charge excitations properly yielding excellent results for electronic bands, hence solving the band gap problem of DFT. However, it fails quite miserably if applied for the computation of optical spectra. Likely, the most notorious case is the frequency- dependent dielectric function

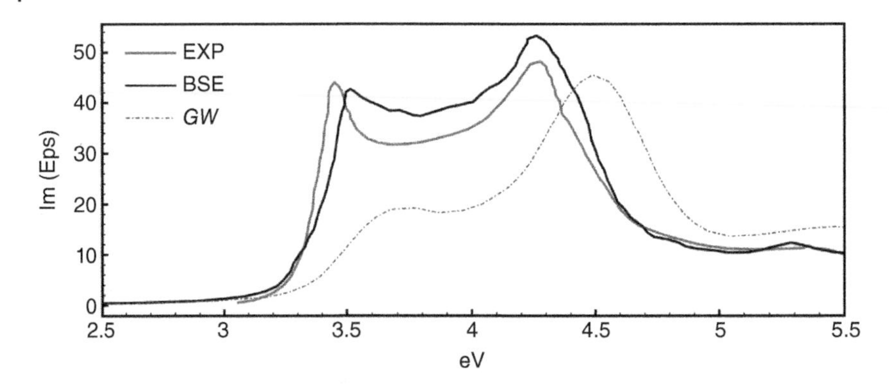

Figure 27.6 Imaginary part of the complex dielectric functional calculated with the *GW* and BSE methods together with experimental results. Data are from Ref. [25]. Source: Umari et al. 2010 [19]. Reproduced with permission of American Physical Society.

of bulk silicon. As displayed in Figure 27.6, *GW* reproduces badly the imaginary part of the dielectric function giving no significant improvement on DFT but for an almost rigid shift of the spectrum. This should not be a surprise for an attentive reader. We have seen in Section 27.3 that *GW* deals with charged excitations. Conversely, the *optical* dielectric function accounts for neutral excitations. This means that we need a theory for addressing the correlation part L^2 of the two-particle Green's functions instead of the one-particle Green's function G^1. The analogous to what done for approximating G^1 through MBPT can be applied to L^2. This leads to a Bethe–Salpeter equation, which determines the latter [4]. Then, an eigenvalue–eigenvector equation can be written for obtaining the neutral excitation energies $\mathcal{E}_s = E_{N,s} - E_{N,0}$ and the corresponding excitonic amplitudes χ_s [26]. It is customary to represent these on a basis of valence (v) conduction (c) transitions:

$$\chi_s(\mathbf{r}_1, \mathbf{r}_2) = \sum_{vc} A^s_{vc}\phi^*_v(\mathbf{r}_1)\phi_c(\mathbf{r}_2) \tag{27.7}$$

where the orbitals ϕ are quasiparticle amplitudes generally approximated with KS orbitals. Hence, the BSE approach gives a Shrödiger- like equation for the matrices A^S_{vc}. At the lowest level of approximation, such equation is not self-consistent:

$$\sum_{v'c'} \left[\delta_{v,v'}\delta_{c,c'}(\epsilon_c - \epsilon_v) + K^d_{vc,v'c'} + K^x_{vc,v'c'}\right] A^s_{v'c'} = \mathcal{E}_s A^S_{vc} \tag{27.8}$$

where the first term inside square brackets is named the diagonal term and simply gives the energy of an unscreened valence–conduction excitation being ϵ_v and ϵ_c quasiparticle energy levels. The direct term K^d accounts for the screened electron–hole interaction, while the exchange term K^x takes care of the Fermi statistics through Pauli's exclusion principle. The details of these two operators can be found in the literature (see, for example, Ref. [26]). Here, it is worth to report that the evaluation of K^d requires the screened Coulomb interaction at zero frequency (static) $W(\omega = 0)$. Then, in case the ground state of the system can be described by doubly occupied orbitals, separate solutions are found for singlet, optically active, and triplet, dark, excitations.

27.7 BSE in Practice

A BSE calculation usually follows a *GW* one that provides with the quasiparticle energy levels and the static screened Coulomb interaction. Therefore, most of the *GW* codes mentioned in Section 27.2 supply with a BSE module. As the scheme outlined in Section 27.6 lacks integrations on frequency, BSE implementations are technically less complex than *GW* ones. However, they have to calculate and manage large matrices such as the operators $K^d_{vc,v'c'}$ and $K^x_{vc,v'c'}$, where the number of valence–conduction transitions is very large but for very tiny systems. As a consequence, BSE requires quite large computational resources. Probably, the main parameter to be checked for assuring convergence is the quality of reciprocal space sampling in case of bulk calculations. The accurate simulation of optical spectra indeed requires very dense (k-points) sampling of the Brillouin's zone [25].

As illustrated in Figure 27.6, the introduction of BSE could finally solve the problem of the dielectric function in bulk silicon as electron–hole interactions are correctly taken into account. This shows also the magnitude and importance of excitonic effects. Similar agreement is generally found in bulk semiconductor and insulators as well as in molecular systems.

The evaluation of the complex dielectric function through BSE permits then to derive a range of optical spectra such as the optical absorption, conductivity, and susceptibility. It is worth noting that Eq. (27.8) gives access to single neutral excitations, permitting to calculate energy levels of excited states (see Eq. (27.2)). This, together with *GW* for the charge energy levels, provides accurate predictions of the energetics of complex systems in particular of interfaces. For example, all the relevant energy levels in dye-sensitized solar cells can be computed, and derived properties such as cell open-circuit voltage and electron injection efficiency are in agreement with available measurements [27].

27.8 Conclusions and Perspectives

After this overview on the *GW* and BSE approaches, it should not be difficult for an attentive reader to catch their potential for modeling catalytic and photocatalytic processes. With the BSE method it is possible to assess the light absorption spectrum and exciton binding energies for determining which part of, and how efficiently, the incoming radiation can be used for activating catalysis reactions. The *GW* method can provide with reliable electron and hole energy levels so that the driving potentials for electron injection in reduction and hole injection in oxidation can be estimated. Therefore, a novel material can be first screened *in silico* before being tested in the laboratory. New *GW* and BSE implementations that allow for faster calculations together with availability of large high-performance computing facilities are expected, in the next years, to lead to highly faithful large datasets of material properties relevant for catalysis as it is already happening at the simpler DFT level of accuracy [28, 29]. Hence the search for a new catalyst or for the optimization of an existing one is expected to start on the computer before moving to the laboratory permitting to examine a much higher number of candidate materials.

References

1 Zou, Z., Ye, J., Sayama, K., and Arakawa, H. (2001). Direct splitting of water under visible light irradiation with an oxide semiconductor photocatalyst. *Nature* 414: 625–627.

2 Hisatomi, T., Kubota, J., and Domen, K. (2014). Recent advances in semiconductors for photocatalytic and photoelectrochemical water splitting. *Chem. Soc. Rev.* 43: 7520–7535.

3 Onida, G., Reining, L., and Rubio, A. (2002). Electronic excitations: density-functional vs. many-body Green's-function approaches. *Rev. Mod. Phys.* 74: 601–659.

4 Strinati, G. (1988). Application of the Green's functions method to the study of the optical properties of semiconductors. *La Riv. Nuovo Cimento (1978–1999)* 11 (12): 1–86.

5 Hedin, L. (1965). New method for calculating the one-particle Green's function with application to the electron-gas problem. *Phys. Rev.* 139: A796–A823.

6 Giannozzi, P., Baroni, S., Bonini, N. et al. (2009). QUANTUM ESPRESSO: a modular and open-source software project for quantum simulations of materials. *J. Phys. Condens. Matter* 21 (39): 395502.

7 Giannozzi, P., Andreussi, O., Brumme, T. et al. (2017). Advanced capabilities for materials modelling with quantum ESPRESSO. *J. Phys. Condens. Matter* 29 (46): 465901.

8 Gonze, X., Amadon, B., Anglade, P.-M. et al. (2009). ABINIT: First-principles approach to material and nanosystem properties. *Comput. Phys. Commun.* 180 (12): 2582–2615. 40 YEARS OF CPC: a celebratory issue focused on quality software for high performance, grid and novel computing architectures.

9 Gonze, X., Jollet, F., Araujo, F.A. et al. (2016). Recent developments in the ABINIT software package. *Comput. Phys. Commun.* 205: 106–131.

10 Hafner, J. (2007). Materials simulations using VASP a quantum perspective to materials science. *Comput. Phys. Commun.* 177 (1): 6–13. Proceedings of the Conference on Computational Physics 2006.

11 Ren, X., Rinke, P., Blum, V. et al. (2012). Resolution-of-identity approach to Hartree–Fock, hybrid density functionals, RPA, MP2 and *GW* with numeric atom-centered orbital basis functions. *New J. Phys.* 14 (5): 053020.

12 Marini, A., Hogan, C., Grüning, M., and Varsano, D. (2009). Yambo: an ab initio tool for excited state calculations. *Comput. Phys. Commun.* 180 (8): 1392–1403.

13 Deslippe, J., Samsonidze, G., Strubbe, D.A. et al. (2012). Berkeley*GW*: a massively parallel computer package for the calculation of the quasiparticle and optical properties of materials and nanostructures. *Comput. Phys. Commun.* 183 (6): 1269–1289.

14 Govoni, M. and Galli, G. (2015). Large scale *GW* calculations. *J. Chem. Theory Comput.* 11 (6): 2680–2696. PMID: 26575564.

15 Hybertsen, M.S. and Louie, S.G. (1986). Electron correlation in semiconductors and insulators: band gaps and quasiparticle energies. *Phys. Rev. B* 34: 5390–5413.

16 Rieger, M.M., Steinbeck, L., White, I.D. et al. (1999). The *GW* space-time method for the self-energy of large systems. *Comput. Phys. Commun.* 117 (3): 211–228.

17 Fleszar, A. and Hanke, W. (1997). Spectral properties of quasiparticles in a semiconductor. *Phys. Rev. B* 56: 10228–10232.

18 Klimeš, J., Kaltak, M., and Kresse, G. (2014). Predictive *GW* calculations using plane waves and pseudopotentials. *Phys. Rev. B* 90: 075125.

19 Umari, P., Stenuit, G., and Baroni, S. (2010). *GW* quasiparticle spectra from occupied states only. *Phys. Rev. B* 81: 115104.

20 Nguyen, N.L., Colonna, N., Ferretti, A., and Marzari, N. (2018). Koopmans-compliant spectral functionals for extended systems. *Phys. Rev. X* 8: 021051.

21 van Schilfgaarde, M., Kotani, T., and Faleev, S. (2006). Quasiparticle self-consistent *GW* theory. *Phys. Rev. Lett.* 96: 226402.

22 Rödl, C., Fuchs, F., Furthmüller, J., and Bechstedt, F. (2009). Quasiparticle band structures of the antiferromagnetic transition-metal oxides MnO, FeO, CoO, and NiO. *Phys. Rev. B* 79: 235114.

23 van Setten, M.J., Caruso, F., Sharifzadeh, S. et al. (2015). *GW*100: benchmarking G_0W_0 for molecular systems. *J. Chem. Theory Comput.* 11(12): 5665–5687. PMID: 26642984.

24 Umari, P., Giacomazzi, L., De Angelis, F. et al. (2013). Energy-level alignment in organic dye-sensitized TiO_2 from *GW* calculations. *J. Chem. Phys.* 139 (1): 014709.

25 Kammerlander, D., Botti, S., Marques, M.A.L. et al. (2012). Speeding up the solution of the Bethe–Salpeter equation by a double-grid method and Wannier interpolation. *Phys. Rev. B* 86: 125203.

26 Rohlfing, M. and Louie, S.G. (2000). Electron–hole excitations and optical spectra from first principles. *Phys. Rev. B* 62: 4927–4944.

27 Marsili, M., Mosconi, E., De Angelis, F., and Umari, P. (2017). Large-scale *GW*-BSE calculations with N^3 scaling: excitonic effects in dye-sensitized solar cells. *Phys. Rev. B* 95: 075415.

28 Jain, A., Ong, S.P., Hautier, G. et al. (2013). Commentary: the materials project: a materials genome approach to accelerating materials innovation. *APL Mater.* 1 (1): 011002.

29 Curtarolo, S., Hart, G.L.W., Nardelli, M.B. et al. (2013). The high-throughput highway to computational materials design. *Nat. Mater.* 12: 191–201.

28

High-Throughput Computational Design of Novel Catalytic Materials

Chenxi Guo, Jinfan Chen, and Jianping Xiao

Dalian Institute of Chemical Physics, Chinese Academy of Sciences, State Key Laboratory of Catalysis, Theoretical Catalysis Group, 457 Zhongshan Road, Dalian, 116023, PR China

28.1 Introduction

In recent decades, there has been an increasing requirement for catalytic materials in our daily life. However, the design of new catalysts by experimental chemistry is still inefficient, considering the cost of time, equipment, and materials. Fortunately, the development of quantum mechanics, which is based on the first-principles theory and computational technologies, makes it possible to rationally analyze a series of surface reactions, which allows computational design for catalysis.

The research on computational catalysis date back to the 1970s with the emergence of density functional theory (DFT), which can calculate the electronic structures with a good trade-off between efficiency and accuracy [1, 2]. A surface reaction can be considered a series of elementary reactions. According to transition state theory [3, 4], an elementary reaction can be described as energies at the initial state, transition state, and final state. Microkinetic modeling [5, 6] correlates the energies in elementary reactions with catalytic reaction rates. More importantly, the fundamental correlations, due to the similar electronic structures on different energy states, have the potential to predict reactive energies and activation barriers with the adsorption energy of a single adsorbate. On the one hand, the correlations were usually reflected as a linear relationship between objective adsorption energies and adsorption energies of a specific adsorbate, known as the descriptor, on a series of catalyst surfaces [7]. As a result, the reaction energies of all elementary reactions can be predicted as a function of the descriptor. On the other hand, activation barriers and associated reaction energies were in a linear correlation for elementary reactions, called the Brønsted–Evans–Polanyi (BEP) relation [8, 9]. Therefore, all the key energies in catalyst design can be linked to a single descriptor.

There is no doubt that a huge amount of data is needed to establish the aforementioned correlations, on diverse catalyst surfaces, including not only metals, alloys, metal oxides, and metal carbides but also zeolites and carbon nanotubes [10–16]. The concept of "high-throughput (HT) computational catalyst design"

Heterogeneous Catalysts: Advanced Design, Characterization and Applications, First Edition.
Edited by Wey Yang Teoh, Atsushi Urakawa, Yun Hau Ng, and Patrick Sit.

was established about 15 years ago, which combines a large database of general characteristics for catalytic materials and the theoretical models for catalyst design [17, 18]. The data and models obtained by DFT calculations can be applied in the screening of catalysts committing to a rational design.

The concept of HT calculations can be applied in many areas, which provides a powerful automatic flow in the material design. In this chapter, we introduce a basic framework of computational catalyst design based on DFT calculations (see Figure 28.1). Generally speaking, the computational catalyst design can be conducted in the following steps:

1. Establish elementary reactions for the objective catalytic reaction over the existing materials in database.
2. Build up the correlations of adsorption energies between all the adsorbates and the descriptor, which is usually the adsorption energy of one or two key adsorbates.
3. Establish the BEP relations between activation energies for all the elementary reactions and the adsorption energy of key adsorbates.
4. Calculate the activity volcano plot based on microkinetic modeling and screening for catalysts with high activity.
5. Performing activity double check and reliability analysis with further data mining for new catalytic materials.

We introduce some examples in the Section 28.3. Some state-of-the-art methods in catalyst design are discussed. A comparison between models, approximations, and automatic HT calculations is discussed.

28.2 The Framework of Computational Catalyst Design

28.2.1 Elementary Reactions and Material Selection

The computational design of catalytic materials always starts with the construction of elementary reactions. In addition to surface reactions, adsorption and desorption should be considered explicitly. Herein, the adsorption energies of all the adsorbates and intermediates are of great importance, which are crucial parameters in calculating the reaction energy for each of the elementary reactions.

The construction of a set of elementary reactions is mainly based on the objective catalytic reaction. For example, the activity of CO_2RR producing methanol ($CO_2 + 3H_2 \leftrightarrow CH_3OH + H_2O$) has been theoretically studied in recent years, in which a set of elementary reactions for a single product of methanol can be established (see Table 28.1). However, a simplified set of elementary reactions, including only eight elementary steps with seven different adsorbates (H*, HCOO*, HCOOH*, H_2COOH*, H_2CO*, OH*, and H_3CO*), was obtained based on a general consideration.

As we know, surface catalytic reactions are much more complex, with more side reactions and by-products (see Figure 28.2 and Table 28.2) [20–24]. Therefore, a more rigorous scenario of catalytic reactions with diverse mechanisms

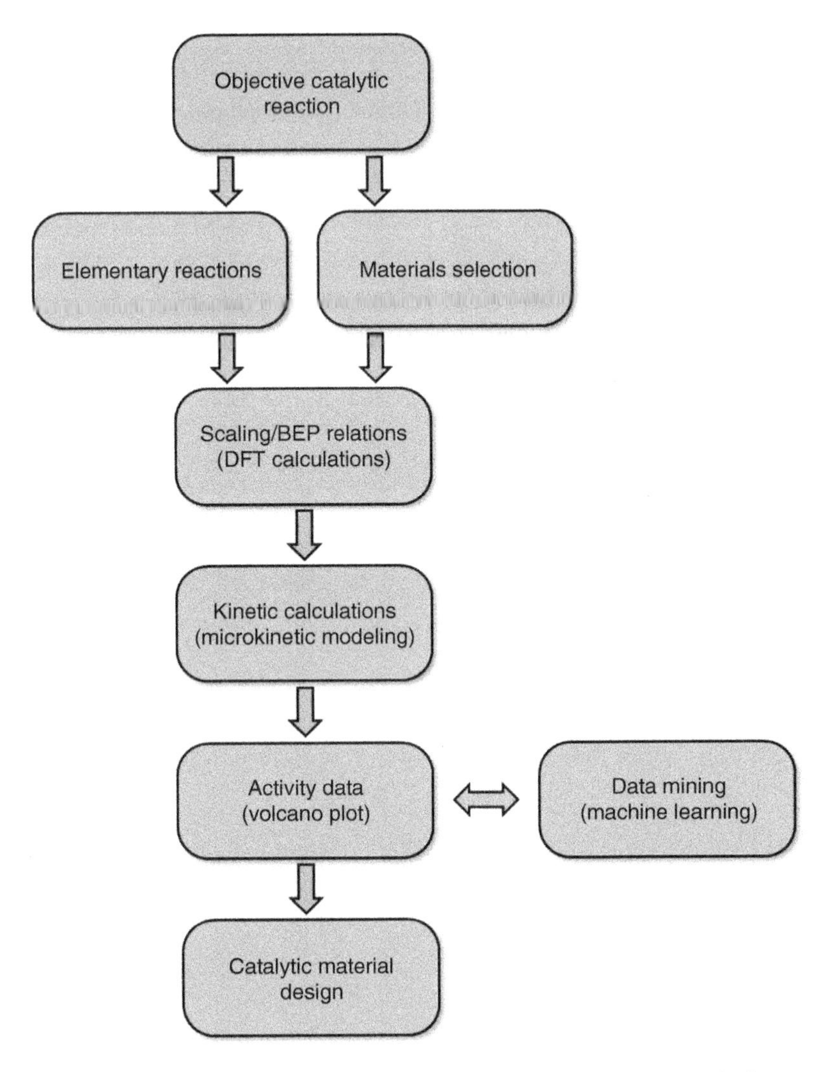

Figure 28.1 Flowchart of computational catalyst design based on DFT calculations and microkinetic modeling.

and products should be considered. The CO_2RR was studied in producing not only methanol (CH_3OH) but also carbon monoxide (CO), formic acid (HCOOH), formaldehyde (CH_2O), and methyl formate ($HCOOCH_3$) [19]. In order to compare the reaction activity with different products and mechanisms, more elementary reactions should be considered. In principle, infinite elementary reactions can be established, as more concerted processes are considered. Therefore, at present, the choice of elementary reactions should be more carefully based on previous researches and experimental results, achieving the compromise between reliability and feasibility.

On the other hand, before establishing scaling correlations, we should choose the general category of catalysts for objective catalytic reactions. Here, the

Table 28.1 Elementary reactions of CO_2RR producing methanol.

No.	Elementary reactions
R1	$H_2(g) + 2^* \leftrightarrow 2H^*$
R2	$CO_2(g) + H^* \leftrightarrow HCOO^*$
R3	$HCOO^* + H^* \leftrightarrow HCOOH^* + ^*$
R4	$HCOOH^* + H^* \leftrightarrow H_2COOH^* + ^*$
R5	$H_2COOH^* + ^* \leftrightarrow H_2CO^* + OH^*$
R6	$H_2CO^* + H^* \leftrightarrow H_3CO^* + ^*$
R7	$H_3CO^* + H^* \leftrightarrow CH_3OH(g) + 2^*$
R8	$OH^* + H^* \leftrightarrow H_2O(g) + 2^*$

Source: Adapted from Grabow and Mavrikakis 2011 [19].

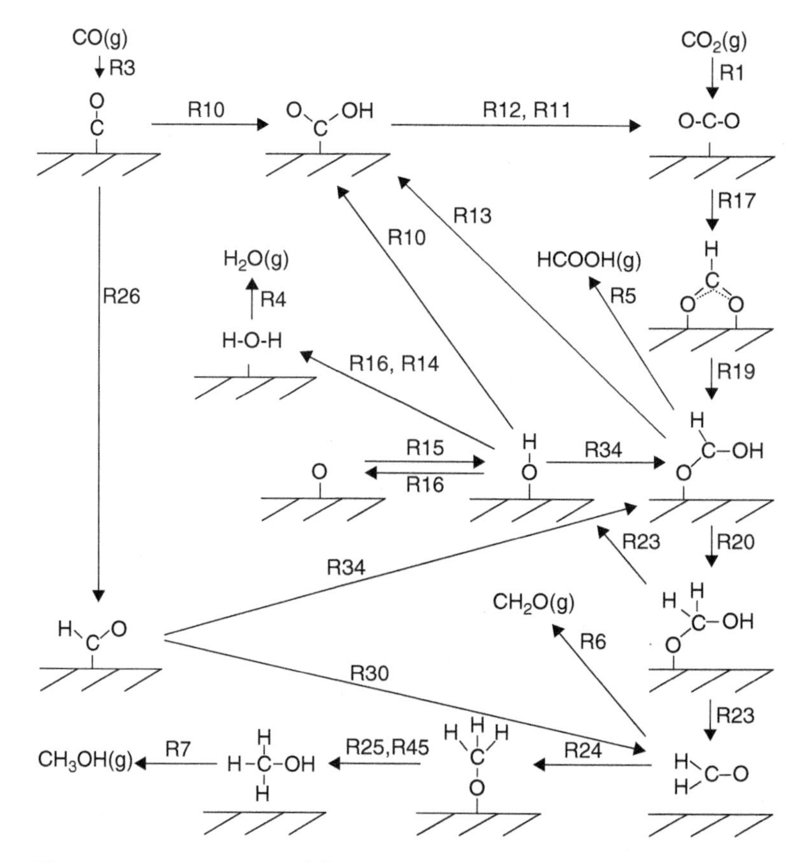

Figure 28.2 Reaction network for CO_2 reduction reaction producing CH_3OH with different by-products and reaction pathways. The number shows the specific elementary reaction associating to Table 28.2. Source: Adapted with permission from Grabow and Mavrikakis [19]. Copyright 2011, American Chemical Society.

Table 28.2 Elementary reactions for CO_2 reduction reaction producing CH_3OH with more by-products and side reactions.

No.	Elementary reactions	No.	Elementary reactions	No.	Elementary reactions
R1	$CO_2(g) + {}^* \leftrightarrow CO_2{}^*$	R18	$HCOO^* + H^* \leftrightarrow H_2CO_2{}^* + {}^*$	R35	$HCOOH^* + {}^* \leftrightarrow HCOH + O^*$
R2	$H_2(g) + 2^* \leftrightarrow 2H^*$	R19	$HCOO^* + {}^* \leftrightarrow HCOOH^* + {}^*$	R36	$CH_3O_2{}^* + {}^* \leftrightarrow CH_2OH^* + O^*$
R3	$CO(g) + {}^* \leftrightarrow CO^*$	R20	$HCOOH^* + {}^* \leftrightarrow CH_3O_2{}^* + {}^*$	R37	$CO_2{}^* + O^* \leftrightarrow CO_3{}^* + {}^*$
R4	$H_2O(g) + {}^* \leftrightarrow H_2O^*$	R21	$H_2CO_2{}^* + H^* \leftrightarrow CH_3O_2{}^* + {}^*$	R38	$CO_3{}^* + H^* \leftrightarrow HCO_3{}^* + {}^*$
R5	$HCOOH(g) + {}^* \leftrightarrow HCOOH^*$	R22	$H_2CO_2{}^* + {}^* \leftrightarrow CH_2O^* + O^*$	R39	$O^* + HCO^* \leftrightarrow OH^* + CO^*$
R6	$CH_2O(g) + {}^* \leftrightarrow CH_2O^*$	R23	$CH_3O_2{}^* + {}^* \leftrightarrow CH_2O^* + OH^*$	R40	$OH^* + HCO^* \leftrightarrow H_2O^* + CO^*$
R7	$CH_3OH(g) + {}^* \leftrightarrow CH_3OH^*$	R24	$CH_2O^* + H^* \leftrightarrow CH_3O^* + {}^*$	R41	$HCOO^* + HCO^* \leftrightarrow HCOOH^* + CO^*$
R8	$HCOOCH_3(g) + {}^* \leftrightarrow HCOOCH_3{}^*$	R25	$CH_3O^* + H^* \leftrightarrow CH_3OH^*$	R42	$HCOO^* + HCO^* \leftrightarrow H_2CO_2{}^* + CO^*$
R9	$CO^* + O^* \leftrightarrow CO_2{}^* + {}^*$	R26	$CO^* + H^* \leftrightarrow HCO^* + {}^*$	R43	$HCOOH^* + HCO^* \leftrightarrow CH_3O_2{}^* + CO^*$
R10	$CO^* + OH^* \leftrightarrow COOH^* + {}^*$	R27	$CO^* + H^* \leftrightarrow COH^* + {}^*$	R44	$CH_2O^* + HCO^* \leftrightarrow CH_3O^* + CO^*$
R11	$COOH^* + {}^* \leftrightarrow CO_2{}^* + H^*$	R28	$HCOO^* + {}^* \leftrightarrow HCO^* + O^*$	R45	$CH_3O^* + HCO^* \leftrightarrow CH_3OH^* + CO^*$
R12	$COOH^* + OH^* \leftrightarrow CO_2{}^* + H^*$	R29	$HCO^* + H^* \leftrightarrow HCOH^* + {}^*$	R46	$CH_3O^* + CH_2O^* \leftrightarrow HCOOCH_3{}^* + O^*$
R13	$COOH^* + H^* \leftrightarrow HCOOH^* + {}^*$	R30	$HCO^* + H^* \leftrightarrow CH_2O^* + {}^*$	R47	$CH_3O^* + CH_2O^* \leftrightarrow H_2COOCH_3{}^* + {}^*$
R14	$H_2O^* + {}^* \leftrightarrow OH^* + H^*$	R31	$CH_2O^* + H^* \leftrightarrow CH_2OH^* + {}^*$	R48	$HCOOCH_3{}^* + H^* \leftrightarrow H_2COOCH_3{}^* + {}^*$
R15	$OH^* + {}^* \leftrightarrow O^* + H^*$	R32	$HCOH^* + H^* \leftrightarrow CH_2OH^* + {}^*$	R49	$2CH_2O^* \rightarrow HCOOCH_3{}^* + {}^*$
R16	$2OH^* \leftrightarrow H_2O^* + O^*$	R33	$CH_2OH^* + H^* \leftrightarrow CH_3OH^* + {}^*$		
R17	$CO_2{}^* + H^* \leftrightarrow HCOO^* + {}^*$	R34	$HCOOH^* + {}^* \leftrightarrow HCO^* + OH^*$		

Source: Adapted with permission from Grabow and Mavrikakis [19]. Copyright 2011, American Chemical Society.

experience from previous research is of eventful value. Nevertheless, pure metal catalysts are generally used to discovering the mechanism, where an initial activity trend and mechanism can be built. It is worth noting that there exist many databases for materials, which are great tools to learn material properties and provide more information for novel catalysts design. For example, a Web application CatApp gives open access to the reaction and activation barriers for a large number of elementary reactions on the surfaces of a series of pure metals [25]. Moreover, many fundamental properties from computations for different materials can be found in the database "The Material Project" [26], including optimized structures, formation energies, band structures, or even phase diagrams. Moreover, the computational parameters are also recorded in the database to help the researchers reproducing the data.

28.2.2 The Scaling Relation and the Reaction Energy

The second step in computational catalyst design is the establishment of scaling relations [7, 27, 28] between all the adsorbates and the descriptor, which should link the basic reactivity of the catalyst to the activity of specific reactions. The scaling relation in heterogeneous catalysis was theoretically discovered using DFT calculations around 10 years ago and was summarized in 2007 [7]. A general linear correlation was found between the adsorption energies of adsorbed A (A = C, N, O, S) and the adsorption energies of AH_x on different kinds of surfaces (see Figure 28.3). Moreover, the similar linear relations were proved to be fitted in many other systems, including metals, alloys, metal oxides, and so on, as well as for diverse adsorbates [29]. In addition, the scaling relations were found valid for different elements in previous works [30–32] (for example, E_O vs. E_H).

With the choice of the descriptor and fitted scaling relations, it is possible to obtain the reaction energies of elementary reactions, which only depends on the parameters in the scaling relations and the specific adsorption energies of the descriptor. In addition, based on the scaling relations, the BEP relations and all the energies used in microkinetic modeling (discussed in Section 28.2.3) can be estimated.

In addition, the reaction energies of each elementary process reflect a basic electronic reactivity to describing the activity of the whole catalytic reaction. The elementary reaction with highest reaction energy can be initially considered the limiting step, which sets the activity limits of the whole catalytic reaction. In what follows, we introduce a succinct method used in a previous work, where the activity limit was established for understanding the general trend in experiments.

As mentioned earlier, the calculations of the thermodynamic trend started with the establishment of elementary reactions [24]. As the all the elementary reactions are reversible, the forward direction needs to be considered carefully, while the backward process should be also allowed. The steps with highest reaction energies are likely to be the rate-determining steps, and the optimum catalyst is always associated with a facile rate-determining step. The activity can also be affected by some other parameters, such as electrochemical field, charge transfer efficiency (Figure 28.4).

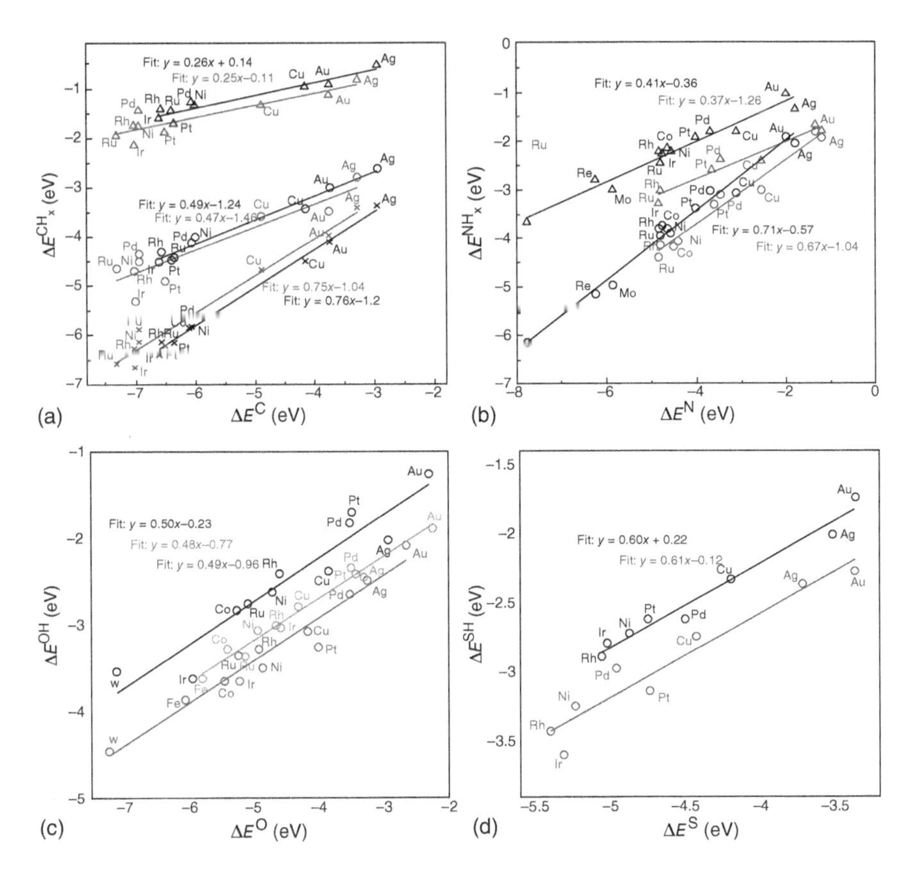

Figure 28.3 General linear scaling relations plotted between adsorption energies of AH_x (A = C, N, O, S) against adsorption energies of A. The subscript "$_x$" shows the number of H binding with adsorbed A ((a) CH_x: crosses $x = 1$, circles $x = 2$, triangles $x = 3$; (b) NH_x: circles $x = 1$, triangles $x = 2$; (c) OH_x and (d) SH_x: circles $x = 1$). The adsorption energy was calculated on the most stable site for flat (111) (black line) and step (211) (red line) surfaces. The scaling relation for adsorbed O and OH was also considered on flat (100) (blue line) surfaces. Source: Adapted with permission from Abild-Pedersen et al. [7]. Copyright 2007, American Physical Society. (See online version for color figure).

The establishment of scaling relations can provide a convenient way to determine the thermodynamic trend; nevertheless, some unreliability in kinetics still needs to be considered carefully later. In previous works, the thermodynamic trend could only be established successfully on simple catalysts, while it was difficult to understand more complex mechanisms and product selectivity. Although the kinetic factors are essential for catalysis, the initial estimation of the thermodynamic trend still shows remarkable values for designing the optimum catalyst.

28.2.3 The BEP Relation and the Activation Barrier

The rate of a reaction can be reasonably described by using microkinetic modeling. Here, transition state theory was used to describe the chemical equilibrium of

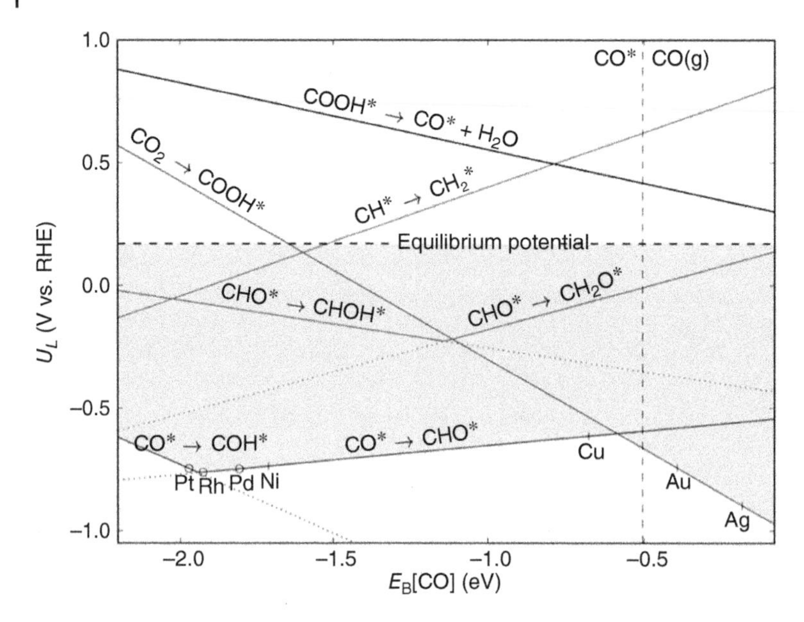

Figure 28.4 Limiting potentials, corrected from reaction free energies, for different elementary reactions in CO_2 electroreduction reaction plotted against the binding energy of CO. The potential lines with different colors show the competitive pathways, and the favored pathways were plotted using the solid lines, while the dotted lines showing less favorable pathways. Source: Adapted with permission from Peterson and Nørskov [24]. Copyright 2012, American Chemical Society. (See online version for color figure).

all the elementary reactions. The transition state was calculated using the climbing image nudged elastic band (NEB) method [33], which has become a generally accepted method in searching transition states in recent years. Some other advanced methods have also been developed, such as constrained optimization [34], which are very efficient in calculating well-defined transition states.

In 2001, a linear relation was found between the dissociative activation barriers and the reaction energies of N_2 on a series of metal surfaces [9]. Later, a linear correlation for many elementary steps over different catalytic materials was found theoretically. A general BEP relation was established, where there was a linear correlation between the dissociative activation barriers and the reaction energies of a specific kind of elementary reaction (see Figure 28.5) [35]. More importantly, all the activation barriers can be also correlated with the descriptor, namely the adsorption energy.

The calculation of barriers in elementary reactions is mainly used to establish the kinetic models of reaction rate. With this in-depth understanding of surface reactions, two theories were developed to describe the mechanism of bimolecular elementary reactions. The Langmuir–Hinshelwood (LH) [36] mechanism describes the coupling of two adsorbates on surfaces, while the Eley–Rideal (ER) [37] mechanism shows the association of adsorbates and gas molecules (see Table 28.3).

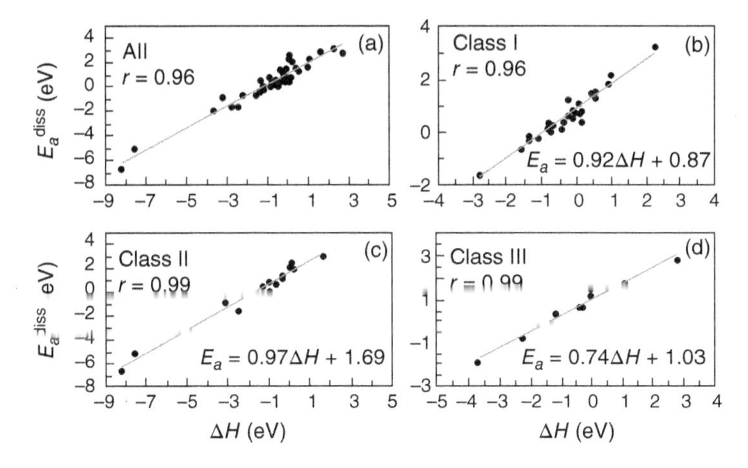

Figure 28.5 (a) The dissociation activation barrier against reaction enthalpy change for more than 50 elementary reaction steps. (b–d) Relation for the systems of dehydrogenation (class I), diatomic activation and hydrocarbon cracking (class II), and triatomic activation (class III). Source: Adapted with permission from Michaelides et al. [35]. Copyright 2003, American Chemical Society.

Table 28.3 Kinetic model of LH and ER mechanisms for the reaction rate calculation.

Mechanism	**LH**	**ER**
Elementary reaction	$A^* + B^* \leftrightarrow AB^* + ^*$	$A^* + B \leftrightarrow AB^*$
Forward rate	$r_f = k_f \cdot \theta_A \cdot \theta_B$	$r_f = k_f \cdot \theta_A \cdot P_B$
Reversed rate	$r_r = k_r \cdot \theta_{AB} \cdot \theta_*$	$r_r = k_r \cdot \theta_{AB}$

k_f and k_r refer to the kinetic constants; θ_A, θ_B, θ_{AB}, and θ_* refer to the surface coverage of adsorbed A, B, AB and free site, respectively; P_B refers to the pressure of B.

The calculation of kinetic constant was also considered in combining the transition state theory and the Arrhenius equation as

$$k = \frac{k_B \cdot T}{h} e^{-\frac{\Delta G}{RT}} \tag{28.1}$$

where k_B and h are Boltzmann's constant and Plank's constant, respectively, and ΔG refers to the standard Gibbs free energy change between the transition state and initial state. Moreover, the adsorption and desorption processes should be considered in the establishment of elementary reactions. Some methods have been developed for describing the activity of adsorption and desorption processes. For example, the adsorption rate can be described using the Hertz–Knudsen equation [38].

The activity simulation of a total reaction is mainly depended on the calculation of kinetic equilibrium, namely, the steady state, for elementary reactions, which

Table 28.4 Kinetic equations for microkinetic modeling.

No.	Elementary reactions	Reaction rate
R1	$H_2(g) + 2^* \leftrightarrow 2H^*$	$r_1 = k_{f_1} \cdot P_{H_2} \cdot \theta_*^2 - k_{r_1} \cdot \theta_H^2$
R2	$CO_2(g) + H^* \leftrightarrow HCOO^*$	$r_2 = k_{f_2} \cdot P_{CO_2} \cdot \theta_H - k_{r_2} \cdot \theta_{HCOO}$
R3	$HCOO^* + H^* \leftrightarrow HCOOH^* + ^*$	$r_3 = k_{f_3} \cdot \theta_{HCOO} \cdot \theta_H - k_{r_3} \cdot \theta_{HCOOH} \cdot \theta_*$
R4	$HCOOH^* + H^* \leftrightarrow H_2COOH^* + ^*$	$r_4 = k_{f_4} \cdot \theta_{HCOOH} \cdot \theta_H - k_{r_4} \cdot \theta_{H_2COOH} \cdot \theta_*$
R5	$H_2COOH^* + ^* \leftrightarrow H_2CO^* + OH^*$	$r_5 = k_{f_5} \cdot \theta_{H_2COOH} \cdot \theta_* - k_{r_5} \cdot \theta_{H_2CO} \cdot \theta_{OH}$
R6	$H_2CO^* + H^* \leftrightarrow H_3CO^* + ^*$	$r_6 = k_{f_6} \cdot \theta_{H_2CO} \cdot \theta_H - k_{r_6} \cdot \theta_{H_3CO} \cdot \theta_*$
R7	$H_3CO^* + H^* \leftrightarrow CH_3OH(g) + 2^*$	$r_7 = k_{f_7} \cdot \theta_{H_3CO} \cdot \theta_H - k_{r_7} \cdot P_{CH_3OH} \cdot \theta_*^2$
R8	$OH^* + H^* \leftrightarrow H_2O(g) + 2^*$	$r_8 = k_{f_8} \cdot \theta_{OH} \cdot \theta_H - k_{r_8} \cdot P_{H_2O} \cdot \theta_*^2$

Table 28.5 Equilibrium conditions for microkinetic modeling.

Equilibrium adsorbate	Equilibrium equations
H	$d\theta_H/dt = 2 \cdot r_1 - r_2 - r_3 - r_4 - r_6 - r_7 - r_8 = 0$
H_2CO	$d\theta_{H_2CO}/dt = r_5 - r_6 = 0$
H_2COOH	$d\theta_{H_2COOH}/dt = r_4 - r_5 = 0$
H_3CO	$d\theta_{H_3CO}/dt = r_6 - r_7 = 0$
HCOO	$d\theta_{HCOO}/dt = r_2 - r_3 = 0$
HCOOH	$d\theta_{HCOOH}/dt = r_3 - r_4 = 0$
OH	$d\theta_{OH}/dt = r_5 - r_8 = 0$
Free site	$d\theta_*/dt = r_3 - 2 \cdot r_1 + r_4 - r_5 + r_6 + 2 \cdot r_7 + 2 \cdot r_8 = 0$

is also called microkinetic modeling (Tables 28.4 and 28.5). Both forward and reverse kinetic equations should be considered for all the elementary steps following the two mechanisms discussed earlier. Note that the activity of adsorption and desorption can also be calculated using the same equation used for the rate calculation for surface reactions. However, it is difficult to locate the transition state in adsorption and desorption using the traditional optimization method, as the barrier in such processes is caused by the change in entropy [39]. In previous research [38, 40], some approximations were used to estimate the rates of adsorption and desorption, where the adsorption barrier was ignored and the desorption energy was used as the desorption barrier. The CatMAP [41] is one of the most popular codes for conducting the numerical calculations of kinetic equilibrium. Coding can also be done with other computer languages [42], such as MATLAB, to solve nonlinear multivariable equations.

Last but not least, the free energy correction must be made in catalyst design. As we all know, the Gibbs free energy should be used in microkinetic modeling because the total energy directly calculated from DFT can only describe the interactions involving nuclei and electrons. A new strategy must be developed to calculate the vibrational motions for adsorbates. The free energy correction can be obtained as a function of vibrational frequency, which can be calculated using the DFT method. In addition, all motions, including transition, rotation, and vibration, should be considered for gas-phase molecules committing to a complete free energy correction [43, 44].

28.2.4 The Activity Volcano Curve

Microkinetic modeling can be used to determine the activity of a specific surface reaction by calculating reaction rates. In catalyst design, the description of activity trend for catalytic reaction plays an important role, as discussed earlier. A series of microkinetic modeling should be conducted with the reaction energies and activation barriers, associated with the adsorption energies of key adsorbates, namely, the descriptor. The activity trend usually shows a "volcano" style [45–48].

The activity "volcano" was first discussed by the Sabatier principle, which states that the binding energies for adsorbates on the catalyst with highest activity should be neither too strong nor too weak. Indeed, it has been proved by DFT calculations on many different systems, where the catalytic activity is usually limited by the effective desorption process with stronger adsorption energies, while the adsorption processes with weaker adsorption energies.

According to the scaling and BEP relations, one-dimensional (1D) and two-dimensional (2D) activity volcano trends were established in the past research [22, 49]. Finding a catalytic material with optimum activity is considered a forever target in catalyst design. As shown in the traditional 1D activity volcano plot (see Figure 28.6a), the activity of oxygen reduction reaction (ORR) was simulated by the correlation with the binding energy of adsorbed O*. Different nanostructures were tested as the design of high-efficiency catalysts aimed to approach the maximum of activity volcano. Moreover, the 2D activity volcano plot provides a further scenario for catalyst design. Indeed, a circumvention of linear relation can be a valid way. The 2D volcano plot performs the activity trend as a function of adsorption energies of two adsorbates (see Figure 28.6b). As a result, a larger area with higher catalytic activity can be obtained, which provides more choices to design efficient catalysts.

28.2.5 Explicit Kinetic Simulations Based on DFT Calculations

It is indispensable to confirm the activity of designed catalysts by other theoretical methods or experiments. More rigorous DFT calculations are recommended to confirm either reaction energies or activation barriers for all the elementary reactions. Moreover, some additional conditions can be considered in this stage, such as adsorbate–adsorbate interactions, electrochemical environment, and confinement effects. The effects from all the extra conditions should be considered regarding the reaction energies and activation barriers for all the elementary processes. Such energies and barriers are the most significant parameters in microkinetic modeling, where the activity and selectivity results can then be rationally predicted.

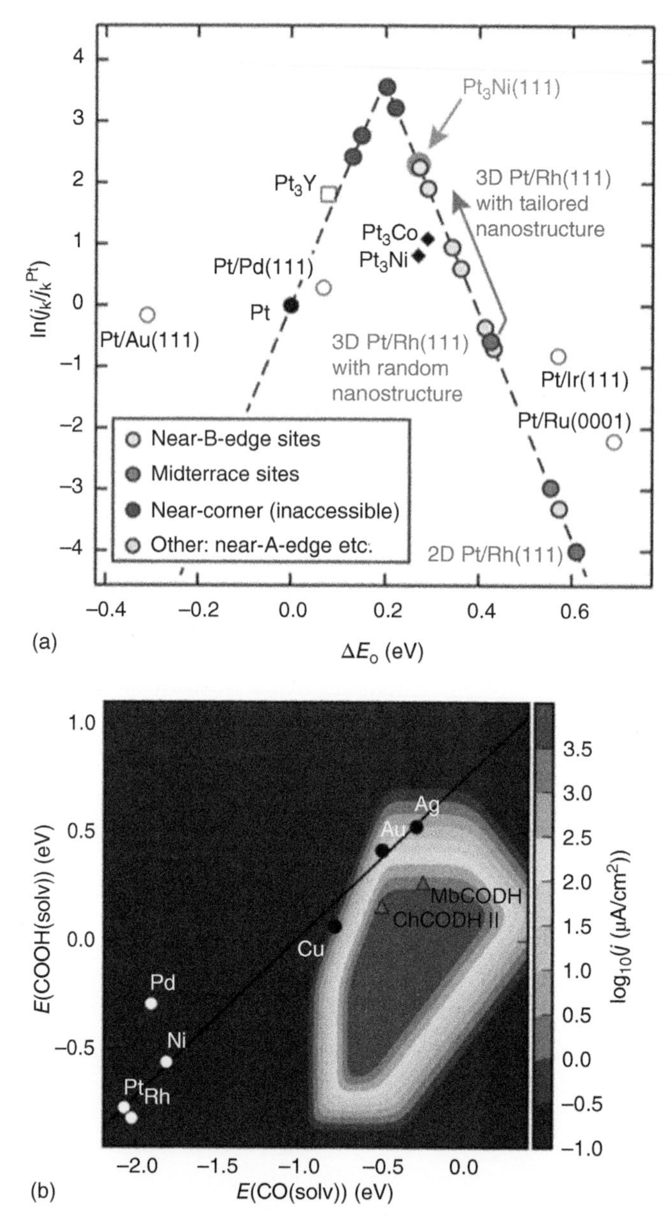

Figure 28.6 (a) Predicted activity of ORR on 1D Pt/Rh(111) surfaces based on different adsorption sites. The dashed line shows the theoretically predicted activity trend for ORR on different material surfaces. Source: Adapted with permission from Friebel et al. [49]. Copyright 2012, American Chemical Society. (b) 2D kinetic volcano for CO evolution on (211) step surfaces of transition metals. The fitted line shows the linear scaling relation between the adsorption energies of CO and COOH. Source: Adapted with permission from Hansen et al. [22]. Copyright 2013, American Chemical Society.

28.2.6 Data Mining and Machine Learning in Catalyst Design

The modern scientific research, both simulations and experiments, has been producing extraordinarily huge amount of data on the structures and properties of materials. Databases (e.g. NOMAD [50], Materials Project [26]) have been built by collecting all the information of materials from quantum mechanical calculations, which are convenient to be read and analyzed in generalized format. Analyzing the data with advanced machine learning algorithms would strengthen our ability to find and predict structures, correlations, and properties of interest, facilitating advanced material design [51]. Data mining with machine learning has been widely used in materials science, which was used to quickly search the objective material with specific physical properties [52–54]. It is more valuable to provide a prediction out of the existing range of knowledge, which effectively promotes the development of material design. Analogously, the DFT-based framework for catalyst design creates a large amount of computational data for the reaction and activation energies of catalytic processes (e.g. CatApp). Numerous thermodynamic and kinetic results from HT computational calculations, as well as existing materials databases, thus constitute a resourceful pool for data mining to identify the real active sites, descriptors, and catalytic mechanisms under certain reaction conditions, acting as a powerful tool for the screening of efficient catalysts [55]. The catalyst design aided with machine learning techniques will be in need considering the extremely complex catalytic processes, where reactions are affected by composition, support, surface morphology, reaction conditions, etc. Moreover, the surface structures and active sites of catalysts could change dynamically during the reactions. With advances in data science and machine learning, it is expected that the development of novel catalytic materials will take place at a much faster pace and more efficiently than ever before.

28.3 Examples for Rational Catalyst Design

The advanced computational technology makes it possible to perform HT calculations on catalyst design. Numerous computations can be used not only to build the fundamental relations in catalyst design but also to screen various kinds of adaptive catalytic materials. In this section, a few examples are discussed from previous research for the application of HT calculations in catalyst design.

28.3.1 Synthesis of Higher Alcohols from Syngas on Alloys

Synthesis of higher alcohols from CO and H_2 was studied as an important topic in heterogeneous catalysis. The computational design of new catalysts for this reaction was also studied, which follows the framework discussed earlier. The catalyst design started with the establishment of elementary reactions. Due to the complicated reaction mechanisms, especially the C–C coupling within numerous pathways, 50 elementary reactions were considered as shown in Table 28.6. Herein, ethanol was considered the only higher carbon alcohol.

Table 28.6 Elementary reactions in alcohol synthesis from syngas.

No.	Elementary reactions	No.	Elementary reactions	No.	Elementary reactions
Rl	$H_2(g) + 2^* \leftrightarrow 2H^*$	R18	$CH_3O^* + {}^* \leftrightarrow CH_3{}^* + O^*$	R35	$CH_2CHOH^* + H^* \leftrightarrow CH_2CH_2OH^* + {}^*$
R2	$CO(g) + {}^* \leftrightarrow CO^*$	R19	$C^* + CO^* \leftrightarrow CCO^* + {}^*$	R36	$CHCH2OH^* + H^* \leftrightarrow CH_2CH_2OH^* + {}^*$
R3	$CO^* + H^* \leftrightarrow COH^* + {}^*$	R20	$CH^* + CO^* \leftrightarrow CHCO^* + {}^*$	R37	$CH_2CH_2OH^* + H^* \leftrightarrow CH_3CH_2OH(g) + 2^*$
R4	$COH^* + {}^* \leftrightarrow C^* + OH^*$	R21	$CH_2{}^* + CO^* \leftrightarrow CH_2CO^* + {}^*$	R38	$CH_3CHOH^* + H^* \leftrightarrow CH_3CH_2OH(g) + 2^*$
R5	$C^* + H^* \leftrightarrow CH^* + {}^*$	R22	$CH_3{}^* + CO^* \leftrightarrow CH_3CO^* + {}^*$	R39	$CCO^* + H^* \leftrightarrow CCOH^* + {}^*$
R6	$CH^* + H^* \leftrightarrow CH_2{}^* + {}^*$	R23	$CCO^* + H^* \leftrightarrow CHCO^* + {}^*$	R40	$CCHO^* + H^* \leftrightarrow CHCHO^* + {}^*$
R7	$CH_2{}^* + H^* \leftrightarrow CH_3{}^* + {}^*$	R24	$CCO^* + H^* \leftrightarrow CCHO^* + {}^*$	R41	$CCHO^* + H^* \leftrightarrow CCHOH^* + {}^*$
R8	$CH_3{}^* + H^* \leftrightarrow CH_4(g) + 2^*$	R25	$CHCO^* + H^* \leftrightarrow CH_2CO^* + {}^*$	R42	$CCOH^* + H^* \leftrightarrow CCHOH^* + {}^*$
R9	$O^* + H^* \leftrightarrow OH^* + {}^*$	R26	$CHCO^* + H^* \leftrightarrow CHCOH^* + {}^*$	R43	$CCOH^* + H^* \leftrightarrow CHCOH^* + {}^*$
R10	$OH^* + H^* \leftrightarrow H_2O(g) + 2^*$	R27	$CHCO^* + H^* \leftrightarrow CHCHO^* + {}^*$	R44	$CCHOH^* + H^* \leftrightarrow CHCHOH^* + {}^*$
R11	$2OH^* \leftrightarrow H_2O(g) + O^* + {}^*$	R28	$CH_2CO^* + H^* \leftrightarrow CH_2COH^* + {}^*$	R45	$CHCHO^* + H^* \leftrightarrow CHCHOH^* + {}^*$
R12	$CO^* + H^* \leftrightarrow HCO^* + {}^*$	R29	$CHCOH^* + H^* \leftrightarrow CH_2COH^* + {}^*$	R46	$CH_2COH^* + H^* \leftrightarrow CH_2CHOH^* + {}^*$
R13	$HCO^* + H^* \leftrightarrow CH_2O^* + {}^*$	R30	$CHCOH^* + H^* \leftrightarrow CHCHOH^* + {}^*$	R47	$CH_3CO^* + H^* \leftrightarrow CH_3COH^* + {}^*$
R14	$CH_2O^* + H^* \leftrightarrow CH_3O^* + {}^*$	R31	$CH_2COH^* + H^* \leftrightarrow CH_2CHOH^* + {}^*$	R48	$CH_3CO^* + H^* \leftrightarrow CH_3CHO^* + {}^*$
R15	$CH_3O^* + H^* \leftrightarrow CH_3OH(g) + 2^*$	R32	$CHCHOH^* + H^* \leftrightarrow CH_2CHOH^* + {}^*$	R49	$CH_3CHO^* + H^* \leftrightarrow CH_3CHOH^* + {}^*$
R16	$HCO^* + {}^* \leftrightarrow CH^* + O^*$	R33	$CHCHOH^* + H^* \leftrightarrow CHCH_2OH^* + {}^*$	R50	$CH_3COH^* + H^* \leftrightarrow CH_3CHOH^* + {}^*$
R17	$CH_2O^* + {}^* \leftrightarrow CH_2{}^* + O^*$	R34	$CH_2CHOH^* + H^* \leftrightarrow CH_3CHOH^* + {}^*$		

The HT calculations were done to obtain all the energies, including adsorption energies and activation barriers for those elementary reactions. Energy results were calculated on several single-metal catalysts (Cu, Pd, Pt, Ag, Ru, Rh, Ni, and Co) to establish the general scaling relations and BEP relations (see Figures 28.7 and 28.8). Herein, the energies on CuCo, which was expected to show high activity, were also calculated. Note that the adsorption energies of C* on different surfaces were used as the main descriptor. Meanwhile, the adsorption energies of O* were applied in some of the scaling relations due to the better linear correlation between the adsorbates with O* binding to the surfaces. In addition, the BEP relations were obtained based on the elementary reactions (see Figure 28.8). Taking account of the cost of computational resources, not all the BEP relations were explicitly calculated in this work, where some of the activation barriers affecting little in the simulation of activity were ignored. It is reasonable to do some simplifications in the DFT calculations; however, it is highly recommended to calculate all the BEP relations with all activation barriers.

In the next stage, the reaction rate was calculated using microkinetic modeling based on the energy results discussed earlier. Microkinetic modeling was conducted using the CatMAP code, where a 2D activity map can be obtained (see Figure 28.9). In the example work, three products were considered CH_4, EtOH, and MeOH, where the material of interest (CuCo) showed high selectivity on EtOH based on the rate simulation from microkinetic modeling. Last but not least, the activity trend based on the activity volcano map also provides great guidelines for catalyst design to achieve both higher activity and specific selectivity. A simple energy calculation of descriptors can be used to predict the performance of new catalysts. Herein, rigorous DFT calculations are recommended to prove the performance of predicted catalysts.

28.3.2 HT Screening for Hydrogen Evolution Reactions (HERs)

Hydrogen evolution reactions (HERs) have been studied as one of the most popular reactions in electrochemistry. On the one hand, HER was experimentally developed as a main method producing H_2. On the other hand, its simple mechanism makes it a valuable classical reaction to investigate theoretical models using the DFT calculations. Besides, the HER is a competitive step in several electrochemical catalytic processes, including electrochemical CO_2 reduction. Therefore, the control of the HER activity plays a crucial role in choosing catalytic materials.

Pure metal surfaces were widely studied both experimentally and theoretically [57, 58] for tuning the HER activity. However, the limited choices of materials in the range of pure metals cannot satisfy the various catalytic requirements. Fortunately, theoretical studies on alloys showed more chances in catalyst design, where different characters of binding energies were considered with a variety of combinations of different elements.

The key adsorption free energy of adsorbed H* was studied on totally 256 pure metals and metal alloy surfaces [59] (see Figure 28.10a). The pure metal materials were separated into two components in the alloy as the host element, which constitutes the slab of the alloy, and the solute element. DFT calculations

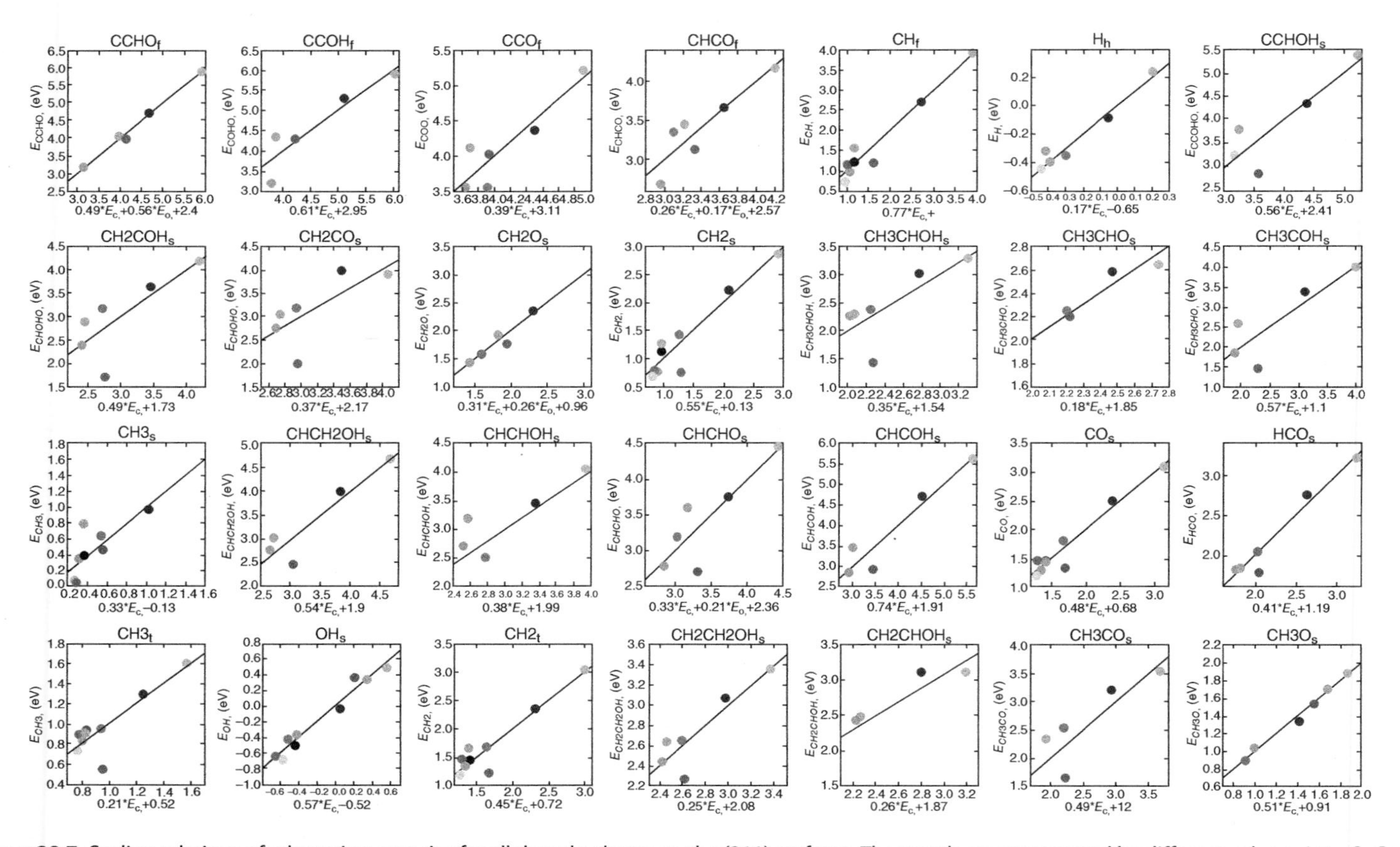

Figure 28.7 Scaling relations of adsorption energies for all the adsorbates on the (211) surfaces. The metals are represented by different color points: CuCo (pink), Cu (blue), Pd (green), Pt (red), Ag (cyan), Rh (orange), Ru (yellow), Ni (black), and Co (greenish black). Source: Adapted with permission from Cao et al. [56]. Copyright 2018, American Chemical Society. (See online version for color figure).

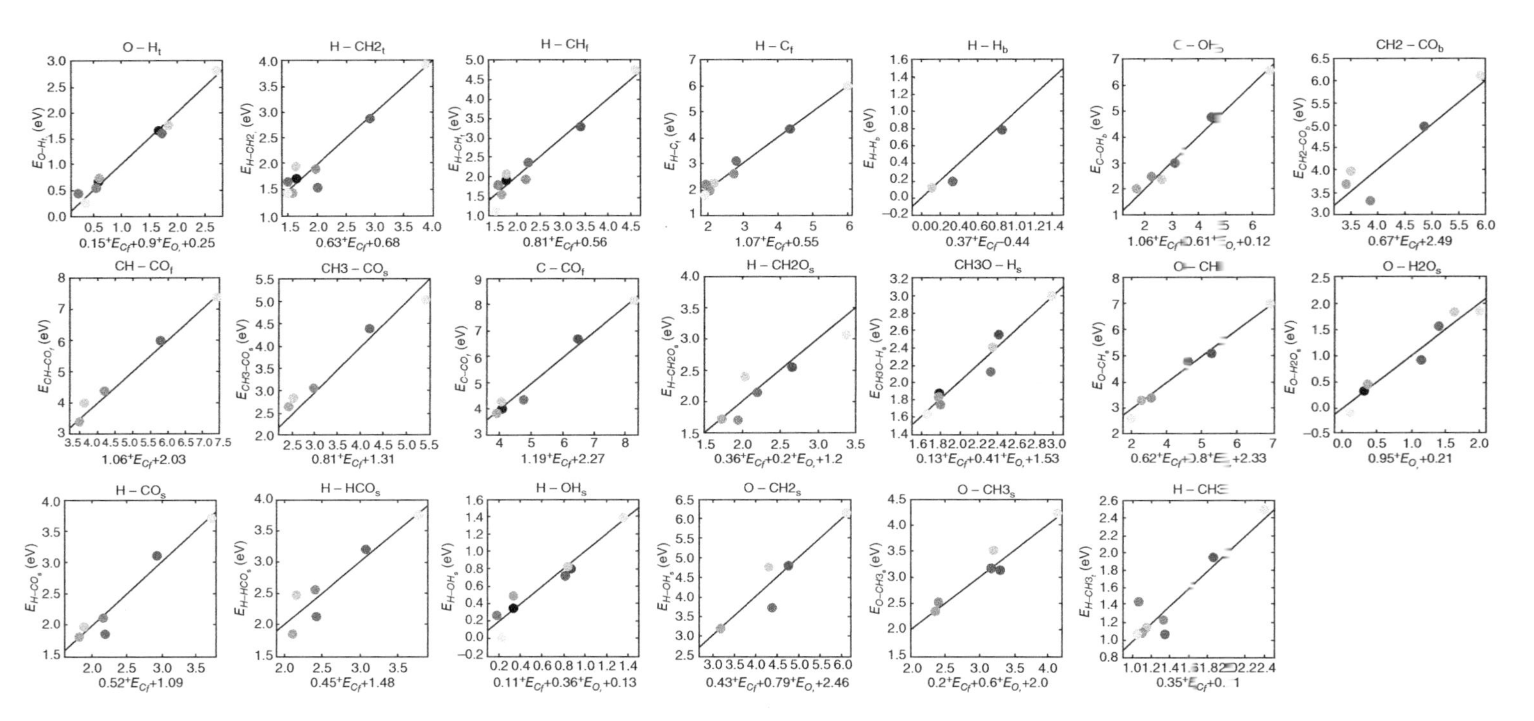

Figure 28.8 BEP relations of adsorption energies for all the adsorbates on the (211) surfaces. The metals are represented by different color points: CuCo (pink), Cu (blue), Pd (green), Pt (red), Ag (cyan), Rh (orange), Ru (yellow), Ni (black), and Co (greenish black). Source: Adapted with permission from Cao et al. [56]. Copyright 2018, American Chemical Society. (See online version for color figure).

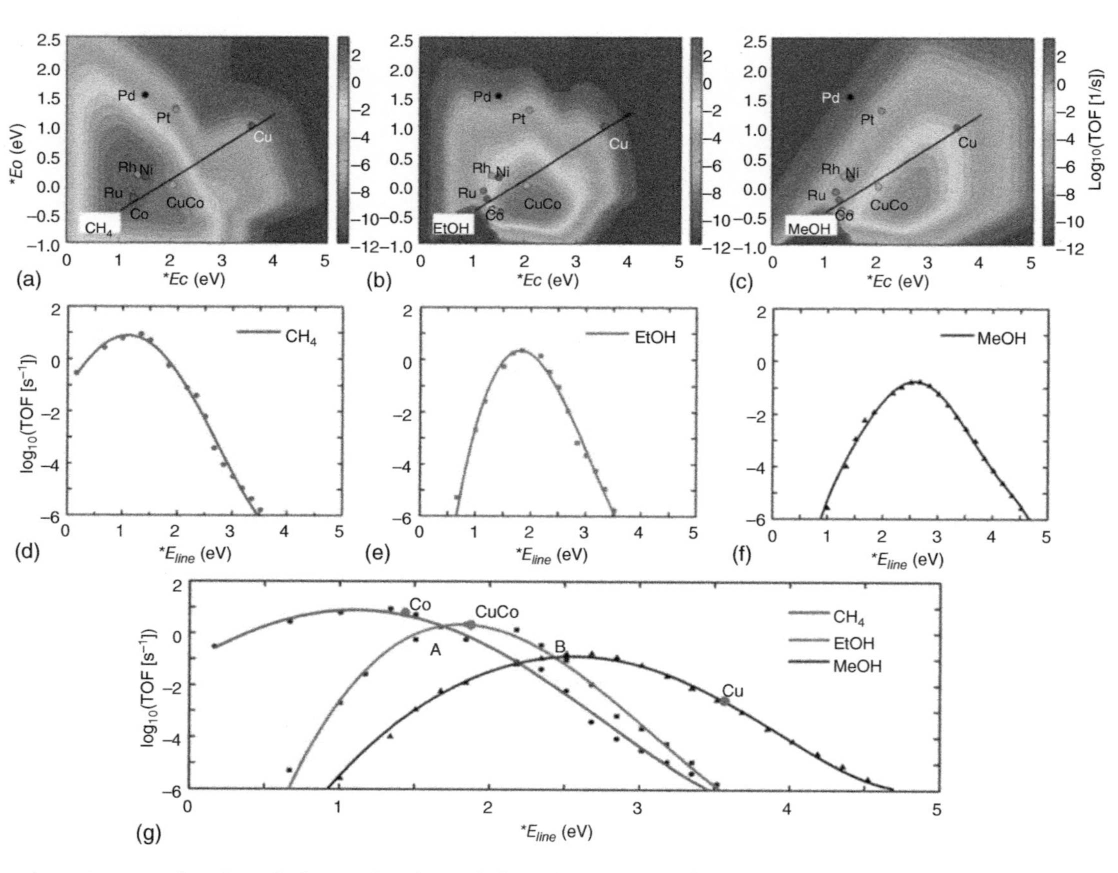

Figure 28.9 Activity volcano map as a function of relevant descriptors (adsorption energies of C* and O*) for different products: CH_4 (a), EtOH (b), and MeOH (c). Reaction activity (TOF) for CH_4 (d), EtOH (e), and MeOH (f) along the line fitted to the points for Co, Cu, and CuCo in (a–c). (g) Comparison of reaction activity for different products. Source: Adapted with permission from Cao et al. [56]. Copyright 2018, American Chemical Society.

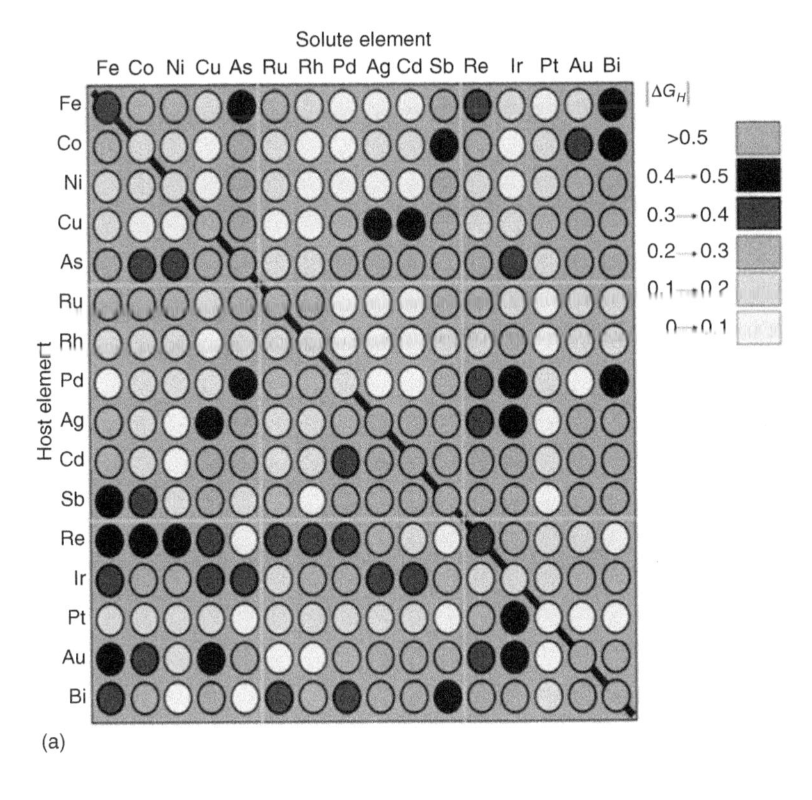

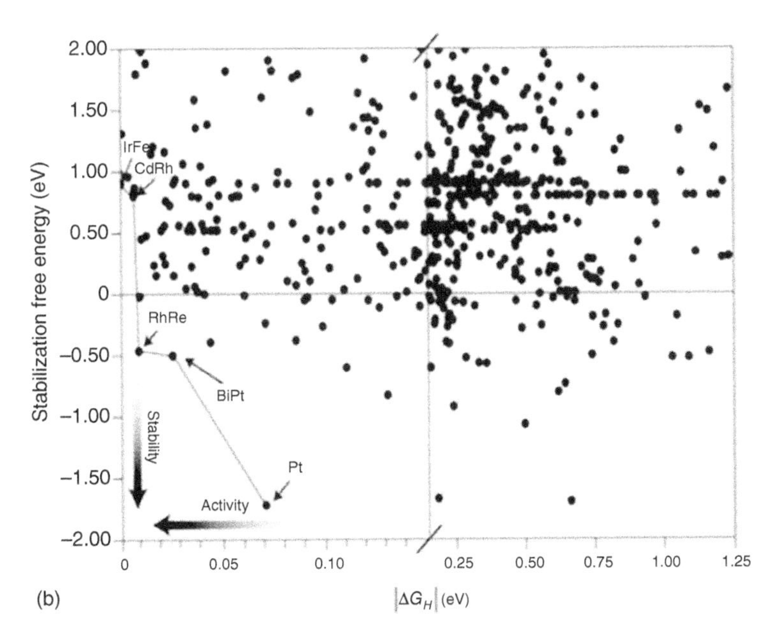

Figure 28.10 (a) Computational high-throughput screening for adsorption free energy of H on 256 pure metals and surface alloys. The slabs were considered to be constructed by the host element with embedded solute element. The solute coverage is 1/3 ML at all cases with the same coverage of adsorbed H. (b) Plot of stability and activity of HER on surface alloys. Source: Adapted with permission from Greeley et al. [59]. Copyright 2006, Nature Publishing Group.

were conducted to determine the strength of H* binding energies. According to the HER mechanisms, the catalytic activity can be simply described using the adsorption free energy of H*, where desorption of H_2 was assumed to be the rate-determining step. In addition, the stability of alloys was considered as another factor in catalyst design. The stabilization free energies for all the materials were calculated, and a 2D catalyst design map was built.

Discovering materials with higher HER activity was set as a main objective in this case. Pt was widely studied in previous research and found to display excellent performance among pure metal materials. Alloy materials better than Pt were found from the HT screening using the computational methods (see Figure 28.10b), including BiPt, which was also tested experimentally in this work to prove its catalytic performance.

28.3.3 Rational Design for CO Oxidation on Multicomponent Alloy Surfaces

The framework of catalyst design in the case of HER generally follows the strategy introduced in Section 28.3.2; however, some reasonable approximations were made due to the complexity of catalytic systems. CO oxidation has been widely studied theoretically in the past 20 years. An in-depth study of the kinetics, namely, transition states, made it another classical system in heterogeneous catalysis.

In the previous example, the alloy was composed of two components. However, "infinite" combinations of alloys can be done numerically with more components of metals. It can be understood that direct screening of adsorption energy on such an infinite number of surfaces will be nearly impossible even through HT computations. In order to overcome this issue, an equation with the bond-counting contribution factor (BCCF) was developed [60, 61], where the parameters in the equation were fitted by DFT calculations, and the equation was then expanded to wider applications (see Figure 28.11a). In their research, four different solute metals were studied with the Pt slab, which also means that the parameters may get changed when the solvent metal is changed and the HT calculation is then of necessity.

Following the framework of catalyst design in the previous part of the chapter, the scaling and BEP relations were built for CO oxidation on pure metal surfaces. A 2D activity volcano surface was then established with microkinetic modeling. The screening of the adsorption energies for both adsorbed CO* and O* on different alloys was done based on BCCFs (see Figure 28.11b). As a result, some surfaces were predicted with better activity than that on pure Pt surface. Last but not least, DFT calculations were conducted on the specific predicted surfaces to check the accuracy of the forecasting approach, and the reaction rate of CO oxidation was calculated by microkinetic modeling.

28.3.4 Adsorbate–Adsorbate Interactions for CO Methanation

There is no doubt that theoretical uncertainty cannot be avoided in the establishment of all the relations in the framework of catalyst design. Indeed, not all

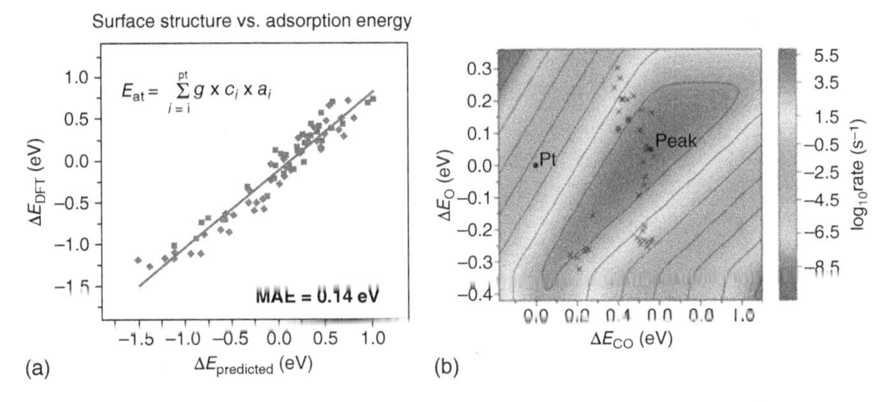

Figure 28.11 (a) Adsorption energy calculated from the bond-counting contribution factor (BCCF) equation plotted with the energies calculated using full DFT calculations. Source: Adapted with permission from Wang and Hu [60]. Copyright 2017, Royal Society of Chemistry. (b) The 2D volcano plot of CO oxidation on close-packed flat surfaces associated with the adsorption of CO and O. The cross points show the predicted activity of different alloys based on the bond-counting contribution factor (BCCF). Source: Adapted with permission from Wang and Hu [61]. Copyright 2017, Royal Society of Chemistry.

the designed catalyst surfaces show predicted activities as tested by either experiments or higher-level theories. Remarkably, the coverage effect was found in significant interactions on the adsorption energies and thus on the predicted activity volcano plot [41, 62–64].

The coverage effect, namely, the adsorbate–adsorbate interactions, was observed in many previous studies and treated by three strategies: (i) The adsorbate–adsorbate interactions were totally ignored when the surface was much less covered. (ii) A specific surface coverage was considered the environment of the system, and all the adsorption energies and activation barriers were calculated under that condition. (iii) Further relations were built to describe the correlation between all the energies, including adsorption energies and activation barriers, and the surface coverage of different kinds of adsorbates.

A self-consistent method with relations between coverages and energies (see Figure 28.12a) has been studied to be a general strategy in kinetic simulations. However, the huge cost of such calculations for the construction of self-interactions and cross-interactions for adsorbates or even activation barriers becomes the main challenge. The improvement of the computational method should make it possible to build complete considerations for adsorbate–adsorbate interactions; however, in fact, previous research mainly focused on the simplification of theoretical models. In this case, only self-interactions were calculated using the DFT calculations, which solely describe the adsorption energies for one specific adsorbate surrounded by itself. In addition, cross-interactions, describing the adsorption energies for one specific adsorbate surrounded by different adsorbates (see Figure 28.12b), were studied by using diverse theoretical methods, and the activation barriers were calculated following the BEP relations. Then, the coverage-dependent activity volcano plot was built, where a larger area with high predicted activity was found, compared with the case without

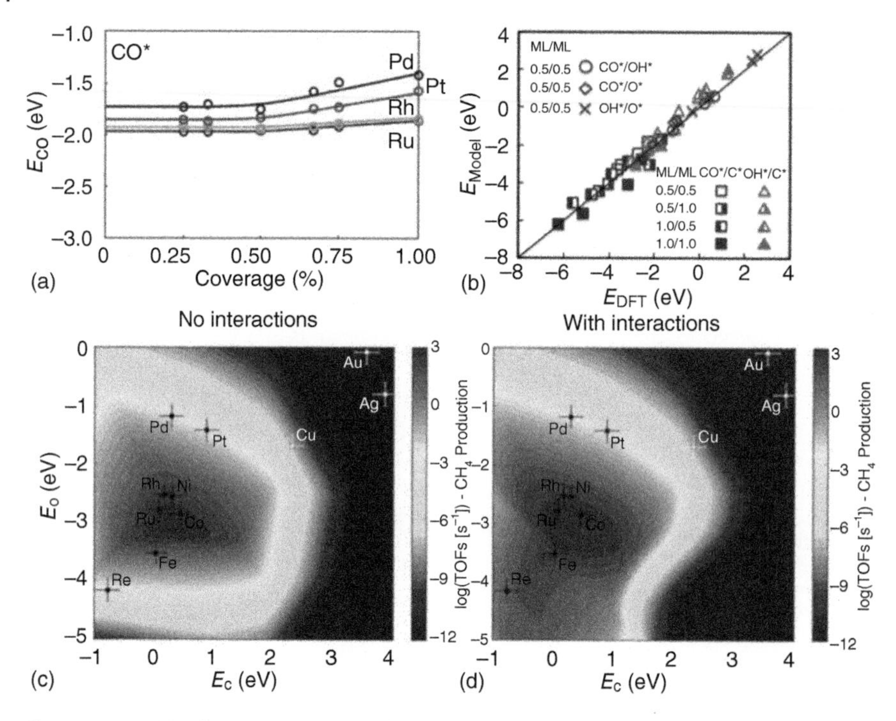

Figure 28.12 (a) Adsorption energies of CO on step (211) surfaces as a function of surface coverage, which describes the self-interactions of coverage effect. (b) Predicted adsorption energies with cross-interactions plotted as a function of DFT-calculated adsorption energies. (c) The activity volcano plot for CO methanation with the descriptors of adsorption energies of C and O. (d) The activity volcano plot for adsorbate–adsorbate interactions. Source: Adapted with permission from Lausche et al. [64]. Copyright 2013, Elsevier Inc.

adsorbate–adsorbate interactions, providing a wider window for the selection of catalytic materials (see Figure 28.12c,d).

28.3.5 RhAu Alloy Nanoparticles for NO Decomposition by Machine Learning

Nanoparticles have been widely used to facilitate chemical reactions. Their high surface-area-to-volume ratio makes them cost saving of novel metals such as gold and platinum. Although nanoparticle surfaces contain varied surface sites, the catalytic activity is often controlled by a few active sites. Identifying the active sites from experiments is still quite costly, regarding the sizes, shapes, and compositions of the nanoparticles, and direct DFT calculations on the nanoparticles would be too expensive. On the other hand, single-crystal surfaces with well-defined sites have been well studied, and their catalytic descriptors such as binding energies of adsorbates can be easily accessible from experiments and DFT calculations or from databases. With the data on single crystals, it is possible to derive a pattern with a machine learning scheme that can predict the catalytic activity of the nanoparticles.

A machine learning algorithm based on a similarity kernel, called the smooth overlap of atomic positions (SOAP), was proposed to study the catalytic performance of $Rh_{(1-x)}Au_x$ alloy nanoparticles for the decomposition of NO ($2NO \rightarrow N_2 + O_2$) [65]. The algorithm is composed of two steps: learning DFT data on single-crystal surfaces and extrapolating the lessons to nanoparticles. The learning data consists of the geometrical information of single-crystal surfaces and binding energies of reaction intermediates on the surfaces. The SOAP evaluates the similarity (noted as K_{IJ}) between the intermediates on different surface sites. SOAP K_{IJ} is the overlap integrals between atomic distributions of surface sites I and J. The binding energy of intermediates on site I is obtained as $E_{b,I} = \sum_J w_J K_{IJ}$. In the learning step, by fitting the existing data of $E_{b,I}$ using a Bayesian linear regression method, regression coefficients of w_J can be obtained. In the extrapolating step, the binding energy of intermediates on site K on nanoparticles is predicted to be $E_{b,K} = \sum_J w_J K_{KJ}$ (see Figure 28.13a). After the active sites and binding energies are found, the catalytic activities on nanoparticles can then be calculated using the binding energies and BEP relations as discussed earlier. As shown in Figure 28.13b, the predicted catalytic activity at 500 K has a volcano-type correlation with respect to the ratio x,

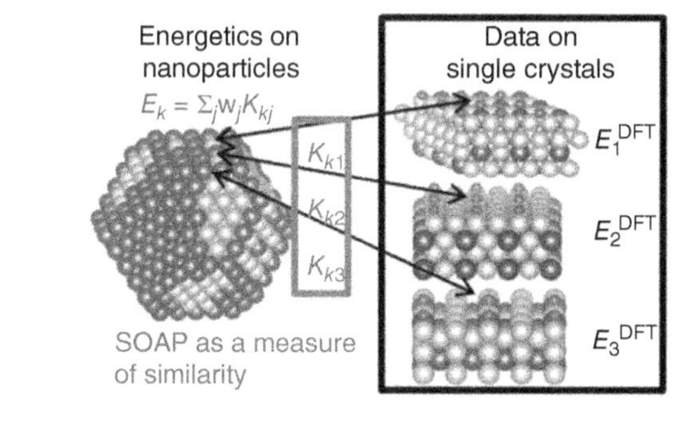

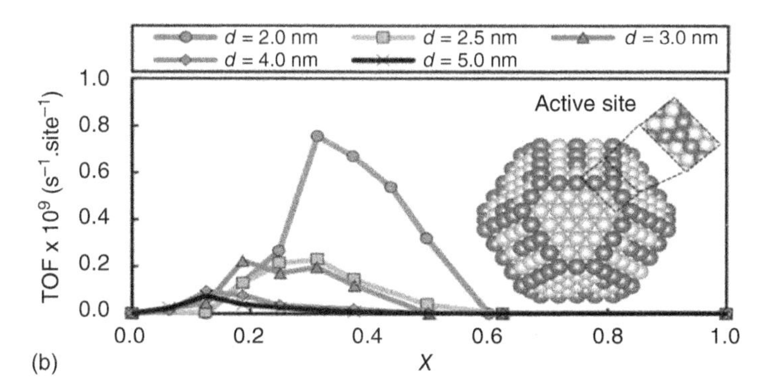

Figure 28.13 (a) Schematic of the machine learning algorithm. (b) Predicted turnover frequencies (TOFs) per surface site at 500 K and the structure of the active site on $Rh_{(1-x)}Au_x$. Source: Adapted with permission from Jinnouchi and Asahi [65]. Copyright 2017, American Chemical Society.

and the maximal activity increases with a decrease in the diameter d of the nanoparticles.

28.4 Summary and Prospects of HT Catalytic Material Design

In this chapter, we introduced one of the most popular strategies as a framework in catalyst design using DFT calculations. A clear and complete process was introduced from the determination of the whole catalytic reaction to the activity prediction of catalysts. Here, the design was more focused on the catalytic reaction based on its fundamental activity on a series of catalytic materials. The HT calculations were involved in many steps of catalytic material design, which are mainly used to determine the activation barriers and the adsorption energies of key adsorbates. Finally, microkinetic modeling was employed to bridge the microscopic energy with the macroscopic reaction rate.

It is obvious that the key point in catalytic material design is the establishment of rules in the prediction of reaction activity based on computational energies. A complicated framework has been now built based on human knowledge of computational catalysis, such as scaling/BEP relations, kinetic model. The improvement of the methods in catalyst design can be divided into two parts. On the one hand, a simpler and more effective model is still the focus of chemists, aiming to obtain a fundamental catalytic activity trend and an in-depth understanding of catalysis with low computational cost. The description and understanding of catalytic reactions using thermodynamic potentials discussed in the chapter have shown great advances in this area. On the other hand, the large amount of data obtained by HT calculations and advanced mathematic algorithms lead to catalytic material design with a blurred understanding. It is expected that the mathematical fitting based on either neutral networks or other machine learning algorithms can bridge the basic property of chemical reactions (such as reactants and products) with their activity rate over a specific catalytic material directly. Moreover, the big data analysis algorithm, or data mining, is rapidly developing to understand materials automatically and create novel materials beyond human knowledge.

References

1 Hohenberg, P. and Kohn, W. (1964). Inhomogeneous electron gas. *Phys. Rev.* 136: B864–B871.
2 Kohn, W. and Sham, L.J. (1965). Self-consistent equations including exchange and correlation effects. *Phys. Rev.* 140: A1133–A1138.
3 Laidler, K.J. and King, M.C. (1983). Development of transition-state theory. *J. Phys. Chem.* 87: 2657–2664.
4 Pechukas, P. (1981). Transition state theory. *Annu. Rev. Phys. Chem.* 32: 159–177.

5 Dumesic, J.A. (1993). *The Microkinetics of Heterogeneous Catalysis*. Washington, DC: American Chemical Society.

6 Chorkendorff, I. and Niemantsve, J.W. (2003). *Concepts of Modern Catalysis and Kinetics*. Weinheim: Wiley-VCH.

7 Abild-Pedersen, F., Greeley, J., Studt, F. et al. (2007). Scaling properties of adsorption energies for hydrogen-containing molecules on transition-metal surfaces. *Phys. Rev. Lett.* 99: 016105. 016105-1–016105-4.

8 Liu, Z.P. and Hu, P. (2001). General trends in CO dissociation on transition metal surfaces. *J. Chem. Phys.* 114: 8244–8247.

9 Logadottir, A., Rod, T.H., Nørskov, J.K. et al. (2001). The Brønsted–Evans–Polanyi relation and the volcano plot for ammonia synthesis over transition metal catalysts. *J. Catal.* 197: 229–231.

10 Yang, B., Burch, R., Hardacre, C. et al. (2014). Mechanistic study of 1,3-butadiene formation in acetylene hydrogenation over the Pd-based catalysts using density functional calculations. *J. Phys. Chem. C* 118: 1560–1567.

11 Sheng, T., Lin, W.F., Hardacre, C., and Hu, P. (2014). Significance of β-dehydrogenation in ethanol electro-oxidation on platinum doped with Ru, Rh, Pd, Os and Ir. *Phys. Chem. Chem. Phys.* 16: 13248–13254.

12 Wang, D., Wang, H., and Hu, P. (2015). Identifying the distinct features of geometric structures for hole trapping to generate radicals on rutile TiO_2(110) in photooxidation using density functional theory calculations with hybrid functional. *Phys. Chem. Chem. Phys.* 17: 1549–1555.

13 Wang, H.F., Li, H.Y., Gong, X.Q. et al. (2012). Oxygen vacancy formation in CeO_2 and $Ce_{1-x}Zr_xO_2$ solid solutions: electron localization, electrostatic potential and structural relaxation. *Phys. Chem. Chem. Phys.* 14: 16521–16535.

14 Mao, Y., Wang, Z., Wang, H.-F., and Hu, P. (2016). Understanding catalytic reactions over zeolites: a density functional theory study of selective catalytic reduction of NO_x by NH_3 over Cu–SAPO-34. *ACS Catal.* 6: 7882–7891.

15 Wang, Y., Li, F., Li, Y., and Chen, Z. (2016). Semi-metallic Be5C2 monolayer global minimum with quasi-planar pentacoordinate carbons and negative Poisson's ratio. *Nat. Commun.* 7: 11488.

16 Giacalone, F., Campisciano, V., Calabrese, C. et al. (2016). Single-walled carbon nanotube-polyamidoamine dendrimer hybrids for heterogeneous catalysis. *ACS Nano* 10: 4627–4636.

17 Norskov, J.K., Bligaard, T., Rossmeisl, J., and Christensen, C.H. (2009). Towards the computational design of solid catalysts. *Nat. Chem.* 1: 37–46.

18 Curtarolo, S., Hart, G.L., Nardelli, M.B. et al. (2013). The high-throughput highway to computational materials design. *Nat. Mater.* 12: 191–201.

19 Grabow, L.C. and Mavrikakis, M. (2011). Mechanism of methanol synthesis on Cu through CO_2 and CO hydrogenation. *ACS Catal.* 1: 365–384.

20 Kortlever, R., Peters, I., Koper, S., and Koper, M.T.M. (2015). Electrochemical CO_2 reduction to formic acid at low overpotential and with high faradaic efficiency on carbon-supported bimetallic Pd–Pt nanoparticles. *ACS Catal.* 5: 3916–3923.

21 Gao, D., Zhou, H., Wang, J. et al. (2015). Size-dependent electrocatalytic reduction of CO_2 over Pd nanoparticles. *J. Am. Chem. Soc.* 137: 4288–4291.

22 Hansen, H.A., Varley, J.B., Peterson, A.A., and Norskov, J.K. (2013). Understanding trends in the electrocatalytic activity of metals and enzymes for CO_2 reduction to CO. *J. Phys. Chem. Lett.* 4: 388–392.
23 Ertem, M.Z., Konezny, S.J., Araujo, C.M., and Batista, V.S. (2013). Functional role of pyridinium during aqueous electrochemical reduction of CO_2 on Pt(111). *J. Phys. Chem. Lett.* 4: 745–748.
24 Peterson, A.A. and Nørskov, J.K. (2012). Activity descriptors for CO_2 electroreduction to methane on transition-metal catalysts. *J. Phys. Chem. Lett.* 3: 251–258.
25 Hummelshøj, J.S., Abild-Pedersen, F., Studt, F. et al. (2012). CatApp: a web application for surface chemistry and heterogeneous catalysis. *Angew. Chem.* 124: 278–280.
26 Jain, A., Ong, S.P., Hautier, G. et al. (2013). Commentary: the materials project: a materials genome approach to accelerating materials innovation. *APL Mater.* 1: 011002.
27 Fernandez, E.M., Moses, P.G., Toftelund, A. et al. (2008). Scaling relationships for adsorption energies on transition metal oxide, sulfide, and nitride surfaces. *Angew. Chem. Int. Ed.* 47: 4683–4686.
28 Montemore, M.M. and Medlin, J.W. (2014). Scaling relations between adsorption energies for computational screening and design of catalysts. *Catal. Sci. Technol.* 4: 3748–3761.
29 Kulkarni, A., Siahrostami, S., Patel, A., and Norskov, J.K. (2018). Understanding catalytic activity trends in the oxygen reduction reaction. *Chem. Rev.* 118: 2302–2312.
30 Seh, Z.W., Kibsgaard, J., Dickens, C.F. et al. (2017). Combining theory and experiment in electrocatalysis: Insights into materials design. *Science* 355: eaad4998.
31 Nie, Y., Li, L., and Wei, Z. (2015). Recent advancements in Pt and Pt-free catalysts for oxygen reduction reaction. *Chem. Soc. Rev.* 44: 2168–2201.
32 Nørskov, J.K., Rossmeisl, J., Logadottir, A. et al. (2004). Origin of the overpotential for oxygen reduction at a fuel-cell cathode. *J. Phys. Chem. B* 108: 17886–17892.
33 Henkelman, G., Uberuaga, B.P., and Jónsson, H. (2000). A climbing image nudged elastic band method for finding saddle points and minimum energy paths. *J. Chem. Phys.* 113: 9901–9904.
34 Wang, H.F. and Liu, Z.P. (2008). Comprehensive mechanism and structure-sensitivity of ethanol oxidation on platinum: new transition-state searching method for resolving the complex reaction network. *J. Am. Chem. Soc.* 130: 10996–11004.
35 Michaelides, A., Liu, Z.P., Zhang, C.J. et al. (2003). Identification of general linear relationships between activation energies and enthalpy changes for dissociation reactions at surfaces. *J. Am. Chem. Soc.* 125: 3704–3705.
36 Eiswirth, M., Bürger, J., Strasser, P., and Ertl, G. (1996). Oscillating Langmuir–Hinshelwood mechanisms. *J. Phys. Chem.* 100: 19118–19123.
37 Weinberg, W.H. (1996). Eley–Rideal surface chemistry: direct reactivity of gas phase atomic hydrogen with adsorbed species. *Acc. Chem. Res.* 29: 479–487.

38 Trinchero, A., Hellman, A., and Grönbeck, H. (2013). Methane oxidation over Pd and Pt studied by DFT and kinetic modeling. *Surf. Sci.* 616: 206–213.

39 Guo, C., Wang, Z., Wang, D. et al. (2018). First-principles determination of CO adsorption and desorption on Pt(111) in the free energy landscape. *J. Phys. Chem. C* 122: 21478–21483.

40 Filot, I.A.W., Broos, R.J.P., van Rijn, J.P.M. et al. (2015). First-principles-based microkinetics simulations of synthesis gas conversion on a stepped rhodium surface. *ACS Catal.* 5: 5453–5467.

41 Medford, A.J., Shi, C., Hoffmann, M.J. et al. (2015). CatMAP: a software package for descriptor-based microkinetic mapping of catalytic trends. *Catal. Lett.* 145: 794–807.

42 Chen, J.-F., Mao, Y., Wang, H.-F., and Hu, P. (2016). Reversibility iteration method to understand reaction networks and to solve micro-kinetics in heterogeneous catalysis. *ACS Catal.* 6: 7078–7087.

43 Cao, X.-M., Burch, R., Hardacre, C., and Hu, P. (2011). An understanding of chemoselective hydrogenation on crotonaldehyde over Pt(111) in the free energy landscape: the microkinetics study based on first-principles calculations. *Catal. Today* 165: 71–79.

44 Wang, Z., Liu, X., Rooney, D.W., and Hu, P. (2015). Elucidating the mechanism and active site of the cyclohexanol dehydrogenation on copper-based catalysts: a density functional theory study. *Surf. Sci.* 640: 181–189.

45 Bligaard, T., Nørskov, J.K., Dahl, S. et al. (2004). The Brønsted–Evans–Polanyi relation and the volcano curve in heterogeneous catalysis. *J. Catal.* 224: 206–217.

46 Cheng, J., Hu, P., Ellis, P. et al. (2008). Brønsted–Evans–Polanyi relation of multistep reactions and volcano curve in heterogeneous catalysis. *J. Phys. Chem. C* 112: 1308–1311.

47 Wang, Z., Wang, H.-F., and Hu, P. (2015). Possibility of designing catalysts beyond the traditional volcano curve: a theoretical framework for multi-phase surfaces. *Chem. Sci.* 6: 5703–5711.

48 Wang, D., Jiang, J., Wang, H.F., and Hu, P. (2016). Revealing the volcano-shaped activity trend of triiodide reduction reaction: a DFT study coupled with microkinetic analysis. *ACS Catal.* 6: 733–741.

49 Friebel, D., Viswanathan, V., Miller, D.J. et al. (2012). Balance of nanostructure and bimetallic interactions in Pt model fuel cell catalysts: in situ XAS and DFT study. *J. Am. Chem. Soc.* 134: 9664–9671.

50 Ghiringhelli, L.M., Carbogno, C., Levchenko, S. et al. (2017). Towards efficient data exchange and sharing for big-data driven materials science: metadata and data formats. *NPJ Comput. Mater.* 3: 46.

51 Beck, D.A.C., Carothers, J.M., Subramanian, V.R., and Pfaendtner, J. (2016). Data science: accelerating innovation and discovery in chemical engineering. *AIChE J.* 62: 1402–1416.

52 Curtarolo, S., Morgan, D., Persson, K. et al. (2003). Predicting crystal structures with data mining of quantum calculations. *Phys. Rev. Lett.* 91: 135503. 135503-1 – 135503-4.

53 Morgan, D., Ceder, G., and Curtarolo, S. (2005). High-throughput and data mining with ab initio methods. *Meas. Sci. Technol.* 16: 296–301.

54 Fischer, C.C., Tibbetts, K.J., Morgan, D., and Ceder, G. (2006). Predicting crystal structure by merging data mining with quantum mechanics. *Nat. Mater.* 5: 641–646.
55 Kitchin, J.R. (2018). Machine learning in catalysis. *Nat. Catal.* 1: 230–232.
56 Cao, A., Schumann, J., Wang, T. et al. (2018). Mechanistic insights into the synthesis of higher alcohols from syngas on CuCo alloys. *ACS Catal.* 8: 10148–10155.
57 Nørskov, J.K., Bligaard, T., Logadottir, A. et al. (2005). Trends in the exchange current for hydrogen evolution. *J. Electrochem. Soc.* 152: J23.
58 Trasatti, S. (1972). Work function, electronegativity, and electrochemical behaviour of metals. *J. Electroanal. Chem. Interfacial Electrochem.* 39: 163–184.
59 Greeley, J., Jaramillo, T.F., Bonde, J. et al. (2006). Computational high-throughput screening of electrocatalytic materials for hydrogen evolution. *Nat. Mater.* 5: 909–913.
60 Wang, Z. and Hu, P. (2017). Formulating the bonding contribution equation in heterogeneous catalysis: a quantitative description between the surface structure and adsorption energy. *Phys. Chem. Chem. Phys.* 19: 5063–5069.
61 Wang, Z. and Hu, P. (2017). A rational catalyst design of CO oxidation using the bonding contribution equation. *Chem. Commun.* 53: 8106–8109.
62 Grabow, L.C., Hvolbæk, B., and Nørskov, J.K. (2010). Understanding trends in catalytic activity: the effect of adsorbate–adsorbate interactions for CO oxidation over transition metals. *Top. Catal.* 53: 298–310.
63 Wu, C., Schmidt, D.J., Wolverton, C., and Schneider, W.F. (2012). Accurate coverage-dependence incorporated into first-principles kinetic models: catalytic NO oxidation on Pt(111). *J. Catal.* 286: 88–94.
64 Lausche, A.C., Medford, A.J., Khan, T.S. et al. (2013). On the effect of coverage-dependent adsorbate–adsorbate interactions for CO methanation on transition metal surfaces. *J. Catal.* 307: 275–282.
65 Jinnouchi, R. and Asahi, R. (2017). Predicting catalytic activity of nanoparticles by a DFT-aided machine-learning algorithm. *J. Phys. Chem. Lett.* 8: 4279–4283.

Section IV

Advancement in Energy and Environmental Catalysis

29

Embracing the Energy and Environmental Challenges of the Twenty-First Century Through Heterogeneous Catalysis

Yun Hau Ng

School of Energy and Environment, City University of Hong Kong, Kong, Kowloon, Hong Kong SAR, PR China

The twenty-first century has introduced great challenges with two of the most significant being the (i) ubiquitous deterioration of environment's quality, in which the causes are highly related to (ii) energy (in)security crisis attributed to the ever-increasing population and the resource consumption pattern of mankind. Driven by the long-established fossil fuel-based processes adopted in all sectors, the currently enjoyable prosperity propelled by industrialization has made detrimental impacts on environment at global scale. As the global pollutions are dominantly due to the emissions or discharge from industrial activities, efforts in tackling environmental problems are therefore logically focused on the reduction of emission at its origin. As elaborated in Sections 1 and 2, heterogeneous catalysis is one of the key pillars in supporting the chemical industries (with their great influences on environment and energy aspects), and it is expected to continue to be the central branch of knowledge in propelling the transition toward carbon-neutral processes. This has driven the rapid development of green chemistry, also known as sustainable chemistry, with the ultimate goals in minimizing, if not eliminating, the use and generation of hazardous substances including carbon emissions [1]. Enabled by insights from fundamental studies, together with ultimate goal for sustainability, heterogeneous catalysis can serve as the center of the chemicals and energy industries, as depicted in Figure 29.1.

One must not be unfamiliar with the term of renewable or alternative energy at this point of time. Emphases have been put forward by many countries over the globe to proactively promote the development of new energy from renewable/clean sources. Besides different types of renewable energy technologies such as photovoltaics and wind energy, hydrogen (H_2) has been regarded a promising clean energy carrier as it has the highest energy density on mass basis (142.2 MJ/kg). Although, on a volume basis, H_2 offers slightly less energy density (12.8 MJ/m^3) than that of natural gas (40.3 MJ/m^3), the only by-product of H_2 consumption to yield water is an irresistible reason to drive its further development. Interest in H_2 energy may have been firstly triggered by the oil crises during the 1970s, but it is now undeniably driven by the need for clean energy. Despite the cleanliness afforded by the utilization of H_2, majority of the H_2

Heterogeneous Catalysts: Advanced Design, Characterization and Applications, First Edition.
Edited by Wey Yang Teoh, Atsushi Urakawa, Yun Hau Ng, and Patrick Sit.

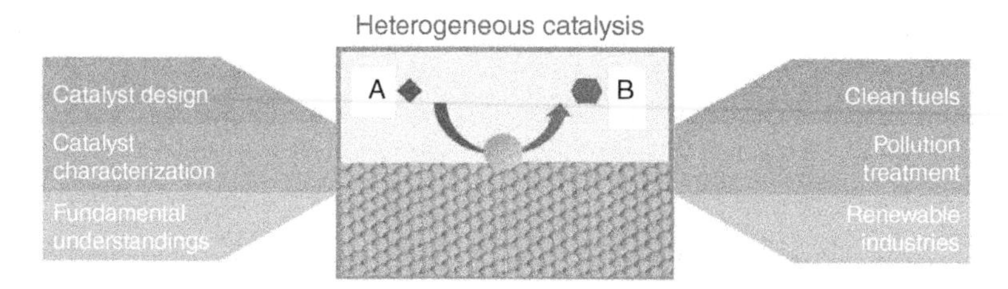

Figure 29.1 Heterogeneous catalysis bridges sciences and energy/environmental industries. (See online version for color figure).

production is still derived from fossil fuels through steam reforming of natural gas and gasification of coal, processes that are carbon emitting [2]. Chapters 30 and 31 discuss the development of robust and active catalytic processes to produce hydrogen solely from water, i.e. water splitting reactions through electrochemical or photo-driven splitting of water. It is the Holy Grail in this area of research to produce clean energy by the cleanest means (water and, perhaps, sunlight) [3]. Upon generation, the most convenient and straightforward usage of H_2 is its conversion into electricity through efficient fuel cells. The working principles of fuel cells rely on electrocatalytic processes [4]. The fundamental understandings on the mechanisms of electron transport and transfer between electrodes and "reactants" are, therefore, critical to designing an active system. Advancements in water splitting and fuel cells are progressing in parallel, with the catalyst as the focal point of research.

Biomass conversion or biomass upgrading is another emerging application for heterogeneous catalysis [5]. Biomass has been regarded as the only substantial source of renewable carbon. Hence, efficiently utilizing biomass as the preferred feedstock to generate renewable chemicals and liquid fuels is of paramount importance in closing the carbon loop in industrial processes. Eliminating waste in the form of biomass (as carbon feedstock) through catalytic processes could have been partially driven by the vision of Anastas and Warner [1], who introduced the 12 principles of green chemistry. Positioned nineth in the postulates, the use of catalysis is emphasized over stoichiometric reagents as the catalysts are not consumed in such reaction. Although it can be a homogeneous catalytic system, the system is often more pH dependent than those in heterogeneous system. In homogeneous catalytic system, the use of large volume of liquid acid (Bronsted and Lewis) could sometimes be less efficient and less environmentally friendly. The replacement with solid acids and zeolites in such reaction is therefore anticipated and desired. In principle, biomass conversion process (or in fact more suitably as an upgrading process) is catalyzed preferably by heterogeneous catalyst to break down a wide range of biomass molecules into fractions of platform compounds prior to their upgrade into desired products. The conversion of biomass using heterogeneous catalysts is discussed in detail in Section 4.

The efforts mentioned earlier are mostly to promote or create a sustainable future for mankind. Current issues, however, should be addressed without further waiting. As current practices in industrial activities could not be replaced or upgraded immediately, it is logical to formulate strategies to tackle the present emissions or discharge. Carbon dioxide (CO_2) emission is undoubtedly correlated to human activities and is often attributed to the large-scale industrial processes and individual's contribution (dominantly through transportation mode). The detrimental impacts of having high atmospheric concentration of CO_2 (or other greenhouse gases) could be catastrophic at global scale and need not to be reiterated in this book. Carbon capture and storage (CCS) has been developed and utilized commercially to capture up to 90% of CO_2 produced from power generation and other industrial processes. A CCS process involves three major steps: (i) capturing CO_2 at the source, (ii) compressing CO_2 for transportation, and (iii) injecting and storing CO_2 into deep underground rock formation (typically at depths of kilometers). The storing process is potentially the most vulnerable and high-cost stage [6]. In the last decade, CO_2 conversion emerges as an alternative solution in addressing excessive CO_2 [7]. As reactions occur on the surface of heterogeneous catalysts, a thoughtful design of catalysts (considering the suitable surface chemistry, electron transfer, energy/kinetics, surface geometry, etc.), a wide range of CO_2 conversion products can potentially be tailored. Ideally, with proper understanding of the reaction mechanism using heterogeneous catalysts, distribution of products can be modulated: carbon monoxide (CO) and methane for gas-phase products and formate, methanol, and longer carbon-chain fuels for liquid products. In principle, CO_2 conversion offers a more effective solution than that of storage in long term. CO_2 conversion using thermal, electrochemical, photocatalytic, and photoelectrochemical methods is covered in Chapters 35 and 36. Though they are yet to be commercially available, these methods hold the key of converting excessive CO_2 and are expected to be adopted in the near future.

In addition to CO_2, nitrogen oxides (NO_x) generated by transportation fuels are major polluting agents in cities. NO_x inhalation triggers a number of respiratory diseases, and it is also an essential component in forming ground-level ozone in the presence of volatile organic compounds and sunlight. Selective catalytic reduction (SCR) process has been formulated to convert NO_x into diatomic nitrogen (N_2) and water [8]. SCR technology was developed in Japan and the United States in the early 1960s with the first large-scale SCR installed in 1978. Though the SCR catalyst has already been widely adopted nowadays, its continued search for less expensive and more durable catalysts is indispensable to meeting the emission standards: Euro I (2003), Euro IV (2005), Euro V (2008/2010), and now Euro VI. Besides gas emission, discharge of pollutant-containing wastewater is another area that benefited from the advancement in heterogeneous catalysis. Existing wastewater treatment processes (such as flocculation–coagulation, zerovalent iron [ZVI], and membrane technology) are, in principle, sufficiently effective in removing bulk pollutants including pesticides, heavy metals, and industrial chemicals. However, these conventional wastewater treatment methods have limited success in treating a number of emerging pollutants such as persistent organic pollutants (POPs), emerging micropollutants, and antibiotics. Heterogeneous catalysis, especially photocatalysis, finds potential applications in these niche fields [9]. Heterogeneous photocatalysts operate on the fundamental basis of semiconductor. Through careful design of the optical band structures and their relevant oxidizing energy, a photocatalyst with proper surface chemistry can help in preferential adsorption of those organic pollutants. Subsequently, in the presence of incident light (could be natural or artificial light), the oxidizing potential of the photocatalyst can mineralize the pollutants completely or partially oxidize them into desired chemicals. Chapters 37 and 38 discussing the development of SCR catalysts for NO_x reduction, as well as photocatalyst for abatement of emerging pollutants are included in the last part of Section 4.

As elaborated above, heterogeneous catalysis is a multidiscipline involving various branches of science and has found great importance in traditional process as well as emerging catalytic applications. From a realistic point of view, excellent properties and performance of a catalyst are results of tedious yet rigorous design, testing, and implementation process ranging from the molecular level all the way up to the large-scale reaction systems engineering. As we look ahead, the multiscale nature in this discipline requires the master or integration of various expertise (from catalyst structure, mechanism, theory, and materials transport); therefore, the education and training of young scientists should have this preparation embedded.

References

1 Anastas, P.T. and Warner, J.C. (1998). *Green Chemistry: Theory and Practice.* New York: Oxford University Press. https://doi.org/10.1159/000143289.

2 Holladay, J.D., Hu, J., King, D.L., and Wang, Y. (2009). An overview of hydrogen production technologies. *Catal. Today* 139 (4): 244–260.

3 (a) Kudo, A. and Miseki, Y. (2009). Heterogeneous photocatalyst materials for water splitting. *Chem. Soc. Rev.* 38 (1): 253–278. (b) Ursua, A., Gandia, L.M., and Sanchis, P. (2012). Hydrogen production from water electrolysis: current status and future trends. *Proc. IEEE* 100 (2): 410–426.

4 Steele, B.C.H. and Heinzel, A. (2001). Materials for fuel-cell technologies. *Nature* 414 (6861): 345–352.

5 Alonso, D.M., Bond, J.Q., and Dumesic, J.A. (2010). Catalytic conversion of biomass to biofuels. *Green Chem.* 12 (9): 1493–1513.

6 Haszeldine, R.S. (2009). Carbon capture and storage: how green can black be? *Science* 325 (5948): 1647–1652.

7 Centi, G., Quadrelli, E.A., and Perathoner, S. (2013). Catalysis for CO_2 conversion: a key technology for rapid introduction of renewable energy in the value chain of chemical industries. *Energy Environ. Sci.* 6 (6): 1711–1731.

8 Shelef, M. (1995). Selective catalytic reduction of NO_x with N-free reductants. *Chem. Rev.* 95 (1): 209–225.

9 (a) Choi, J., Lee, H., Choi, Y. et al. (2014). Heterogeneous photocatalytic treatment of pharmaceutical micropollutants: effects of wastewater effluent matrix and catalyst modifications. *Appl. Catal., B* 147: 8–6. (b) Pi, Y., Li, X., Xia, Q. et al. (2018). Adsorptive and photocatalytic removal of persistent organic pollutants (POPs) in water by metal-organic frameworks (MOFs). *Chem. Eng. J.* 337: 351–371. (c) Elmolla, E.S. and Chaudhuri, M. (2010). Photocatalytic degradation of amoxicillin, ampicillin and cloxacillin antibiotics in aqueous solution using UV/TiO_2 and UV/H_2O_2/TiO_2 photocatalysis. *Desalination* 252 (1–3): 46–52.

30

Electrochemical Water Splitting

Guang Liu[1,2], *Kamran Dastafkan*[1], *and Chuan Zhao*[1]

[1] *The University of New South Wales, School of Chemistry, Sydney, NSW 2052, Australia*
[2] *Taiyuan University of Technology, Research Institute of Special Chemicals, College of Chemistry and Chemical Engineering, Shanxi Key Laboratory of Gas Energy Efficient and Clean Utilization, Taiyuan, Shanxi 030024, PR China*

As a crucial reactant in industry, hydrogen is considered an expensive consumable mostly used in petroleum refining and ammonia synthesis [1]. Such extensive utilization demands large-scale and highly efficient hydrogen production technologies. Hydrogen mass production is largely dependent on the reforming and gasification of fossil fuels which entail high cost, environmental pollution, and product's low purity [2]. Besides, hydrogen is a sustainable energy carrier for renewable energy and regarded as an alternative to fossil fuels for the growing economy [3]. Electrochemical water splitting that uses the electricity generated from the intermittent waste heat, wind, and solar energies is a facile approach to efficiently produce pure hydrogen and oxygen [4]. In this chapter, we expand on the fundamentals, ongoing challenges, and progress, as well as the outlook for water electrolysis technology.

30.1 Fundamentals of Electrochemical Water Splitting

Use and development of electrochemical water splitting date back to more than two centuries ago. Hydrogen production can be produced by employing the electricity output from any generator. However, it is most efficient and environmentally friendly when the processes are derived from renewable sources. Conventional electrochemical water splitting consists of an archetypal electrolysis cell, a conducting electrolyte, and two redox electrodes for oxidation–reduction reactions as shown in Figure 30.1. At anode, the half reaction known as water oxidation or oxygen evolution reaction (OER) proceeds. Depending on the kind and strength of the ionic medium, the following equations are dominant [5, 6]:

$$\text{Acidic medium:}\ \ 2H_2O \rightarrow O_2 + 4H^+ + 4e^- \quad (E^0 = 1.23\,\text{V vs. SHE}) \tag{30.1}$$

Heterogeneous Catalysts: Advanced Design, Characterization and Applications, First Edition.
Edited by Wey Yang Teoh, Atsushi Urakawa, Yun Hau Ng, and Patrick Sit.

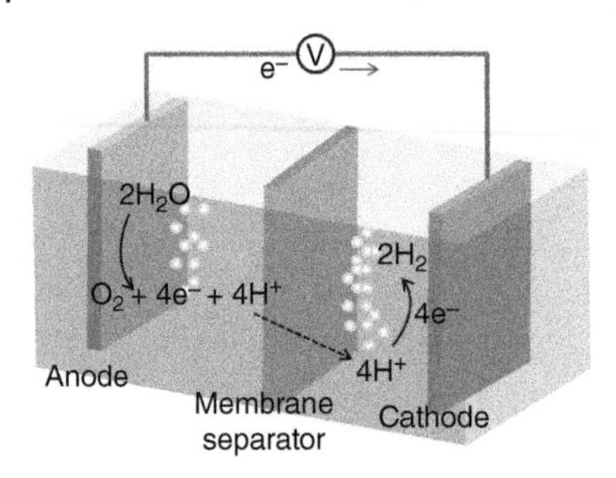

Figure 30.1 Illustration of the electrolysis cell for electrochemical water splitting. (See online version for color figure).

$$\text{Alkaline medium:}\quad 4OH^- \rightarrow 2H_2O + O_2 + 4e^- \quad (E^0 = 0.40\ \text{V vs. SHE}) \tag{30.2}$$

At cathode, the reduction of water, or more commonly referred to as hydrogen evolution reaction (HER), occurs, in which this reaction proceeds has different pathways in different electrolytes [5, 6]:

$$\text{Acidic medium:}\quad 4H^+ + 4e^- \rightarrow 2H_2 \quad (E^0 = 0.00\ \text{V vs. SHE}) \tag{30.3}$$

$$\text{Alkaline medium:}\quad 4H_2O + 4e^- \rightarrow 4OH^- + 2H_2 \quad (E^0 = -0.83\ \text{V vs. SHE}) \tag{30.4}$$

Combining both half-reactions balanced with acid or base results in the overall water splitting into oxygen and hydrogen, where hydrogen ions migrate toward the cathode while hydroxide ions toward the anode:

$$\text{Overall reaction:}\quad 2H_2O(l) \rightarrow 2H_2(g) + O_2(g) \quad (E^0_{\text{cell}} = 1.23\ \text{V vs. SHE}) \tag{30.5}$$

The total number of transmitted electrons in water is twice the number of produced hydrogen molecules and four times the number of evolved oxygen molecules at ambient conditions. The oxidation and reduction potentials required to thermodynamically drive OER and HER occur at 1.23 and 0.00 V vs. a standard hydrogen electrode (SHE), respectively, at standard temperature and pressure (25 °C, 1 atm). However, both half-reactions are kinetically sluggish due to the complex charge and mass transfer processes at the interface of the electrode and electrolyte. As a result, activation energy barriers exist due to the need of additional voltage beyond the thermodynamic potentials (1.23 V) that are termed as overpotential (η_{OER} and η_{HER}). In addition, more energy barriers evolve during water electrolysis leading to the appearance of other types of overpotential, namely, surface blockage by the evolved O_2 (η_{bubble}, O_2) and H_2 (η_{bubble}, H_2) gas bubbles, electrolyte resistance ($\eta_{\text{electrolytes}}$), and electrical resistance of the external circuit (η_{circuit}). For practical applications, an effective interface between electrode and electrolyte phases must be developed to overcome the aforementioned sources of poor efficiency [7].

However, the utilization of renewable energy for hydrogen production requires appropriate technologies. Practical aspects of electrochemical water splitting involve development of both energy- and cost-effective water electrolyzers since the efficiency and high purity of hydrogen gas determine a commercially advantageous technology, specifically when the subsequent energy conversion to electricity by fuel cell technology is considered [8]. The elaboration and development of electrocatalysts for water splitting in this chapter is focused on the low-temperature liquid electrolysis. Development of efficient electrocatalyst materials is a straightforward option to reduce anodic and cathodic overpotentials. To date, oxide compounds of noble metals of ruthenium (RuO_2) and iridium (IrO_2) are the current benchmark OER electrocatalysts, where platinum is known as the best electrocatalyst for HER. However, supply and application of these precious metals for large-scale water splitting are costly and yet not sustainable. Accordingly, the design of novel compounds based on earth-abundant metals is focused not only to improve the thermodynamically unfavorable and kinetically limited energy consumption but also to achieve a cost-effective water electrolysis system.

30.2 Technological and Practical Considerations

Considering the intermittent nature of renewable energy sources, the advancement in the practical aspects of water electrolysis is greatly envisaged where durability and effectiveness of water electrolyzers are the most crucial factors. In principle, water electrolyzers operate similarly, but different conditions necessitate different requirements. From the technological perspective, water splitting can be performed in different electrolytic and temperature conditions. Both liquid and solid media can serve as electrolyte while having different ion species that are electrochemically transferred during the process, such as OH^-, H^+, and O^{2-}. The range of operation temperature can be between room temperature and 1000 °C. Low operating temperatures allow for a facile design and safe functioning, while high operating temperatures lead to higher efficiency. Based on these factors, water electrolyzers have been categorized into three major types: liquid electrolyte, polymer electrolyte membrane (PEM), and solid oxide electrolyte (SOE) cells (Figure 30.2) [10–12]. Specific conditions, advantages, and disadvantages exist for each type, making them the main sustainable technologies in industry.

30.2.1 Liquid Electrolyte Water Electrolysis

Liquid electrolyzers operate at low temperatures and depending on the acidity of electrolyte. Two kinds of acid liquid electrolyte water electrolyzer (ACIWE) and alkaline liquid electrolyte water electrolyzer (ALKWE) exist. ALKWE has found industrial applications due to more durability of electrode materials in alkaline solution. Water splitting in ALKWE occurs between 20 and 80 °C and at pressures below 3 MPa either in potassium or sodium hydroxide solution, generally 20–40 wt%, and has become the most mature among the three technologies [13]. ALKWE is recognized as a practical and sustainable hydrogen generation technology thanks to its flexibility in cell design especially when the input energy

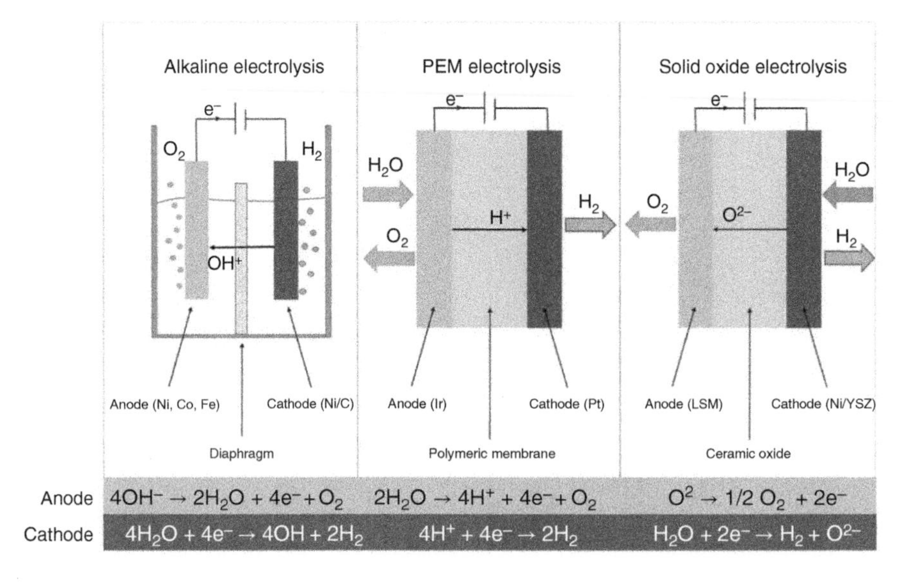

Figure 30.2 Operation principles of ALKWE, PEMWE, and SOEWE. The overall water splitting reaction in all electrolyzers is $H_2O \rightarrow H_2 + 1/2O_2$. Source: Reprinted with permission from Sapountzi et al. [9]. Copyright 2017, Elsevier. (See online version for color figure).

is provided by renewable sources [14]. The charge carrier is OH^- whose transfer through electrolyte triggers gas evolution reactions at anode and cathode. Besides electrode, electrolyte, and electrocatalyst, cell components in ALKWE include diaphragm, which is a permeable membrane for OH^- to separate the evolved gases on electrodes. Older types were composed of asbestos, but due to safety concerns, asbestos has been replaced by composite materials such as polyethersulfone (PES), polyphenylene sulfide (PPS), and nickel oxide-coated titanium mesh [15]. Anodic and cathodic reactions have already been mentioned in Section 30.1. Regarding the commercial availability, Section 30.3 is devoted to the electrode materials in liquid electrolyte technology. Considering the ohmic losses in the electrical circuit, slow ion conduction is one of the major disadvantages. Furthermore, diaphragm is also a potential barrier to OH^- transport, thereby leading to low current densities in ALKWE. Gas permeation through diaphragm is also another source for low efficiency, typically between 59% and 70% in this electrolyzer. Accordingly, relatively high energy is required which adds up to the operation cost [9, 13, 16]. Lastly, due to the corrosive nature of electrolytes, electrode durability is an ongoing problem. In this regard, application of additives such as ionic liquids has been proposed to reduce the corrosive nature and to increase the ionic strength of electrolytes [17].

However, the total cost for commercialization is still low for ALKWE compared to other technologies because of the use of non-noble metal catalysts and their long-term stability. In the cell design, thermodynamic studies have shown that water splitting could be pursued in ALKWE under high pressure as well, which, indeed, meets lower energy consumption. Typical investigations in this direction revealed that the pressure difference in two half-cells leads to hydrogen

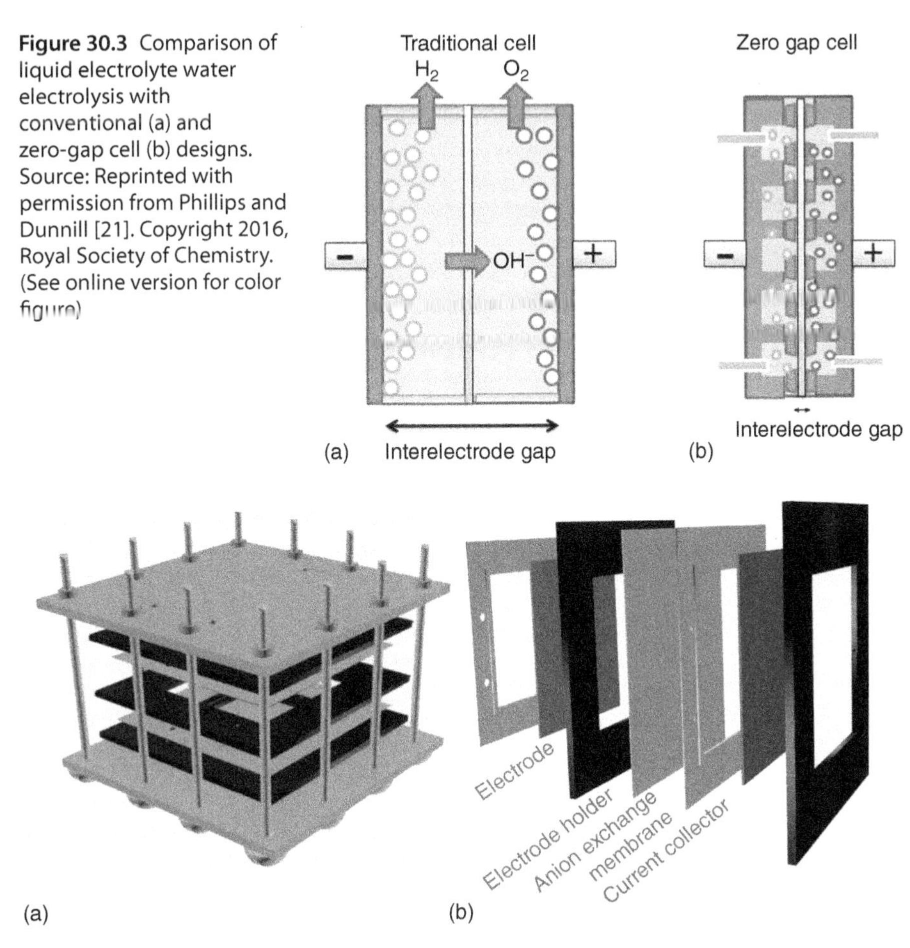

Figure 30.3 Comparison of liquid electrolyte water electrolysis with conventional (a) and zero-gap cell (b) designs. Source: Reprinted with permission from Phillips and Dunnill [21]. Copyright 2016, Royal Society of Chemistry. (See online version for color figure)

Figure 30.4 Schematic representation of a water splitting electrolyzer. (a) 3D model of a cell stack and (b) zoom-in view of the redox electrode's assembly. Source: Reprinted with permission from Peng et al. [22]. Copyright 2019, American Chemical Society. (See online version for color figure).

storage with high purity [18–20]. Furthermore, recent innovations have introduced state-of-the-art electrolyzers such as zero-gap electrode and stacked cell designs. Zero-gap electrode cell eliminates high ohmic losses in conventional electrolyzers with a minimum distance among the electrodes and diaphragm, resulting in a decreased voltage input (Figure 30.3) [21]. Stacked cell design has been proposed for practical considerations and is an assembly of a bipolar series of cells. A single unit of a stacked design is shown in Figure 30.4 [22].

In another perspective, four innovative strategies have been developed for ALKWE, and the schematic designs are depicted in Figure 30.5 [23].

30.2.1.1 Overall Water Electrolysis (OWE)

Although OER and HER reactions, respectively, favor alkaline and acidic environments, whole water splitting should be performed in one medium. Besides, regarding the sluggish kinetics of OER and weak stability of most metallic

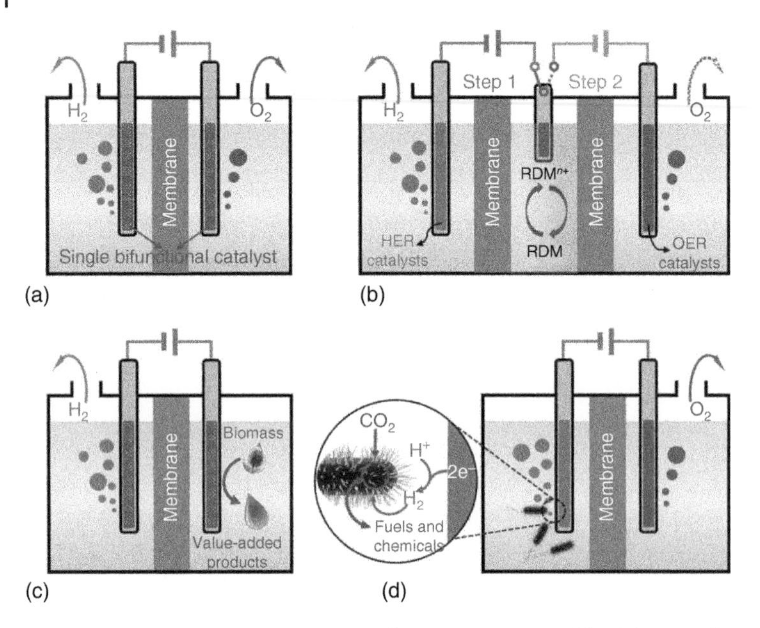

Figure 30.5 Four innovative strategies for nonconventional liquid water electrolysis. (a) Overall water electrolysis; (b) decoupled water electrolysis; (c) hybrid water electrolysis; (d) tandem water electrolysis. RDM = redox mediator. Source: Reprinted with permission from You and Sun [23]. Copyright 2018, American Chemical Society. (See online version for color figure).

materials in acid, the overall efficiency would improve in alkaline solution. Therefore, overall water electrolysis (OWE) operating in alkaline condition has been recognized as the simplest practical technology.

30.2.1.2 Doubled Water Electrolysis (DWE)

In this design, redox species are used between anode and cathode chambers. These are reactive materials such as ferrocene derivatives that behave as reversible electron donor–acceptor pairs. Accordingly, the first anodic reaction is oxidation of these species, and then with their consecutive reduction, the coupled OER on a second electrode would occur. This way, not only the mixing of the evolved product gases is avoided, but also higher HER kinetics would be rendered.

30.2.1.3 Hybrid Water Electrolysis (HWE)

In doubled water electrolysis (DWE), oxygen gas still evolves on a second anode. Being the bottleneck reaction in water splitting because of its large overpotential and the fact that oxygen is not the target product, OER causes a major energy loss in the whole electrolysis. It has been demonstrated that organic oxidation reactions are more thermodynamically feasible than OER and thus ideal as an anodic counterpart. Hybrid water electrolysis (HWE) exhibits several privileges:

(i) Overall cell voltage is reduced owing to more favorable organic oxidation, which in turn leads to improved efficiency.
(ii) More valuable anodic products are obtained.

(iii) Formation of reactive oxygen species along with O_2/H_2 mixture is inhibited.
(iv) Membrane-free cells could be developed due to high selectivity in hydrogen generation.

Organic species, i.e. biomass, should be highly soluble in water, form non-gaseous products, possess lower onset potential, and have minimum competitiveness with HER. One ideal biomass organic is 5-hydroxymethylfurfural (HMF) as it can be oxidized into 2,5-furandicarboxylic acid (FDCA), a valuable chemical required for synthesizing polyamides, polyesters, and polyurethanes [23, 24].

30.2.1.4 Tandem Water Electrolysis (TWE)

In this strategy, the generated hydrogen gas is simultaneously used by biological catalysts, e.g. enzymes to form valuable fuels such as methane and ammonia. Hence, costly requirements in terms of storage and transport in the form of hydrogen are eliminated.

Notwithstanding the lower efficiency, ALKWE is being the most pragmatic approach in water electrolysis technology owing to developments and the utilization of non-precious catalysts.

30.2.2 Polymer Electrolyte Membrane Water Electrolysis

Polymer electrolyte membrane water electrolyzer (PEMWE) is also one of the current practical technologies in water splitting. In this design, a membrane electrode assembly (MEA) consisting of a solid polymer, two redox electrodes, and coated electrocatalyst materials is the main component of the electrolyzer. Solid polymer membrane is conductive and transports ionic agents such as protons to form an acidic environment. Deionized water without any electrolytic species is passed through the membrane, which acts as both electrolyte and gas separator. Other parts include a gas diffusion layer (GDL) or the so-called current collector and a gasket for smooth current flow between bipolar plates and electrodes. Bipolar plates are electrically conductive. They carry the deionized water through, and convey the product gases out of electrodes. These plates and GDL are usually made of titanium or protected stainless steel [15]. The backbone membrane can be made of perfluorosulfonic acid (PFSA) polymers where interesting chemistry, namely, good oxidative behavior, mechanical robustness, and dimensional stability with the changes in temperature and pressure as well as ionic conductivity, results in high efficiency. Membrane thickness is a determining factor for controlling gas permeability and usually varies between 100 and 200 μm. But to date, the most utilized polymer in PEMWE has been Nafion [15]. OER and HER reactions are shown in Figure 30.2.

Over time, efforts have been made to merge the principles in ALKWE and PEMWE, and a hybrid system called solid alkaline water electrolyzer (SAWE) or anion-exchange membrane water electrolyzer (AEMWE) is developed. The ionic agent OH^- is selectively conducted through PEM, creating a local alkaline environment. This new technology uses the advantages of the both electrolyzers described. SAWE is still operating at laboratory scale but presents some merits such as lower capital cost and the use of non-noble metallic electrocatalysts since

OER could proceed better in basic electrolytes. However, the insufficient selectivity of PEMs for OH^- poses serious obstacles to further electrolyte developments [25, 26]. Anodic and cathodic reactions in SAWE are as follows:

$$\text{Anodic:}\ \ 4OH^- \rightarrow 2H_2O + O_2 + 4e^-$$

$$\text{Cathodic:}\ \ 2H_2O + 4e^- \rightarrow 4OH^- + 2H_2$$

Both types of PEMWE cells operate between 20 and 200 °C, but the cell efficiency of the protonic PEMWE is high and can be up to 82% [9]. Generally, the compact design of PEMWE leads to rapid responses, and protonic conductivity decreases the ohmic loss. Therefore, PEMWE can achieve higher current densities. These merits together with higher hydrogen purity result in an overall improved efficiency with respect to ALKWE where the diffusion rates are slow [10]. As in ALKWE, gas shift reactions can happen in PEMWE as well. However, optimization in the membrane structure and thickness effectively reduces this issue. Solid electrolyte also favors the development of stacked cell design with differential pressure in PEMWE [27]. Nonetheless, PEM materials are not as diverse as electrocatalysts, and the local acidity in protonic PEMWE specifically inhibits the utilization of abundant metal catalysts. Therefore, total cost for the cell design in PEMWE is high [11].

30.2.3 Solid Oxide Electrolyte Water Electrolysis

In solid oxide electrolyte water electrolysis (SOEWE) technology, electrolyte is in the form of solid, typically an inorganic solid oxide, and conductive to oxygen anion (O^{2-}) as the ionic carrier. The main operational characteristic in SOEWEs is the broad range and high temperature of the cell, expanded between 500 and 1000 °C. Thus, basically SOEWE is a steam electrolyzer where the kinetics of water splitting is significantly improved compared with those of low-temperature electrolyzers, i.e. ALKWE and PEMWE. Thermodynamic studies have shown that the total energy demand (ΔH) in SOEWE undergoes small variations with high temperature, while heat demand ($T\Delta S$) increases significantly especially after the intermediate temperature range (500–700 °C). Therefore, the electrical energy demand (ΔG) required to fulfill the electrolysis is relatively low with the increment in temperature [28]. This largely contributes to the reduction in operation cost. When coupled with renewable energy sources, i.e. solar, hydrogen production efficiency can reach 100% [9]. SOEWEs mainly use ceramic materials as solid oxide electrolyte, which in addition to the high O^{2-} conductivity surpassing the gas permeation inside the cell. In recent years, composite materials have been prepared and used in solid electrolytes where the ceramic oxide phase is mixed with the electrocatalyst phase. This strategy can enhance the magnitude of the three-phase boundary and increase the contact between the electrode and electrolyte. Hence, both electrocatalytic and mechanical stability could be improved [29, 30]. Ni-based composites and yttria-stabilized zirconia (YSZ) are typical materials for cathodes, while perovskite and the related oxides such as lanthanum strontium manganite (LSM) are common candidates for anodes [9]. OER and HER reaction routes occur as shown in Figure 30.2. Overall, excellent ion conductivity and high temperature result in significant ohmic losses for SOEWE.

In addition to anion oxygen conductive solid electrolytes, proton conductors that are mainly composed of ceramic oxide have been developed and found applications in steam electrolysis. The reasons are better ion conductivity of ceramic materials and higher cell efficiency at the intermediate temperature range (500–700 °C). Also, ceramic proton conductors make better composites with Ni materials for cathode [31]. OER and HER reactions in this design are as follows:

$$\text{Anodic:}\quad 2H_2O \rightarrow 4H^+ + 4e^- + O_2$$

$$\text{Cathodic:}\quad 4H^+ + 4e^- \rightarrow 2H_2$$

Both oxygen anion and protonic SOEWEs operate between 500 and 1000 °C. They exhibit advanced kinetics and thermodynamics as well as high efficiency up to 100%. However, protonic SOEWE is still at laboratory scale. Another advantage is that solid electrolyte assists the reaction dynamics and favors the compact design, e.g. stacked cell resulting in faster cell response with respect to ALKWE [32]. High temperature in anion oxygen conductive SOEWEs has led to co-electrolysis of H_2/CO_2. Related developments in this area pave the way to produce syngas ($H_2 + CO$) with SOE technology with the same energy demand as that of steam electrolysis at temperatures between 700 and 900 °C. This is beneficial since syngas could be transformed into value-added fuels such as methanol via the Fischer–Tropsch method [33]. Anodic and cathodic reactions in this type of steam electrolyzers are as follows:

$$\text{Anodic:}\quad O^{2-} \rightarrow 1/2O_2 + 2e^-$$

$$\text{Cathodic:}\quad H_2O + 2e^- \rightarrow H_2 + O_2{}^-$$

$$CO_2 + 2e^- \rightarrow CO + O_2{}^-$$

In this regard, co-electrolytic SOEWE exhibits both energy storage and CO_2 reduction features. However, SOEWE technology has its own problems. Even with reduced gas shift reactions owing to the use of solid electrolyte, successful sealing of the electrode chambers is hard to attain. Prolonged operations at high temperatures put mechanical stability and safe reaction cycles at risk. On this account, electrode degradation often narrows down the range of electrocatalyst materials [34].

30.3 Electrocatalyst Materials in Liquid Electrolyte Water Splitting

30.3.1 Oxygen Evolution Reaction Electrocatalysts

30.3.1.1 Metal Oxides

Regarding the four-electron transfer through the electrical double layer, the charge and ion transport from bulk electrolyte to anode surface during OER

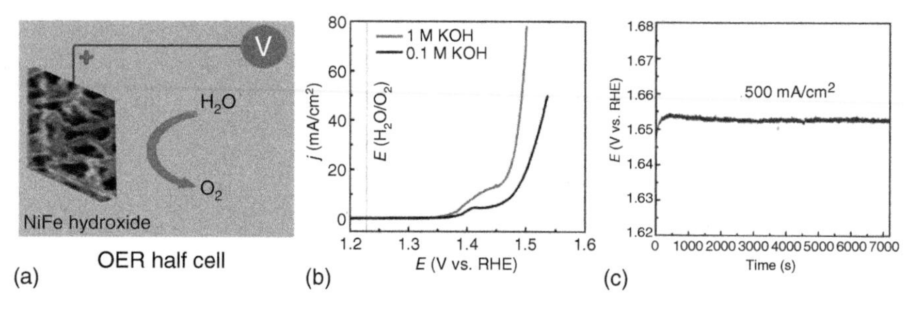

Figure 30.6 (a) Water oxidation schematic diagram based on NiFe hydroxide supported on Ni foam in alkaline electrolyte. (b) OER polarization curves of the NiFe/NF oxygen electrode in 0.1 and 1 M KOH solutions at 5 mV/s with 95% iR compensations. (c) The chronopotentiometry curve of NiFe/NF in 10 M KOH with a constant current density of 500 mA/cm^2 (vs., versus). Source: Reprinted with permission from Lu and Zhao [36]. Copyright 2015, Nature Publishing Group. (See online version for color figure).

is more complicated than HER, and hence higher overpotential is required. Recent investigations on the first-row transition metals have demonstrated that Fe-based atoms (Fe, Co, and Ni) and their compounds such as spinel oxides, perovskite oxides, and layered (oxy)hydroxides could be effective substituents to RuO_2/IrO_2. For example, Boettcher and coauthors reported the OER performances of NiO_x, CoO_x, FeO_x, $Ni_yCo_{1-y}O_x$, $Ni_{0.9}Fe_{0.1}O_x$, IrO_x, and MnO_x films [35] and found that $Ni_{0.9}Fe_{0.1}O_x$ with a low iron loading in nickel oxide matrix exhibited the highest OER activity of about an order of magnitude higher than that of IrO_x control film. To date, NiFe hydroxides have been the most efficacious system based on non-precious metals. Zhao and Lu [36] reported 3D NiFe hydroxide thin nanosheets that possess amorphous, mesoporous, hierarchical, and interconnected structures, making them quite a robust, stable, and yet inexpensive electrocatalyst for alkaline electrolytic splitting of water. The proposed NiFe hydroxide-based oxygen electrode have become a new benchmark by outperforming noble Ir and Ru electrodes and meeting industrially required current densities (up to 1 A/cm^2) in concentrated alkaline electrolytes (Figure 30.6). Highly active NiFe oxides/hydroxides in alkaline electrolytes have been also reported by Zhou et al. [37], Liu et al. [2], and Long et al. [38]. The effect of Fe sites in NiFe oxyhydroxide catalyst was investigated by Bell's group [39] who found a 500-fold enhancement in OER activity at an optimal 25% Fe loading over those of pure Ni and Fe parent composites. The authors suggested that Fe^{3+} centers in oxyhydroxide structure tend to occupy octahedral sites with lower overpotential that are considered to be the active sites for OER, while Ni^{2+} centers reveal larger overpotentials.

Although there are still some issues and mechanisms to be addressed for binary metal systems, new directions in OER activity were found after a comprehensive combinatorial screening of nearly 3500 trimetallic oxides. In this trend, $A_xB_yC_zO_q$ mixed oxide structures were assessed by a red fluorescence quenching approach associated with oxygen evolution ability. The sensitive fluorescence combinatorial screening method provided valuable information about these mixed oxides and their superiority over well-investigated pure oxide compounds,

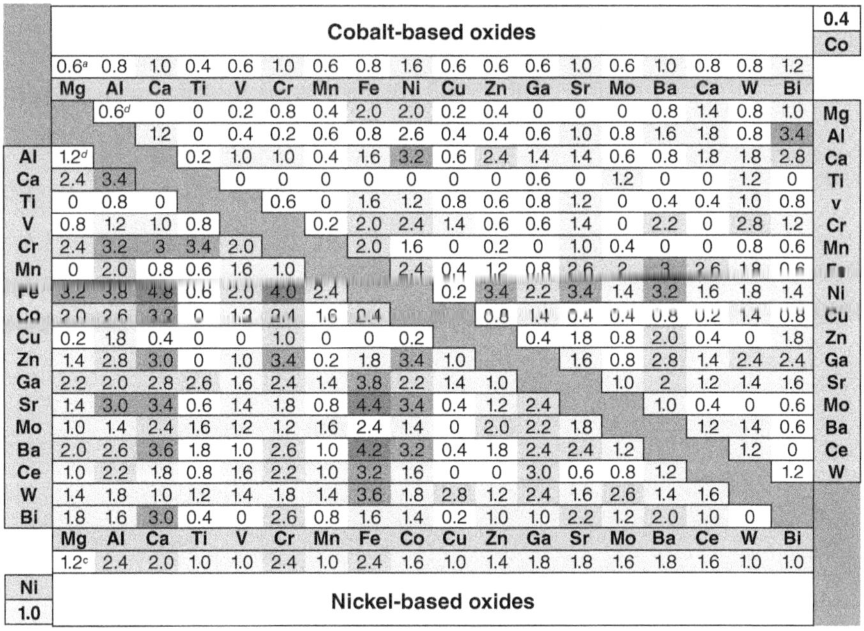

	Cobalt-based oxides																		0.4 Co
	0.6[a]	0.8	1.0	0.4	0.6	1.0	0.6	0.8	1.6	0.6	0.6	0.6	1.0	0.6	1.0	0.8	0.8	1.2	
	Mg	Al	Ca	Ti	V	Cr	Mn	Fe	Ni	Cu	Zn	Ga	Sr	Mo	Ba	Ca	W	Bi	
		0.6[d]	0	0	0.2	0.8	0.4	2.0	2.0	0.2	0.4	0	0	0	0.8	1.4	0.8	1.0	Mg
			1.2	0	0.4	0.2	0.6	0.8	2.6	0.4	0.4	0.6	1.0	0.8	1.6	1.8	0.8	3.4	Al
Al	1.2[d]			0.2	1.0	1.0	0.4	1.6	3.2	0.6	2.4	1.4	1.4	0.6	0.8	1.8	1.8	2.8	Ca
Ca	2.4	3.4			0	0	0	0	0	0	0	0.6	0	1.2	0	0	1.2	0	Ti
Ti	0	0.8	0			0.6	0	1.6	1.2	0.8	0.6	0.8	1.2	0	0.4	0.4	1.0	0.8	V
V	0.8	1.2	1.0	0.8			0.2	2.0	2.4	1.4	0.6	0.6	1.4	0	2.2	0	2.8	1.2	Cr
Cr	2.4	3.2	3	3.4	2.0			2.0	1.6	0	0.2	0	1.0	0.4	0	0	0.8	0.6	Mn
Mn	0	2.0	0.8	0.6	1.6	1.0			2.4	0.4	1.2	0.8	2.6	2	3	2.6	1.8	0.6	Fe
Fe	3.2	3.8	4.8	0.6	2.0	4.0	2.4			0.2	3.4	2.2	3.4	1.4	3.2	1.6	1.8	1.4	Ni
Co	2.0	2.6	3.2	0	1.2	2.1	1.6	2.4			0.8	1.4	0.4	0.4	0.8	0.2	1.4	0.8	Cu
Cu	0.2	1.8	0.4	0	0	1.0	0	0	0.2			0.4	1.8	0.8	2.0	0.4	0	1.8	Zn
Zn	1.4	2.8	3.0	0	1.0	3.4	0.2	1.8	3.4	1.0			1.6	0.8	2.8	1.4	2.4	2.4	Ga
Ga	2.2	2.0	2.8	2.6	1.6	2.4	1.4	3.8	2.2	1.4	1.0			1.0	2	1.2	1.4	1.6	Sr
Sr	1.4	3.0	3.4	0.6	1.4	1.8	0.8	4.4	3.4	0.4	1.2	2.4			1.0	0.4	0	0.6	Mo
Mo	1.0	1.4	2.4	1.6	1.2	1.2	1.6	2.4	1.4	0	2.0	2.2	1.8			1.2	1.4	0.6	Ba
Ba	2.0	2.6	3.6	1.8	1.0	2.6	1.0	4.2	3.2	0.4	1.8	2.4	2.4	1.2			1.2	0	Ce
Ce	1.0	2.2	1.8	0.8	1.6	2.2	1.0	3.2	1.6	0	0	3.0	0.6	0.8	1.2			1.2	W
W	1.4	1.8	1.0	1.2	1.4	1.8	1.4	3.6	1.8	2.8	1.2	2.4	1.6	2.6	1.4	1.6			
Bi	1.8	1.6	3.0	0.4	0	2.6	0.8	1.6	1.4	0.2	1.0	1.0	2.2	1.2	2.0	1.0	0		
	Mg	Al	Ca	Ti	V	Cr	Mn	Fe	Co	Cu	Zn	Ga	Sr	Mo	Ba	Ce	W	Bi	
	1.2[c]	2.4	2.0	1.0	1.0	2.4	1.0	2.4	1.6	1.0	1.4	1.8	1.8	1.6	1.8	1.6	1.0	1.0	
Ni 1.0	Nickel-based oxides																		

Figure 30.7 Maximal average activity of cobalt- and nickel-containing triads and dyads. Source: Reprinted with permission from Gerken et al. [40]. Copyright 2014, Royal Society of Chemistry. (See online version for color figure).

i.e. ABO_3 and/or AB_2O_4 structures in alkaline environment. This sheds light on the composition–activity relationship for abundant metal catalysts. Reports indicate that the incorporation of redox active and Lewis acidic elements such as Fe, Al, Ga, and Cr dramatically improves the activity of single Ni and Co oxides (Figure 30.7) [40]. Particularly, ternary systems based on NiFe mixed oxides have been shown to greatly boost OER efficiency. Various Fe-family-based OER catalysts are listed in Table 30.1. Gregoire and coworkers surveyed the effectiveness of cerium-rich quaternary systems through high-throughput screening of 5456 discrete Fe-family oxides where pseudo-ternary $Ni_{0.2}Co_{0.3}Ce_{0.5}O_x$ was reported to be the most efficient composition with the ability of delivering 10 mA/cm^2 current density at 310 mV [52]. Theoretical studies have shown that high-valence non-3d metals such as tungsten are able to modulate the adsorption energy on the surface of 3d metal oxides [42]. Experimentally, this was testified by augmented activity and stability of gelled FeCoW oxyhydroxides upon an overpotential as small as 191 mV at 10 mA/cm^2 and prolong OER for more than 500 hours. Similarly, high-valence molybdenum (Mo^{6+}) was also found to have influence over CoFe oxide [53]. The other verified mixed-metal oxide system is perovskites among which BSCF family, e.g. $Ba_{0.5}Sr_{0.5}Co_{0.8}Fe_{0.2}O_{3-\delta}$, has emerged as active OER catalysts in both acid and base media [54].

Computational methods provide two paradigms for effective design of these compounds: (i) stronger covalent interaction between oxygen and transition metal and (ii) near-unity occupancy of e_g orbital both yielding a promoted

Table 30.1 OER performances of earth-abundant OER electrocatalysts.

Catalyst	Substrate	Mass loading (mg/cm^2)	Overpotential @10 mA/cm^2 (mV vs. reversible hydrogen electrode [RHE])	Tafel slope (mV/dec)	Elec-trolyte	Refer-ences
NiFe-layered double hydroxide (LDH)	Fe foam	2.78	200	NA [a]	1 M KOH	[2]
$Ni_{0.9}Fe_{0.1}O_x$	Quartz crystal microbalance (QCM) electrodes	~0.012	336	30	1 M KOH	[35]
$NiFeO_x$	Ni foam	NA	215	28	1 M KOH	[36]
NiFeOOH	Ni foam	4.0	174 [b]	41.5	1 M KOH	[37]
FeNi-rGO LDH	Ni foam	0.25	195	39	1 M KOH	[38]
$CoFe_2O_4$	GC [b]	0.285	314	30.69	1 M KOH	[41]
Gelled FeCoW	Au foam	0.21	190	NA	1 M KOH	[42]
$CoFe_2O_4$	Ni foam	0.82	253	44	0.1 M KOH	[43]
Fe-mCo_3O_4	GC	NA	380	60	1 M KOH	[44]
CoMn LDH	GC	0.142	324	43	1 M KOH	[45]
NiFeCu	Ni foam	10.2	180	33	1 M KOH	[46]
Ni_2P	GC	0.14	290	47	1 M KOH	[47]
Ni_3S_2	Ni foam	37	187	159.3	0.1 M KOH	[48]
Ni_3N	Carbon cloth	NA	256	41	1 M KOH	[49]
NiCo-MOF	GC	0.2	189	42	1 M KOH	[50]
Co_3O_4-C	Cu foil	0.2	290	70	0.1 M KOH	[51]

a) NA represents not available.
b) GC denotes glassy carbon.

active phase and surface adsorption energy, hence greater OER activity. On this account, moderate oxygen adsorption energy in perovskites such as $LaNiO_3$ and $LaCoO_3$ is more optimal than either too strong adsorption energy in $LaCuO_3$ or too weak adsorption energy in $LaMnO_3$. This suggests the importance of the presence of Fe, Ni, and Co atoms in almost all prominent catalysts through a synergistic effect on OER (Table 30.1). Thus, combinatorial screening and computational methods are promising approaches to develop ultra-efficient catalysts in the future.

30.3.1.2 Metal Chalcogenides

This category consists of transition metal (basically Fe-family) sulfides, selenides, and tellurides that were, foremost, exploited for their HER activities as they offer higher electroconductivity compared with their oxide counterparts and excellent chemical stability in acidic/alkaline solutions. Later they were also tested for OER in search for potential bifunctional electrocatalysts to perform overall water splitting and demonstrated robust activity with comparable overpotentials in the range of 200–300 mV at 10 mA/cm^2 with respect to oxide structures [48, 55, 56]. In this regard, Ni_3S_2 nanorods/Ni foam and $CoSe_2$ nanosheets/glassy carbon electrode (GCE) with exceptional O_2 turnover in diluted alkaline electrolyte were among the first efforts to evaluate OER behavior of nanostructured single metallic chalcogenides [48, 56]. Atomically thick $CoSe_2$ nanosheets revealed Co vacancies as opposed to their bulk counterpart that could serve as active sites for efficacious water adsorption. FeS is another practical chalcogenide. Very recently, mackinawite-type nanosheet arrays of FeS on Fe foam exhibited substantial performance via *in situ* transformation into porous amorphous FeO_x phase during OER to modulate the number and activity of active sites [57]. This opens a new window for investigating hybrid materials based on the synergy between metal chalcogenides and oxides (e.g. Fe_3O_4 [58], NiFeOOH [59]).

Considering the great performance of unmodified metal chalcogenides, reinforced OER response is expected upon further doping of heteroatoms. Binary NiFe sulfide supported on Ni foam was a recent case study that produced both higher current density and lower overpotential compared with single Ni sulfide [60]. While Ni sulfide was still better than NiFe hydroxide, it was stated that introduction of Fe increases the population of active sites and electrochemically active surface area (ECSA) in addition to improving OER stability. Similarly, the addition of Ni to $CoSe_2$ was reported by Zhu et al. [61] to prepare a diselenide material that indicated enlarged surface area for mass transfer, suitable surface roughness with populated active sites, enhanced electroconductivity, and 3D nanonetwork for facilitating detachment of O_2 bubbles. *In situ* transformation is also viable for hybrid chalcogenides and plays a dual role in OER activity. Xu et al. [62] found that $Ni_xFe_{1-x}Se_2$ acts as a pre-electrocatalyst and a conductive core that augments the charge transfer upon conversion into oxide/hydroxide form as the real active phase during oxidative sweeps. Owing to such phase transformation, further mechanistic pathways and increased active site density are expected for this category.

30.3.1.3 Metal Pnictides, Carbides, and Borides

The pnictides comprised of metal nitrides and phosphides usually possess good electroconductivity, and similar to chalcogenides, they are susceptible to phase change during anodic potentials. *In situ* formation of an oxide/hydroxide thin-layer active phase is prevalent at the electrode–electrolyte interface with facilitated electron transfer kinetics by metal pnictide as the conductive core. Numerous synthesis strategies have been developed to exploit these features. Specifically, when deposited on a 3D scaffold like microporous metal foams or conductive carbon cloth, both electroconductivity and the number of active sites over metal pnictides are increased dramatically. Liu and coworkers [63] observed that FeP nanorods on carbon paper evince a superior OER activity compared with many transition metal–based OER catalysts. On the other hand, surface nanostructuring is another approach to boost the correlation between electroconductivity and activity. Nickel nitrides and phosphides have been extensively studied in this regard. It is demonstrated that Ni_3N nanosheets own higher electroconductivity and density of states around the Fermi level than bulk Ni_3N, leading to a twofold increment in the charge transfer kinetics and a significant shift in the onset potential [49]. A similar observation was derived upon modifying the morphology of Ni_2P from nanoparticles to nanowires in which a overpotential shift of 100 mV was reported [64].

Cationic/anionic manipulation is also recognized as a feasible strategy for expanding the active sites over surface. As a proof of concept, Fe and O doping in Co (oxy)phosphide have been stated to optimize the electrocatalytic activity [65] similar to anion regulation and surface electronic polarization mechanism for OER onset potential shift in NiFeS where gradual variation of S/O atomic ratio optimizes water adsorption, electron transfer, and gas (O_2) bubble detachment [66]. Furthermore, metal carbides (e.g. Ni_3C [67]) and borides (Co_2B [68], NiB [69], NiFeB [70]) have attracted interest recently with consistent mechanism pathways to metal oxides and other groups, while their application for whole water splitting is currently a growing field. Comparison of OER activity of recent electrocatalysts based on carbide and boride is illustrated in Table 30.1.

30.3.1.4 Metal–Organic Frameworks and Related Materials

Metal–organic frameworks (MOFs) composed of metallic centers and coordination organic ligands have emerged as new OER catalysts [71]. Inherited with merits of both heterogeneous and homogeneous catalysts, MOFs offer a great potential in chemical robustness for electrolysis. However, bulk MOFs have limited electrical conductivity, and the active metal centers are blocked by vulcanization organic ligands. Therefore, novel design of the periodic structural units and nanostructured MOFs is an effective strategy to overcome these obstacles. Ultrathin NiCo bimetal–organic framework nanosheets were synthesized for this purpose and exhibited outstanding responses including diminished onset potential and overpotential values [50]. It was verified that the coordinately unsaturated Ni and Co atom nodes exhibit a coupling effect, making them dominating active OER centers. Zhao and coworkers [72] investigated 2D NiFe-based bimetallic organic frameworks through a dissolution–crystallization mechanism and found that the corresponding ultrathin arrays possess excellent

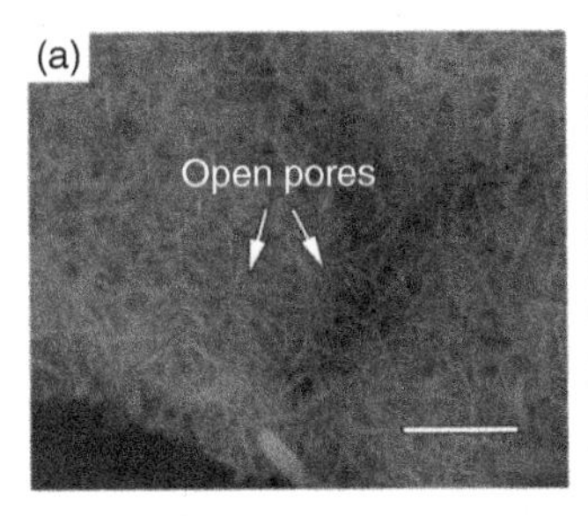

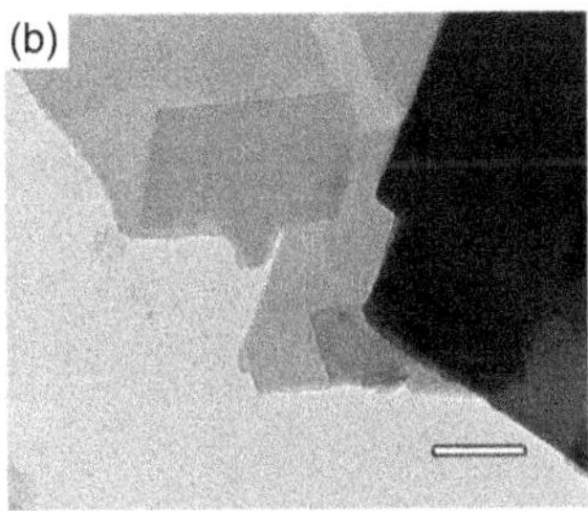

Figure 30.8 (a) Scanning electron microscopy (SEM) and (b) transmission electron microscopy (TEM) images of ultrathin NiFe-MOFs. (Scale bars are 1 μm for (a) and 100 nm for (b)). Source: Reprinted with permission from Duan et al. [72]. Copyright 2017, Nature Publishing Group. (See online version for color figure).

catalytic activity with low overpotential and long-term stability (Figure 30.8). Very recently, water-stable NH_2-MIL-88B(Fe_2Ni) MOF materials were reported to be highly active bifunctional electrocatalysts for both OER and HER in alkaline whole water splitting [73]. Nevertheless, most MOFs still suffer from inferior electroconductivity and mass permeability.

Alternatively, MOF-derived functional electrocatalysts such as nanocarbon and metal–carbon hybrid materials have been fabricated by controllable calcination at high temperatures with MOFs as precursors. MOF-derived carbon (nano)materials gain the inherent advantages from parent MOFs such as adjustable size, morphology, and porosity, easy functionalization with heteroatoms, or combination with metal–metal oxides. For example, MOF-derived hybrid Co_3O_4–carbon porous nanowire array was prepared *in situ* on Cu foil. Owing to the enlarged active surface area, promoted mass transport, remarkable durability originated from *in situ* carbon incorporation, and unique nanowire array configuration, the resulted hybrid material was directly used as OER electrode. Such a novel design afforded stable and higher current densities in alkaline solution than that of commercial IrO_2/C electrodes with the same carbon content [51]. Self-sacrificial template is another platform to transmit MOF advantages to final catalyst materials. Sun and coauthors used ZIF-67 as a template to prepare porous Co-P/NC nanopolyhedra. The observed superior OER activity of Co-P/NC was ascribed to the high conductivity and large active surface area attained from the synergistic effects from Co-P and N-doped carbon matrices [74].

30.3.2 Hydrogen Evolution Reaction Electrocatalysts

30.3.2.1 Non-noble Metals and Noble Metal–Free Alloys

Noble metals in platinum group including Pt, Ir, Ru, Rh, and Pd are known as the most active HER electrocatalysts, but their utilization is hindered by scarcity and high cost. Non-noble transition metals mainly including W, Mo, Co, Cu, Ni, and Fe (which are also six of the most earth-abundant metals with increasing order [75] illustrated in Figure 30.9) along with their alloys have been in the center of interests for cost-effective H_2 production. Miles and Thomason [76] measured the HER activity of these six transition metals and found the following trend:

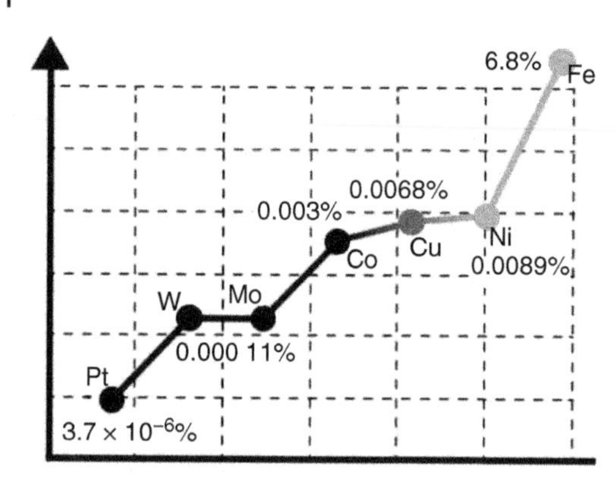

Figure 30.9 Crustal abundance of most used metals for HER electrocatalysts. Source: Reprinted with permission from Zou and Zhang [75]. Copyright 2015, Royal Society of Chemistry. (See online version for color figure).

Cu < Fe < W < Co < Mo < Ni. Since HER is a surface reaction, a nanostructured configuration would be preferred over a dense microstructured surface. Asefa and coauthors [77] showed that co-embedded N-doped carbon nanotubes (CNTs) effectively drive the HER process with a comparable activity to that of Pt. Tang and coworkers [78] observed that porous graphene decorated with single-atom Ni has the potential for substantial HER activity in acidic electrolyte due to sp-d orbital charge transfer between carbon and Ni atoms. However, alkaline water electrolysis with these metals is challenging as they do not show prolonged stability due to the reversible formation of hydride species on their surface in the basic solutions [79].

To date, electronic structure tuning, noble metal–free alloying, and hybridizing the resulting alloys with other compounds have been recognized as key approaches to improve both HER activity and alkaline stability of non-precious metals. Considering the higher intrinsic activity of Ni and Mo among other non-precious metals, development of Ni- and Mo-based HER catalysts has been extensively investigated. Ni-based alloys display good corrosion resistance in high-pH environment and hence are the most applied HER electrode in the industrially interested alkaline water electrolysis. Early efforts revealed the structure–activity relationship for several Ni-based binary alloys in a decreasing trend of NiMo > NiZn > NiCo > NiW > NiFe > NiCr, which were prepared on steel substrate by electrodeposition [80]. NiMo possesses not only the most HER active sites but also excellent durability, and hence its binary or ternary alloys are among the best HER catalysts. Such a merit originates from the hypo-hyper-d-electronic coupling effects induced by two unpaired and half-filled d-electron orbitals of Ni and Mo, respectively. Accordingly, electronic adjustment by a third element in NiMo ternary alloys has been followed recently. Huang and coworkers [81] identified the essential role of Zn in boosting NiMo intrinsic electrocatalytic behavior. They identified that 2 atom% as the optimum Zn content facilitates the formation of high-valent Ni and Mo, thus featuring a noble-metal-like HER activity for NiMoZn alloy. Ternary NiMoTi alloy synthesized by mechanical alloying and heat treatment also exhibited a positive influence of Ti doping on the alkaline HER efficiency of NiMo alloy [82]. Synergistic effect between NiMo intermetallic phase and Ni(Ti) solid solution was mostly promoted in $Ni_{50}Mo_{40}Ti_{10}$ system. Moreover, hybridizing with conductive carbon materials as in the case of recently reported alloys, e.g. CoNi@NC [83] and CoFe@N-rGO [84], is another perspective to get advanced HER catalysts.

30.3.2.2 Non-precious Metal Composites

In addition to intermetallic compounds, nonmetal species such as Se, S, P, N, C, and B are potential additives to noble metal–free-based composites. Like OER electrocatalysts, transition metal chalcogenides, pnictides, carbides, and borides have been extensively evaluated for their electrocatalytic HER behavior. In this section, recent improvements on nonmetal composites are discussed, and the corresponding HER performances are listed in Table 30.2. Nanostructured MoS_2 has been one of the well-studied transition metal chalcogenides and is considered a promising alternative to noble Pt for electrochemical reduction of water. Bulk MoS_2 possesses a graphite-like hexagonally packed layered structure (hexagonal close-packed [HCP] crystal system), and crystallinity brings a variety of surface sites. Distinct polytypes exist in bulk MoS_2 crystallites where electron–hole mobility happens with much faster kinetics along the basal planes (1T-MoS_2 polytype) compared with the perpendicular transfer between sheets (2H-MoS_2 polytype). Much of the surface of bulk MoS_2 is composed of these

Table 30.2 HER performances of earth-abundant nonmetal electrocatalysts.

Catalyst	Substrate	Overpotential @10 mA/cm^2 (mV vs. RHE)	Tafel slope (mV/dec)	Electrolyte	References
MoS_2/reduced graphene oxide (RGO)	GC[a)]	~150	41	0.5 M H_2SO_4	[85]
MoS_2	Fluorine-doped tin oxide (FTO)	260	50	0.5 M H_2SO_4	[86]
MoS_2	Au	170	60	pH = 0.2 H_2SO_4	[87]
CoS_2	GC	145	51	0.5 M H_2SO_4	[88]
Mo_2C	Carbon	240[b)]	54	1 M KOH	[89]
Mo_2C	GC	151	59	1 M KOH	[90]
Ni_2P	Ti foil	130	46	0.5 M H_2SO_4	[91]
Co-P	Ti foil	85[b)]	50	0.5 M H_2SO_4	[92]
FeP	Ti plate	55	38	0.5 M H_2SO_4	[93]

a) GC denotes as glassy carbon.
b) Overpotential obtained at 20 mA/cm^2.

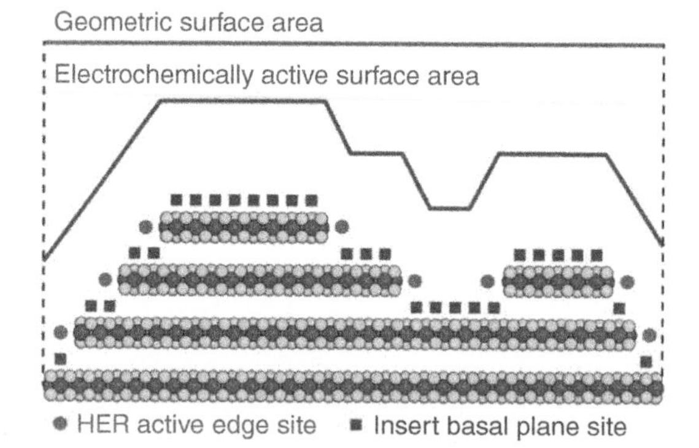

Figure 30.10 Two-dimensional representation of crystalline 2H polytype MoS_2 catalyst electrochemically active surface area and projected geometric surface area. The edge sites, indicated by green dots, are active for HER. Basal plane sites, indicated by blue squares, are not active for HER. Source: Reprinted with permission from Benck et al. [94]. Copyright 2014, American Chemical Society. (See online version for color figure).

basal plate sites that are favored thermodynamically but inert catalytically [94]. In contrast, edge sites in each layer of MoS_2 octahedral coordination show high activity for HER by having optimal hydrogen binding energies close to zero that could be observed in noble metals such as Pt. Considerably large number of active edge sites could be brought into effect through nanostructuring [95]. Hence, unlike bulk structure, nanostructured MoS_2 is ideal for HER, especially since it offers enhanced surface roughness. Generally, active edge sites are more available in a roughened surface, and by taking ECSA as an intrinsic activity measurement into account, higher electrocatalytic activity could be projected for nanocrystalline MoS_2 per unit of geometric surface area. Distinctive active edge and inert basal sites in MoS_2 are illustrated in Figure 30.10. A considerable amount of research has been conducted on exposing more active sites in MoS_2, and the results proposed several strategies for excellent Pt-like HER activity, namely, (i) design of active site-rich nanostructures with considerable porosity such as reticular nanosheets [86]; (ii) doping of active heteroatoms such as Co, Ni, V, and Li [96]; (iii) functionalization with reactive anions containing O, Se, and N [97]; (iv) addition of adjunct conductive structures such as graphene [98] and CNTs [85]; as well as (v) amorphization [99].

Mo_2C is among the best metal carbides with preferable electronic structure and a Pt-like electrocatalytic performance. The cumulative effects of Mo activity, C conductivity, combinatorial durability, and favorable structural phase were reported by Vrubel and Hu for the substantial HER performance of β-Mo_2C nanoparticles in both acidic and alkaline media [89]. Shortly after that, multiple phases of MoC (α-MoC_{1-x}, η-MoC, γ-MoC, and β-Mo_2C) were thoroughly examined, and a decreasing trend of β-$Mo_2C > \gamma$-MoC $> \eta$-MoC $> \alpha$-MoC_{1-x} was obtained for their HER ability [100], evincing the significance of structural optimization for nanoscale Mo_2C. Morphology also has a great influence, which was demonstrated by Mo_2C nanowires prepared via simple heat treatment under inert conditions by Liao et al. [101]. Having a porous, aggregation-free structure along with large specific surface area, the as-prepared Mo_2C nanowires yielded small overpotentials at high current densities about 8 and 50 times higher than those of bulk and commercial Mo_2C catalysts, respectively. MOF-template strategy was examined by Lou and coworkers [90] for Mo_2C catalyst, and the results verified its electrocatalytic HER capability with small overpotentials in both acidic and basic solutions.

In pnictide group, one of the most promising HER catalysts is Ni_2P. Initial theoretical studies by Rodriguez's group in 2005 suggested an ensemble effect where P sites participate in HER and promote a moderate bonding between intermediate species and products with the catalyst surface, placing Ni_2P among the best practical HER catalysts. Inspired by their work, Popczun et al. [91] confirmed Ni_2P exceptional HER activity experimentally in acid electrolyte. However, the behavior was not satisfactory in basic conditions because of the formed Ni phase from Ni_2P. Herein, structural changes such as nanostructuring, heteroatom doping, anion regulation, and affixing conductive nanocarbons are the plausible options to boosting Ni_2P alkaline HER efficiency and stability, and numerous recent research reports signify their efficacy.

30.3.2.3 Metal-Free Electrocatalysts

Nonmetallic compounds have attracted interests recently toward economical H_2 production. Certain features seem necessary, namely, large surface area, excellent charge transfer, facile mass transport, and chemical stability in corrosive electrolytes that potentially were supplied by nanoscale carbonaceous compounds such as CNTs and graphene as support materials. However, further improvement of the electronic structure of these nanocarbon materials can be achieved by heteroatom doping, leading to a dramatic increase in their intrinsic HER activity. Accordingly, metal-free carbon-based electrocatalysts emerged as alternative candidates such as O-, S-, N-, P-, and B- (dual) doped graphene [102], N-doped activated carbon [103], B-doped multiwalled carbon nanotubes (MWCNTs) [104], and N-doped hexagonal carbon [105]. Using theoretical computations, Qiao and coworkers reported that the valence orbital energy level of heteroatom-doped graphene could be originated from the N-/P-co-activated adjacent carbon atoms in graphene matrix providing higher activity compared with single-atom doped graphene. Carbon nitride (C_3N_4) also possesses a 2D structure like graphene and is capable of showing a low HER onset overpotential in pure phase but with low current density due to poor electroconductivity. Hence, combining C_3N_4 with conductive heteroatom-doped graphene could be an effective resolution. C_3N_4-N graphene introduced by Zheng et al. [106] shows an inherent chemical–electronic synergy suitable for optimum proton adsorption and reduction kinetics, which makes it comparable to transition metal catalysts.

30.4 Conclusions and Outlook

In this chapter, the main technologies and promising electrocatalysts used in electrochemical water splitting to date were summarized. Practical considerations necessitate ongoing development in cell design and different cell components to reach maximum efficiency. ALKWE is a mature and the simplest unit currently utilized in the industrial sector. PEMWE benefits from the compact design and improved kinetics, but considering the capital costs due to membrane and electrode materials, it is at near-term commercialization. SOE water electrolyzer depicts the highest efficiency thanks to high-temperature electrolysis; however, different issues in cell components have kept it at laboratory scale. Although considerable achievements have been made based on non-precious metal and nonmetal compounds for both OER and HER procedures, developed systems still face several issues for the use in large-scale applications: (i) Prolonged mechanical durability and electrochemical stability often fall behind the state-of-the-art electrode advances though some have demonstrated comparable activity with commercial electrodes. While the lack of or poor connecting interface between the current collector and catalyst is a continuing problem for the electrodes prepared via binder-assisted electrocatalysts coating, 3D conductive scaffolds are not completely robust for long-term use. Hence, engineering advancement in electrode configuration is highly desired. (ii) Insufficient catalytic activity is usually accompanied with poor stability in corrosive acidic solutions. Current electrochemical water splitting technology utilizes alkaline and acidic electrolytes for OER and HER, respectively. But this entails complexity and increased cost in electrode preparation and cell design. Therefore, both efficient catalytic and electrolytic evaluations in various electrolytes should be expanded. (iii) Industrial electrolyzers often require large current densities of 1 or 2 A/cm^2 for both anodic and cathodic reactions, while the majority of reported electrocatalysts either were studied below 100 mA/cm^2 or unable to deliver such amounts. Besides, in common three-electrode systems, evolved O_2 and H_2 may diffuse to the opposite area and react with each other to form water, which occurs more sharply at high currents. In any case, mass transport effect is crucial for the electrodes with high current densities, and such systems should be evaluated in novel electrolytic cells with functional separators and more advanced techniques beyond cyclic voltammetry and Tafel analysis. Accordingly, the urge for more active and stable materials, bifunctional electrocatalysts, along with pragmatic approaches in catalyst and cell design, is envisaged for future electrochemical water splitting.

References

1 Chen, J.G., Crooks, R.M., Seefeldt, L.C. et al. (2018). *Science* 360: eaar6611.
2 Liu, Y., Liang, X., Gu, L. et al. (2018). *Nat. Commun.* 9: 2609.
3 Schlapbach, L. and Zuttel, A. (2001). *Nature* 414: 353.
4 Cook, T.R., Dogutan, D.K., Reece, S.Y. et al. (2010). *Chem. Rev.* 110: 6474.
5 Roger, I., Shipman, M.A., and Symes, M.D. (2017). *Nat. Rev. Chem.* 1: 0003.
6 Vanýsek, P. (2012). Electrochemical series. In: *CRC Handbook of Chemistry and Physics*, 93e (ed. M.W. Haynes), 5–80. Chemical Rubber Company. ISBN: 9781439880494.
7 Li, X., Hao, X., Abudula, A., and Guan, G. (2016). *J. Mater. Chem. A* 4: 11973.

8 Barbir, F. (2005). *Sol. Energy* 78: 661.
9 Sapountzi, F.M., Gracia, J.M., Weststrate, C.J. et al. (2017). *Prog. Energy Combust. Sci.* 58: 1.
10 Zeng, K. and Zhang, D. (2010). *Prog. Energy Combust. Sci.* 36: 307.
11 Carmo, M., Fritz, D.L., Mergel, J., and Stolten, D. (2013). *Int. J. Hydrogen Energy* 38: 4901.
12 Hauch, A., Ebbesen, S.D., Jensen, S.H., and Mogensen, M. (2008). *J. Mater. Chem.* 18: 2331.
13 Acar, C. and Dincer, I. (2014). *Int. J. Hydrogen Energy* 39: 1.
14 Kim, S.W., Kim, H., Yoon, K.J. et al. (2015). *J. Power Sources* 280: 630.
15 Mamoon Rashid, M.D., Mohammed, K., Mesfer, A. et al. (2015). *Int. J. Eng. Adv. Technol.* 4: 80.
16 Manabe, A., Kashiwase, M., Hashimoto, T. et al. (2013). *Electrochim. Acta* 100: 249.
17 Pool, D.H., Stewart, M.P., O'Hagan, M. et al. (2012). *Proc. Natl. Acad. Sci. U.S.A.* 109: 15634.
18 Todd, D., Schwager, M., and Merida, W. (2014). *J. Power Sources* 269: 424.
19 Ganley, J.C. (2009). *Int. J. Hydrogen Energy* 34: 3604.
20 Parrondo, J., Arges, C.G., Niedzwiecki, N. et al. (2014). *RSC Adv.* 4: 9875.
21 Phillips, R. and Dunnill, C.W. (2016). *RSC Adv.* 6: 100643.
22 Peng, Y., Jiang, K., Hill, W. et al. (2019). *ACS Appl. Mater. Interfaces* 11: 3971–3977. https://doi.org/10.1021/acsami.8b19251.
23 You, B. and Sun, Y. (2018). *Acc. Chem. Res.* 51: 1571.
24 Han, G., Jin, Y.H., Burgess, R.A. et al. (2017). *J. Am. Chem. Soc.* 139: 15584.
25 Pan, J., Chen, C., Li, Y. et al. (2014). *Energy Environ. Sci.* 4: 354.
26 Varcoe, J.R., Atanassov, P., Dekel, D.R. et al. (2014). *Energy Environ. Sci.* 7: 3135.
27 Siracusano, S., Baglio, V., Di Blasi, A. et al. (2010). *Int. J. Hydrogen Energy* 35: 5558.
28 Ebbesen, S.D., Jensen, S.H., Hauch, A., and Mogensen, M.B. (2014). *Chem. Rev.* 114: 10697.
29 Fabbri, E., Bi, L., Pergolesi, D., and Traversa, E. (2011). *Energy Environ. Sci.* 4: 4984.
30 Li, S.S. and Xie, K. (2013). *J. Electrochem. Soc.* 160: F224.
31 Bi, L., Boulfrad, S., and Traversa, E. (2014). *Chem. Soc. Rev.* 43: 8255.
32 Udagawa, J., Aquiar, P., and Brandon, N.P. (2007). *J. Power Sources* 166: 127.
33 van de Loosdrecht, J. and Niemantsverdriet, J.W. (2013). Synthesis gas to hydrogen, methanol and synthetic fuels. In: *Chemical Energy Storage* (ed. R. Schoegl), 443–457. Berlin: De Gruyter.
34 Rostrup-Nielsen, J. and Christiansen, L.J. (2011). *Concepts in Syngas Preparation*, Catalytic Science Series. London: Imperial College Press.
35 Trotochaud, L., Ranney, J.K., Williams, K.N., and Boettcher, S.W. (2012). *J. Am. Chem. Soc.* 134: 17253.
36 Lu, X. and Zhao, C. (2015). *Nat. Commun.* 6: 6616.
37 Zhou, H., Yu, F., Zhu, Q. et al. (2018). *Energy Environ. Sci.* 11: 2858.
38 Long, X., Li, J., Xiao, S. et al. (2014). *Angew. Chem. Int. Ed.* 53: 7584.

39 Friebel, D., Louie, M.W., Bajdich, M. et al. (2015). *J. Am. Chem. Soc.* 137: 1305.

40 Gerken, J.B., Shaner, S.E., Masse, R.C. et al. (2014). *Energy Environ. Sci.* 7: 2376.

41 Liu, Y., Li, J., Li, F. et al. (2016). *J. Mater. Chem. A* 4: 4472.

42 Zhang, B., Zheng, X., Voznyy, O. et al. (2016). *Science* 352: 333.

43 Liu, L., Zhang, H., Mu, Y. et al. (2016). *J. Power Sources* 327: 599.

44 Xiao, C., Lu, X., and Zhao, C. (2014). *Chem. Commun.* 50: 10122.

45 Song, F. and Hu, X. (2014). *J. Am. Chem. Soc.* 136: 16481.

46 Zhang, P., Li, L., Nordlund, D. et al. (2018). *Nat. Commun.* 9: 381.

47 Stern, L.A., Feng, L., Song, F., and Hu, X. (2015). *Energy Environ. Sci.* 8: 2347.

48 Zhou, W., Wu, X.J., Cao, X. et al. (2013). *Energy Environ. Sci.* 6: 2921.

49 Xu, K., Chen, P., Li, X. et al. (2015). *J. Am. Chem. Soc.* 137: 4119.

50 Zhao, S., Wang, Y., Dong, J. et al. (2016). *Nat. Energy* 1: 16184.

51 Ma, T.Y., Dai, S., Jaroniec, M., and Qiao, S.Z. (2014). *J. Am. Chem. Soc.* 136: 13925.

52 Haber, J.A., Cai, Y., Jung, S. et al. (2014). *Energy Environ. Sci.* 7: 682.

53 Liu, P.F., Yang, S., Zheng, L.R. et al. (2017). *Chem. Sci.* 8: 3484.

54 Suntivich, J., May, K.J., Gasteiger, H.A. et al. (2011). *Science* 334: 1383.

55 Chen, G.F., Ma, T.Y., Liu, Z.Q. et al. (2016). *Adv. Funct. Mater.* 26: 3314.

56 Liu, Y., Cheng, H., Lyu, M. et al. (2014). *J. Am. Chem. Soc.* 136: 15670.

57 Zou, X., Wu, Y., Liu, Y. et al. (2018). *Chem* 4: 1139.

58 Yang, J., Zhu, G., Liu, Y. et al. (2016). *Adv. Funct. Mater.* 26: 4712.

59 Zou, X., Liu, Y., Li, G.D. et al. (2017). *Adv. Mater.* 29: 1700404.

60 Dong, B., Zhao, X., Han, G.Q. et al. (2016). *J. Mater. Chem. A* 4: 13499.

61 Zhu, H., Jiang, R., Chen, X. et al. (2017). *Sci. Bull.* 62: 1373.

62 Xu, X., Song, F., and Hu, X. (2016). *Nat. Commun.* 7: 12324.

63 Xiong, D., Wang, X., Li, W., and Liu, L. (2016). *Chem. Commun.* 52: 8711.

64 Han, A., Chen, H., Sun, Z. et al. (2015). *Chem. Commun.* 51: 11626.

65 Duan, J., Chen, S., Vasileff, A., and Qiao, S.Z. (2016). *ACS Nano* 10: 8738.

66 (a) Li, B.Q., Zhang, S.Y., Tang, C. et al. (2017). *Small* 13: 1700610. (b) Hou, Y., Qiu, M., Zhang, T. et al. (2017). *Adv. Mater.* 29: 1701589.

67 Fan, H., Yu, H., Zhang, Y. et al. (2017). *Angew. Chem. Int. Ed.* 56: 12566.

68 Masa, J., Weide, P., Peeters, D. et al. (2016). *Adv. Energy Mater.* 6: 1502313.

69 Masa, J., Sinev, I., Mistry, H. et al. (2017). *Adv. Energy Mater.* 7: 1700381.

70 Liu, G., He, D., Yao, R. et al. (2018). *Nano Res.* 11: 1664.

71 Miner, E.M. and Dincă, M. (2016). *Nat. Energy* 1: 16186.

72 Duan, J., Chen, S., and Zhao, C. (2017). *Nat. Commun.* 8: 15341.

73 Senthil Raja, D., Chuah, X.-F., and Lu, S.-Y. (2018). *Adv. Energy Mater.* 8: 1801065.

74 You, B., Jiang, N., Sheng, M. et al. (2015). *Chem. Mater.* 27: 7636.

75 Zou, X. and Zhang, Y. (2015). *Chem. Soc. Rev.* 44: 5148.

76 Miles, M.H. and Thomason, M.A. (1976). *J. Electrochem. Soc.* 123: 1459.

77 Zou, X., Huang, X., Goswami, A. et al. (2014). *Angew. Chem. Int. Ed.* 53: 4372.

78 Qiu, H.J., Ito, Y., Cong, W. et al. (2015). *Angew. Chem. Int. Ed.* 54: 14031.

79 Soares, D.M., Teschke, O., and Torriani, I. (1992). *J. Electrochem. Soc.* 139: 98.
80 Raj, I.A. (1993). *J. Mater. Sci.* 28: 4375.
81 Wang, X., Su, R., Aslan, H. et al. (2015). *Nano Energy* 12: 9.
82 Panek, J., Kubisztal, J., and Bierska-Piech, B. (2014). *Surf. Interface Anal.* 46: 716.
83 Deng, J., Ren, P., Deng, D., and Bao, X. (2015). *Angew. Chem. Int. Ed.* 54: 2100.
84 Wu, Z., Li, P., Qin, Q. et al. (2018). *Carbon* 139: 35.
85 Li, Y., Wang, H., Xie, L. et al. (2011). *J. Am. Chem. Soc.* 133: 7296.
86 Kibsgaard, J., Chen, Z., Reinecke, B.N., and Jaramillo, T.F. (2012). *Nat. Mater.* 11: 963.
87 Li, H., Tsai, C., Koh, A.L. et al. (2015). *Nat. Mater.* 15: 48.
88 Faber, M.S., Dziedzic, R., Lukowski, M.A. et al. (2014). *J. Am. Chem. Soc.* 136: 10053.
89 Vrubel, H. and Hu, X. (2012). *Angew. Chem. Int. Ed.* 124: 12875.
90 Wu, H.B., Xia, B.Y., Yu, L. et al. (2015). *Nat. Commun.* 6: 6512.
91 Popczun, E.J., McKone, J.R., Read, C.G. et al. (2013). *J. Am. Chem. Soc.* 135: 9267.
92 Popczun, E.J., Read, C.G., Roske, C.W. et al. (2014). *Angew. Chem. Int. Ed.* 53: 5427.
93 Jiang, P., Liu, Q., Liang, Y. et al. (2014). *Angew. Chem. Int. Ed.* 53: 12855.
94 Benck, J.D., Hellstern, T.R., Kibsgaard, J. et al. (2014). *ACS Catal.* 4: 3957.
95 Hinnemann, B., Moses, P.G., Bonde, J. et al. (2005). *J. Am. Chem. Soc.* 127: 5308.
96 Wang, H., Lu, Z., Xu, S. et al. (2013). *Proc. Natl. Acad. Sci. U.S.A.* 110: 19701.
97 Zhou, W., Hou, D., Sang, Y. et al. (2014). *J. Mater. Chem. A* 2: 11358.
98 Firmiano, E.G.S., Cordeiro, M.A.L., Rabelo, A.C. et al. (2012). *Chem. Commun.* 48: 7687.
99 Ge, X., Chen, L., Zhang, L. et al. (2014). *Adv. Mater.* 26: 3100.
100 Wan, C., Regmi, Y.N., and Leonard, B.M. (2014). *Angew. Chem. Int. Ed.* 126: 6525.
101 Liao, L., Wang, S., Xiao, J. et al. (2014). *Energy Environ. Sci.* 7: 387.
102 Jiao, Y., Zheng, Y., Davey, K., and Qiao, S.Z. (2016). *Nat. Energy* 1: 16130.
103 Zhang, B., Wen, Z., Ci, S. et al. (2014). *RSC Adv.* 4: 49161.
104 Cheng, Y., Tian, Y., Fan, X. et al. (2014). *Electrochim. Acta* 143: 291.
105 Liu, Y., Yu, H., Quan, X. et al. (2014). *Sci. Rep.* 4: 6843.
106 Zheng, Y., Jiao, Y., Zhu, Y. et al. (2014). *Nat. Commun.* 5: 3783.

31

New Visible-Light-Responsive Photocatalysts for Water Splitting Based on Mixed Anions

Kazuhiko Maeda

Tokyo Institute of Technology, Department of Chemistry, School of Science, 2-12-1-NE-2 Ookayama, Meguro-ku, Tokyo 152-8550, Japan

31.1 Introduction

As discussed in Chapter 11, band engineering of a semiconductor is very important for the development of a heterogeneous photocatalyst. Scaife pointed out that the flat band potential (E_{FB}) of various metal oxides with d^0 or d^{10} electronic configuration follows an empirical relation:

$$E_{FB}\ (\text{NHE}) \approx 2.94 - E_g$$

where E_g is the band gap of an oxide semiconductor [1]. This means that if a metal oxide has a band gap smaller than 2.94 eV, the E_{FB} that is nearly equal to the conduction band minimum (CBM) lies at a potential more positive than the water reduction potential (i.e. 0 V vs. normal hydrogen electrode [NHE] at pH 0). A typical example is WO_3, which has a band gap of 2.7 eV and does not have the ability to reduce proton to give H_2.

Mixed anion compounds that contain more than two anionic species in the same phase are good candidates of band-engineered photocatalysts, in particular, to produce a visible-light-responsive narrow-gap photocatalyst for overall water splitting [2]. The key concept is to utilize p orbitals of less electronegative anion than oxygen, thereby forming a new valence band at more negative positions while keeping the conduction band potential, as shown in Figure 31.1.

Oxynitrides (oxide nitrides) are in general stable, nontoxic materials and can be readily obtained by nitriding a corresponding metal oxide powder. Domen et al. have developed oxynitrides and nitrides as photocatalysts for water splitting under visible light [3–5]. Figure 31.2a shows ultraviolet (UV)–visible diffuse reflectance spectra of Ta_2O_5, TaON, and Ta_3N_5. The absorption edge is red shifted from Ta_2O_5 to Ta_3N_5 via TaON. The band structures of Ta_2O_5, TaON, and Ta_3N_5 determined by UV photoelectron spectroscopy and electrochemical analysis are schematically drawn in Figure 31.2b [6], which clearly shows that the valence band maximum (VBM) is shifted negatively in the order $Ta_2O_5 < TaON < Ta_3N_5$ without significant change in the CBM.

Heterogeneous Catalysts: Advanced Design, Characterization and Applications, First Edition.
Edited by Wey Yang Teoh, Atsushi Urakawa, Yun Hau Ng, and Patrick Sit.

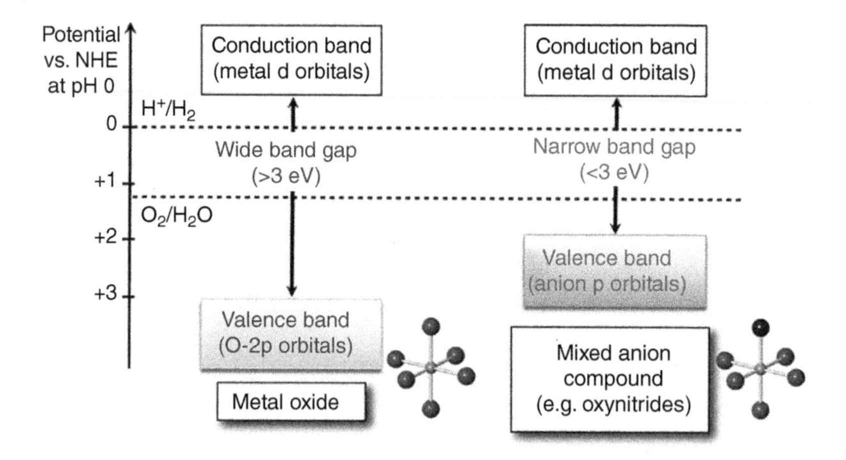

Figure 31.1 Basic concept of band gap narrowing of mixed anion compounds, as compared with metal oxides. (See online version for color figure).

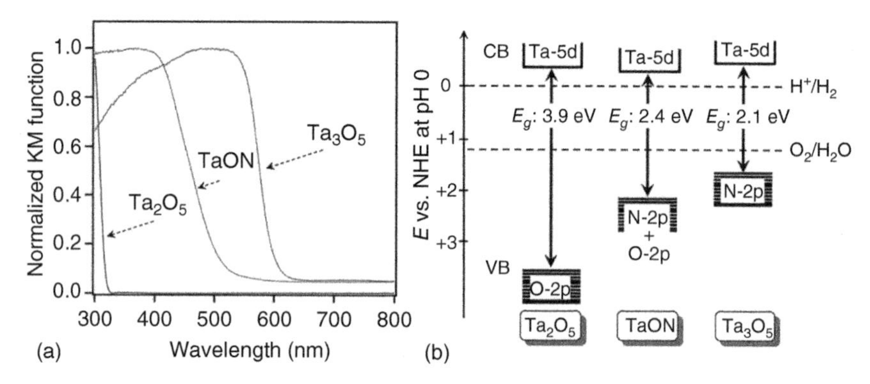

Figure 31.2 (a) UV–visible diffuse reflectance spectra of Ta_2O_5, TaON, and Ta_3N_5 and (b) their band-edge potentials. Source: Reproduced with permission from Maeda and Domen [3]. Copyright 2007, American Chemical Society.

Overall water splitting to form H_2 and O_2 using a heterogeneous photocatalyst is difficult to achieve because of the uphill nature of the reaction involving a large positive change in Gibbs energy ($\Delta G° = 237$ kJ/mol). Therefore, electron donors/acceptors that are more susceptible to oxidation/reduction than water are used for test reactions of water splitting. For example, WO_3 works as a good photocatalyst for water oxidation into O_2 in the presence of an electron donor (e.g. Ag^+, Fe^{3+}, or IO_3^-) whose reduction potential is more positive than the CBM of WO_3. While Ta_2O_5 exhibited little photocatalytic activity under visible light ($\lambda > 400$ nm), both TaON and Ta_3N_5 served as stable photocatalyst for water reduction/oxidation in the presence of suitable electron donor/acceptor, meaning that both have a potential to split pure water under visible light [7, 8]. Since the initial reports on TaON and Ta_3N_5, a variety of (oxy)nitrides have been reported as candidates for photocatalytic water splitting under visible light.

GaN–ZnO and $ZnGeN_2$–ZnO solid solutions with d^{10} electronic configurations are active photocatalysts to split pure water into H_2 and O_2 under visible light without any sacrificial reagents [9, 10]. In particular, GaN–ZnO solid solutions have been demonstrated to be the first reproducible example of overall water splitting among semiconductors having a band gap smaller than 3 eV. Significant improvement of water splitting activity was accomplished upon surface modification of GaN–ZnO with RuO_2 or Rh–Cr mixed oxide nanoparticles that served as H_2 evolution cocatalysts [9, 11], with stable operation even as long as three months [12]. Followed by these d^{10}-based oxynitrides, visible-light overall water splitting using d^0 oxynitrides has recently been achieved. The key to accomplish overall water splitting is proper surface modification. For example, ZrO_2-grafted TaON became an active photocatalyst for overall water splitting under visible light when modified with core/shell-structured RuO_x/Cr_2O_3 nanoparticles and colloidal IrO_2, which work as water reduction and oxidation sites, respectively [13]. A similar surface modification strategy was applied for $LaTaO_2N–LaMg_{2/3}Ta_{1/3}O_3$ solid solutions [14] and $CaTaO_2N$ [15] to achieve overall water splitting. One of the drawbacks of oxynitride photocatalysts is their instability against oxidation reactions by photogenerated holes. Even with benchmarking GaN–ZnO photocatalyst, degradation of photocatalytic activity was observed, because the photogenerated holes oxidize the N^{3-} anion in the material [12]. The stability issue is a common challenge in not only oxynitrides but also other non-oxide-type photocatalysts such as sulfides and nitrides.

While visible-light-driven overall water splitting using a single semiconductor photocatalyst still remains a difficult subject, a two-step photoexcitation scheme, so-called Z-scheme shown in Figure 31.3, using two different semiconductors allows one to apply a variety of mixed anion materials, which are not limited to oxynitrides, to one (or both) side of the system in the presence of a suitable shuttle redox mediator (e.g. IO_3^-/I^- and Fe^{3+}/Fe^{2+}) or with the aid of intimate interfacial contact between the two semiconductors [16–23].In the Z-scheme system, a wider range of visible light can be utilized because the energy to drive each reaction is reduced. It is also possible to apply a semiconductor material that satisfies either water oxidation or reduction potential to one side of the system. Recent technological advancement to assemble both H_2 and O_2 evolution photocatalysts based on metal oxides in a conductive sheet enables efficient Z-scheme water splitting, with a solar energy conversion efficiency of 1.1% [24]. This photocatalyst sheet strategy could be applicable to oxynitrides, although the efficiency still remains low [25].

Thus, significant progress has been made in photocatalytic water splitting community in the past 20 years. Meanwhile, some of mixed anion photocatalysts, most of which have been developed for overall water splitting, have been applied to CO_2 fixation scheme, with the aid of a functional metal complex [26–29]. In this chapter, the latest development of new mixed anion photocatalysts is described not only for water splitting but also CO_2 reduction.

31.2 New Doped Rutile TiO_2 Photocatalysts for Efficient Water Oxidation

Rutile TiO_2 having a band gap of 3.0 eV is a good photocatalyst for water oxidation under UV irradiation (<400 nm) [30]. Even in the presence of an electron

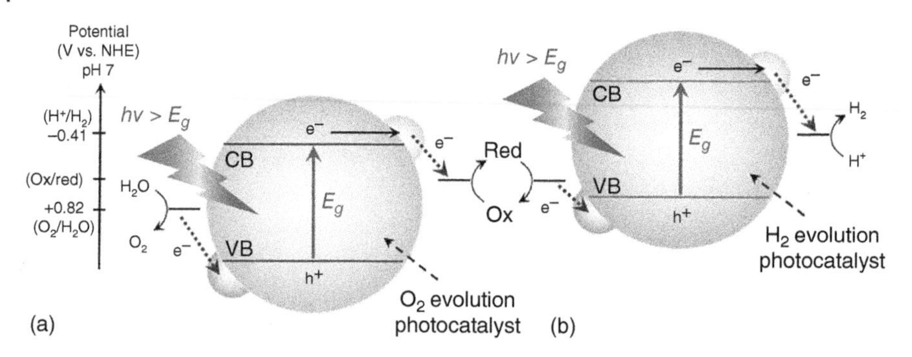

Figure 31.3 Basic principle of overall water splitting using two different semiconductor photocatalysts through two-step photoexcitation. Source: Reproduced with permission from Maeda and Domen [4]. Copyright 2010, American Chemical Society.

donor that is more susceptible to oxidation than water, rutile TiO_2 is capable of selectively oxidizing water into O_2. While this property is very useful as a water oxidation photocatalyst for the construction of a Z-scheme water splitting system (see Figure 31.3b), the band gap of rutile TiO_2 is too wide to effectively absorb visible light.

Nitrogen-doped TiO_2 is a visible-light-responsive photocatalyst that has been mostly used for the degradation of harmful organic compounds [31]. However, the use of anion-doped TiO_2 for light energy conversion scheme such as water splitting and CO_2 fixation is very rare. The key to produce an active doped TiO_2 photocatalyst is to compensate the charge imbalance that is usually caused by aliovalent ion doping into TiO_2 (e.g. O^{2-}/N^{3-} exchange). Such charge imbalance in the crystal lattice leads to the generation of vacancies to compensate the charge balance, which potentially works as recombination centers of electrons and holes. For example, Cr-doped rutile TiO_2 is able to absorb an entire range of visible light, but does not work as a photocatalyst for individual H_2/O_2 evolution in the presence of sacrificial regents under visible light due to the formation of Cr^{6+} [32]. By contrast, rutile TiO_2 codoped with Cr and Sb shows photocatalytic activities, because the generation of Cr^{6+} is suppressed by the Sb^{5+} codopant. The codoping strategy can be applied to anion-doped TiO_2, as exemplified by N- and F-codoped anatase TiO_2 [33, 34].

Recently, the codoping strategy was applied to cation–anion combinations to synthesize a new doped rutile TiO_2 photocatalyst [35]. Ta- and N-codoped rutile TiO_2 (TiO_2:Ta,N) powder was synthesized by nitriding Ta-doped rutile TiO_2 that had been in prior synthesized by a microwave-assisted solvothermal method. X-ray diffraction and elemental analyses indicated that TiO_2:Ta,N (1.0 mol% Ta and <0.1 wt% N) having a single-phase rutile structure was successfully obtained. Figure 31.4a shows UV–visible diffuse reflectance spectra of TiO_2:Ta,N, along with TiO_2:Ta, TiO_2:N, and TaON for comparison. The absorption edge of TiO_2:Ta is located at around 400 nm, which is very close to that of rutile TiO_2 [30]. This is reasonable because the Ta^{5+} dopant having higher orbital energy does not form any energy level in the band gap of rutile TiO_2.

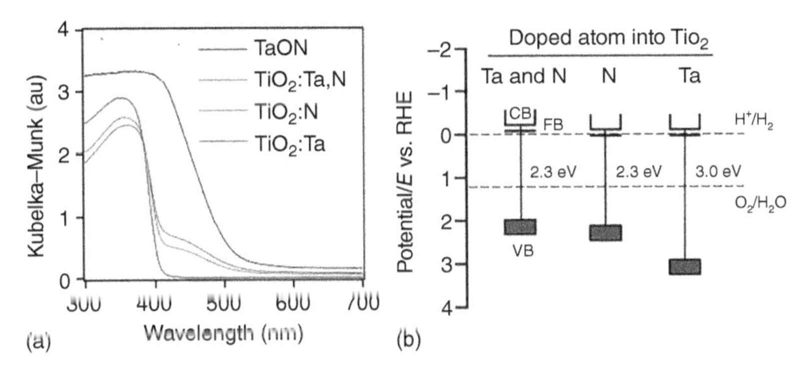

Figure 31.4 (a) UV–visible diffuse reflectance spectra and (b) band-edge potentials of TiO_2:Ta,N and related compounds. Source: Reproduced with permission from Nakada et al. [35]. Copyright 2017, Royal Society of Chemistry. (See online version for color figure).

A schematic illustration of the band gap structures of the synthesized materials is displayed in Figure 31.4b, which was obtained from the results of (photo)electrochemical analysis and UV–visible diffuse reflectance spectroscopy. It indicates that the VBMs of TiO_2:N and TiO_2:Ta,N are located at more negative potential than that of TiO_2:Ta, while the CBMs are almost the same. This feature is identical to that observed in tantalum (oxy)nitrides (Figure 31.2), as the N-2p state contributes to the negative shift of VBM.

Although there was no significant change in the light absorption profile and the band-edge potentials between TiO_2:Ta,N and TiO_2:N, photocatalytic activities of the two were quite different. With modification by RuO_2 nanoparticles as bifunctional cocatalysts for reduction/oxidation reactions [20], TiO_2:Ta,N became photocatalytically active for water oxidation to form O_2 in the presence of reversible electron acceptors (IO_3^- or Fe^{3+}). However, TiO_2:N did not show activity regardless of the RuO_2 modification.

In order to investigate the reason for the big difference in activity between TiO_2:Ta,N and TiO_2:N, transient absorption spectroscopy measurements were performed. Transient absorption spectroscopy in the visible to mid-infrared region is a powerful method that can visualize the photogenerated charge carrier dynamics in a semiconductor material [36, 37]. As shown in Figure 31.5, TiO_2:N exhibited characteristic signals in the range of 20 000 to 3000 cm^{-1}, which are attributed to trapped electrons and/or holes in the mid-gap states (most likely oxygen vacancies) [36, 37]. On the other hand, this absorption is very weak in TiO_2:Ta,N, indicating that the density of oxygen vacancies is reduced upon codoping with Ta. The Ta/N codoping into TiO_2 also resulted in improving the lifetime of photogenerated free electrons that are influential in photocatalytic reactions. The relatively low density of trap states and prolonged lifetime of mobile electrons are both strong contributors to the higher activity of TiO_2:Ta,N.

The RuO_2/TiO_2:Ta,N photocatalyst could be applied as an O_2 evolution photocatalyst in Z-scheme water splitting, in combination with Ru-loaded $SrTiO_3$:Rh as a H_2 evolution photocatalyst. Nearly stoichiometric H_2 and O_2 evolution was

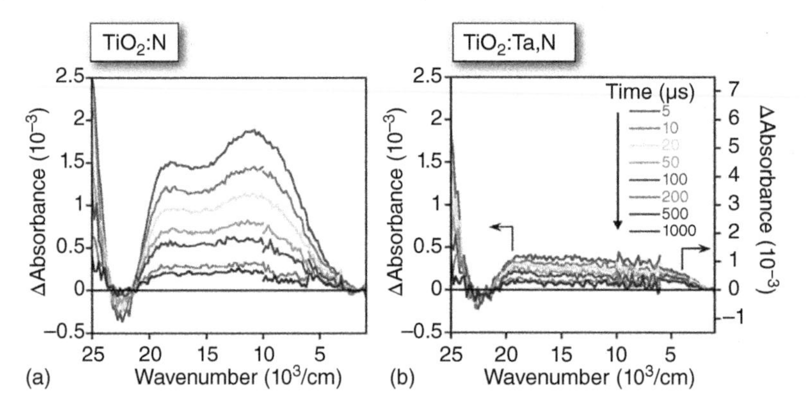

Figure 31.5 Transient absorption spectra for TiO_2:N and TiO_2:Ta,N excited with 355 nm laser pulses under vacuum. Source: Reproduced with permission from Nakada et al. [35]. Copyright 2017, Royal Society of Chemistry. (See online version for color figure).

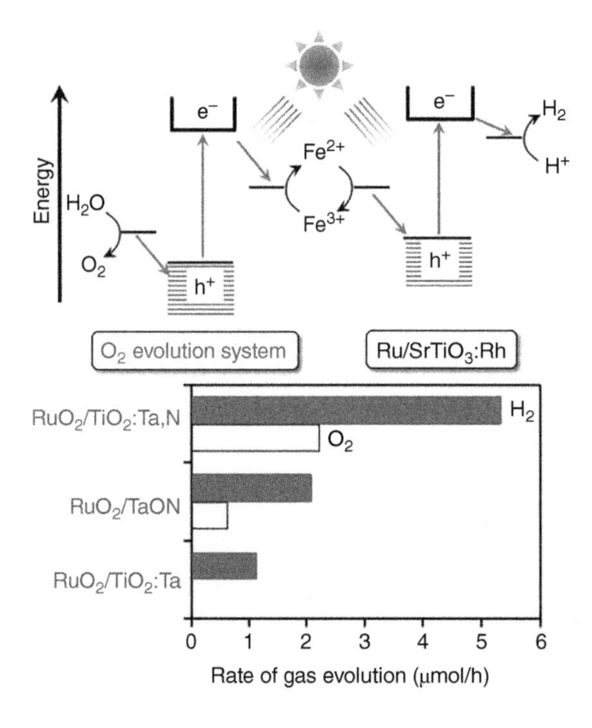

Figure 31.6 Rates of solar-driven H_2 and O_2 evolution from mixtures of an oxygen evolution photocatalyst (RuO_2/TiO_2:Ta,N, RuO_2/TiO_2:Ta, or RuO_2/TaON; 50 mg) and Ru/$SrTiO_3$:Rh (25 mg) dispersed in an aqueous solution (100 ml) containing $FeCl_3$ (1 mM) under simulated sunlight irradiation (AM 1.5G, 100 mW/cm^2). Source: Reproduced with permission from Nakada et al. [35]. Copyright 2017, Royal Society of Chemistry.

achieved using this combination in the presence of an Fe^{3+}/Fe^{2+} redox couple under visible-light irradiation ($\lambda > 420$ nm) and under AM 1.5G simulated sunlight. The solar-to-hydrogen energy conversion efficiency achieved in this system was 0.02%. Although this value is low, it is almost double compared with that using TaON (Figure 31.6), which is one of the most active oxynitride photocatalysts for visible-light water oxidation [3], even though the visible-light absorption of TiO_2:Ta,N is obviously inferior to that of TaON (see Figure 31.4a).

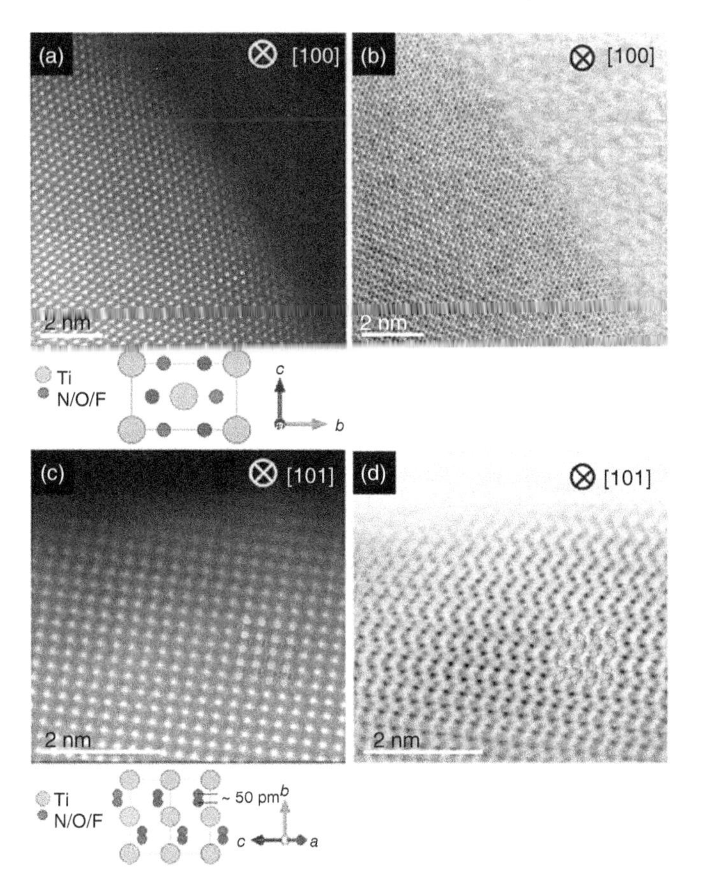

Figure 31.7 Results of STEM observations for TiO_2:N,F. (a, b) High-angle annular dark-field (HAADF) and annular bright-field (ABF) images taken along the [100] and (c, d) [101] direction. As the signal intensity in HAADF imaging is approximately proportional to Z^2 (where Z represents the atomic number), Ti atomic columns in the HAADF images can be seen as bright dots. Source: Reproduced with permission from Miyoshi et al. [38]. Copyright 2018, Royal Society of Chemistry. (See online version for color figure).

Another interesting example of doped rutile TiO_2 is N/F-codoped rutile TiO_2 (TiO_2:N,F) [38]. This material could be synthesized by heating a mixture of rutile TiO_2 and $(NH_4)_2TiF_6$ at 773 K under a flow of ammonia gas. Scanning transmission electron microscopy (STEM) observations indicated the production of the rutile-type structure of TiO_2:N,F with good crystallinity and no disorder of anion arrangement (Figure 31.7). Visible-light absorption capability of TiO_2:N,F was dependent on the concentration of nitrogen in the material, as determined by the relative concentration of $(NH_4)_2TiF_6$ in the starting mixture. As shown in Figure 31.8, the visible-light absorption band was more pronounced with increasing the relative concentration of $(NH_4)_2TiF_6$.

Photocatalytic activity of TiO_2:N,F for visible-light water oxidation was also influenced by the $(NH_4)_2TiF_6$ concentration. It is noted that rutile TiO_2 nitrided at the same temperature without $(NH_4)_2TiF_6$ showed negligible activity, due

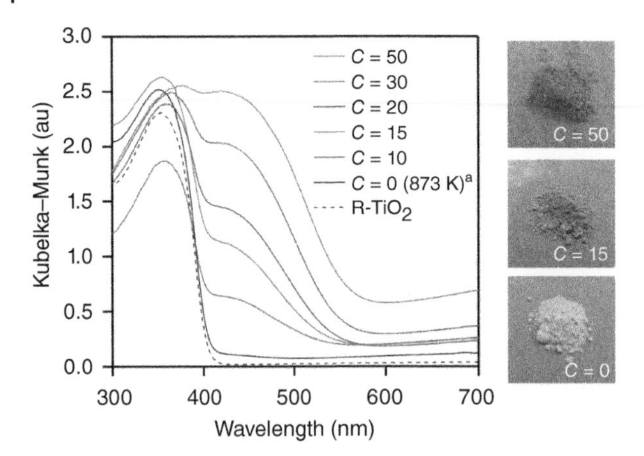

Figure 31.8 UV–visible diffuse reflectance spectra of TiO_2:N,F obtained by nitriding the mixture of TiO_2 and $(NH_4)_2TiF_6$ with various concentrations of $(NH_4)_2TiF_6$ at 773 K for one hour. [a]Nitrided at 873 K. Here "C" indicates the molar concentration of $(NH_4)_2TiF_6$ in the mixture. Source: Reproduced with permission from Miyoshi et al. [38]. Copyright 2018, Royal Society of Chemistry. (See online version for color figure).

primarily to its poor visible-light absorption. The highest activity for water oxidation into O_2 from aqueous $AgNO_3$ solution was obtained with the sample prepared at the $(NH_4)_2TiF_6/TiO_2$ ratio of 15/85. Similar to the TiO_2:Ta,N case, transient absorption spectroscopy revealed that appropriate N/F codoping reduces the density of mid-gap states working as deep traps of photogenerated electrons and increases the number of free electrons compared with only N-doped rutile TiO_2. Thus, photocatalytic activity of the TiO_2:N,F for water oxidation could be enhanced by improving visible-light absorption capability through the N/F codoping while suppressing the density of deep trap sites.

The optimized TiO_2:N,F, further modified with RuO_2 cocatalyst, was applied as the water oxidation component in Z-scheme water splitting in combination with a H_2 evolution photocatalyst Ru/$SrTiO_3$:Rh in the presence of $[Co(bpy)_3]^{3+/2+}$ (bpy = 2,2′-bipyridine) as a shuttle redox mediator under visible light. It is known that most of mixed anion photocatalysts, more or less, suffer from oxidative degradation by photogenerated holes, which results in low stability [3]. Interestingly, the optimized RuO_2/TiO_2:N,F provided stable H_2 and O_2 evolution during the 35 hours of reaction, as shown in Figure 31.9.

31.3 Unprecedented Narrow-Gap Oxyfluoride

As discussed above, oxynitrides are promising materials as visible-light-driven photocatalysts. This is because N-2p orbitals can form a valence band that has a more negative potential, as compared with O-2p orbitals. In this regard, oxyfluorides (oxide fluorides) seem unsuitable because of the higher electronegativity of fluorine. Very recently, the oxyfluoride $Pb_2Ti_2O_{5.4}F_{1.2}$ that has an anion-ordered $A_2B_2X_6X'_{0.5}$-type pyrochlore structure (Figure 31.10a) [40] was found to possess an unprecedented small band gap (c. 2.4 eV), absorbing visible light of up to

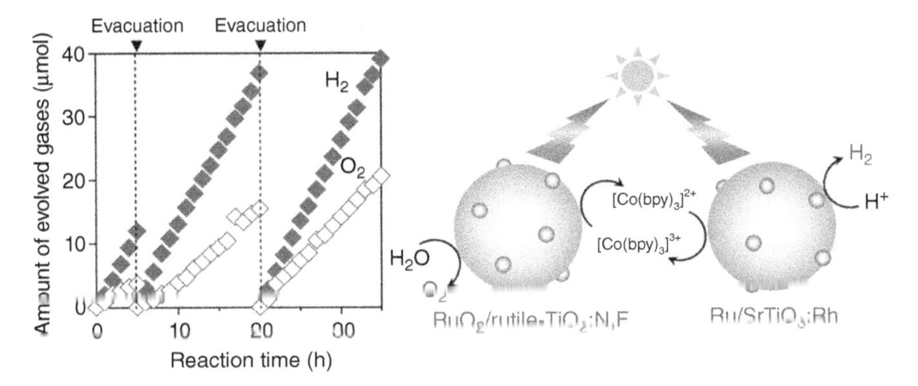

Figure 31.9 Time course of H_2 and O_2 evolution from mixtures of RuO_2/TiO_2:N,F (50 mg) and Ru/$SrTiO_3$:Rh (25 mg) dispersed in an aqueous solution (120 ml) containing tris(2,2′-bipyridyl)cobalt(II) sulfate (0.5 mM) under visible-light irradiation ($\lambda > 420$ nm). Source: Reproduced with permission from Miyoshi et al. [38]. Copyright 2018, Royal Society of Chemistry.

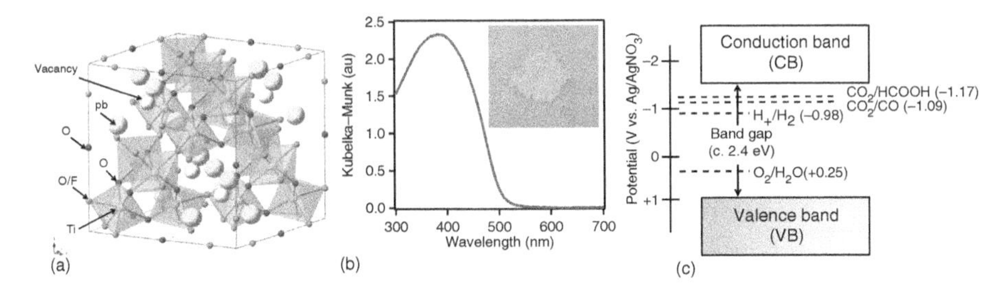

Figure 31.10 (a) Crystal structure of $Pb_2Ti_2O_{5.4}F_{1.2}$. The annotations indicate elements, sites, and Wyckoff positions. In this space group, the X and X′ sites are divided into two sites, which are denoted as X1/X2 and X′1/X′2, respectively. (b) UV–visible diffuse reflectance spectrum of $Pb_2Ti_2O_{5.4}F_{1.2}$. The inset shows a picture of $Pb_2Ti_2O_{5.4}F_{1.2}$. (c) Schematic band structure diagram of $Pb_2Ti_2O_{5.4}F_{1.2}$, along with some redox potentials. Source: Reproduced with permission from Kuriki et al. [39]. Copyright 2018, American Chemical Society. (See online version for color figure).

510 nm (see Figure 31.10b) [39]. Electrochemical impedance spectroscopy and UV–visible diffuse reflectance spectroscopy indicated that $Pb_2Ti_2O_{5.4}F_{1.2}$ has band-edge potentials suitable for water reduction/oxidation and CO_2 reduction, as illustrated in Figure 31.10c. Actually, $Pb_2Ti_2O_{5.4}F_{1.2}$ worked as a stable photocatalyst for visible-light-driven H_2 evolution and CO_2 reduction when modified with a suitable promoter unit such as Pt nanoparticles (for H_2 evolution) or a Ru(II) binuclear complex (for CO_2 reduction), respectively, in the presence of triethanolamine (TEOA) as an electron donor. In addition, water oxidation to form O_2 was possible using $AgNO_3$ as an electron acceptor with the aid of RuO_2 cocatalyst.

Density functional theory (DFT) calculations were performed to investigate the origin of the visible light response of $Pb_2Ti_2O_{5.4}F_{1.2}$. Figure 31.11a shows the total and partial density of state (DOS) of the material; the CBM of $Pb_2Ti_2O_{5.4}F_{1.2}$ consists mainly of Ti-3d orbitals with some hybridization of Pb-6p (in particular,

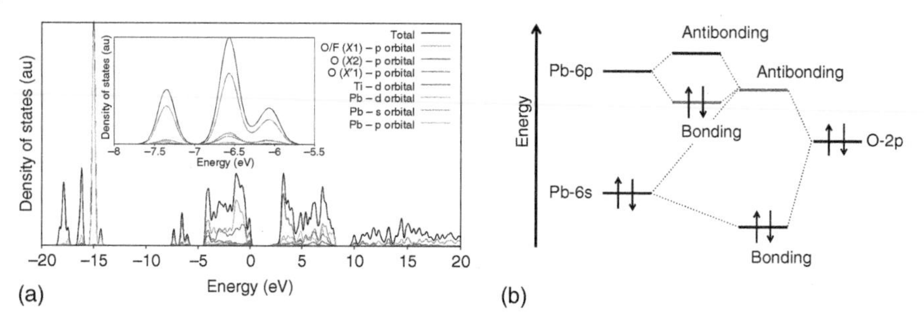

Figure 31.11 (a) Total and partial DOS of $Pb_2Ti_2O_{5.4}F_{1.2}$. In $Pb_2Ti_2O_{5.4}F_{1.2}$, the X′1 (4b) and X′2 (4d) sites are occupied only by oxide anion, with the occupation factors of 0.9 and 0.16, respectively. DFT calculations were performed with a simple model of X′1/X′2 = 1.00/0.00, assuming that the X′1 site is fully occupied by oxide anion, while the X′2 site is vacant. The inset shows an enlarged view of the lower valence band region. (b) Orbital interaction between Pb-6s and O-2p in $Pb_2Ti_2O_{5.4}F_{1.2}$ along with an energy level diagram based on the RLP model. Source: Reproduced with permission from Kuriki et al. [39]. Copyright 2018, American Chemical Society. (See online version for color figure).

higher-energy side), whereas the VBM is formed by O-2p orbitals with certain contributions of Pb-6s orbitals. In lower-energy side (from −8 to −5 eV) of the valence band, it is clear that Pb-6s orbitals are hybridized with O-2p orbitals (Figure 31.11a inset). It was also found that the O-2p contribution to Pb-6s band (per atom) is largest in the X′1 site, which is fivefold those in X1 and X2 sites, on the basis of integral calculation for each partial DOS.

The valence band character can be interpreted in terms of the revised lone pair (RLP) model, proposed by Walsh et al. [41]. In this model (schematically shown in Figure 31.11b), the antibonding orbitals formed by Pb-6s and O-2p orbitals are stabilized through the interaction with the empty Pb-6p orbitals, accounting for the elevated O-2p orbitals in $Pb_2Ti_2O_{5.4}F_{1.2}$. It is noted that some Pb(II)-based oxides such as PbO and $PbTiO_3$ are known to exert the RLP effect [41]. However, the band gaps of such known Pb(II)-containing oxides are wider than that of $Pb_2Ti_2O_{5.4}F_{1.2}$, indicating that the Pb-6s/O-2p hybridization in $Pb_2Ti_2O_{5.4}F_{1.2}$ is much stronger. This is at least in part due to the unique coordination environment around the A-site in the pyrochlore structure, where in the ideal $A_2B_2O_7$ case there are two short A—O bonds and six long A—O bonds. The length of the shortest Pb—O bond in $Pb_2Ti_2O_{5.4}F_{1.2}$ is 2.248 Å [40], which is much shorter than that in perovskite $PbTiO_3$ (2.510 Å). This difference in the local coordination environment could qualitatively explain the narrower band gap of $Pb_2Ti_2O_{5.4}F_{1.2}$ (c. 2.4 eV) than $PbTiO_3$ (c. 2.8 eV). A follow-up study on another oxyfluoride of $Pb_2Ti_4O_9F_2$ supported the idea [42].

Thus, the unprecedented visible light response of $Pb_2Ti_2O_{5.4}F_{1.2}$ is concluded to originate from strong interaction between Pb-6s and O-2p orbitals, which is caused by a short Pb—O bond in the pyrochlore lattice due to the fluorine substitution. The low coordination number of the A-site of pyrochlore (or defect-pyrochlore) structure is preferable for strengthening the electronic interaction between M-6s (M = Pb and Bi) and O-2p, thereby producing more prominent visible-light absorption.

31.4 Conclusion and Future Perspective

Mixed anion materials such as oxynitrides are promising as photocatalysts for water splitting. In this chapter, several examples of recently reported mixed anion materials were discussed. Rutile TiO_2 codoped with Ta/N or N/F served as an effective O_2 evolution photocatalyst in visible-light Z-scheme water splitting systems, in combination with $SrTiO_3$:Rh as a H_2 evolution photocatalyst in the presence of a suitable shuttle redox mediator. In such doped TiO_2 materials, the suppression of oxygen vacancies, which are generated due to aliovalent cation and/or anion doping (e.g. Ti^{4+}/Rh^{3+} or O^{2-}/N^{3-} exchange) and work as deep traps of photogenerated charge carriers, is essential to obtain measurable photocatalytic activity for water oxidation, as clearly demonstrated by transient absorption spectroscopy. Ta/N-codoped rutile TiO_2 exhibited higher performance as a photocatalyst for O_2 evolution in solar-driven Z-scheme water splitting with $SrTiO_3$:Rh than TaON, which is one of the most active non-oxide photocatalysts for visible-light water oxidation. Very recently, an electrode material consisting of the rutile TiO_2:Ta,N has been shown to work as a stable water oxidation photoanode [43]. With post-modification of the rutile TiO_2:Ta,N electrode by conductive TiO_x layer and a RuO_x cocatalyst, the optimized photoanode produced O_2 upon simulated sunlight over five hours of operation with a faradaic efficiency of 94% and no sign of deactivation. This is distinct from the ordinary oxynitride-based photoanodes that usually suffer from self-oxidative degradation, even with the aid of a water oxidation cocatalyst. The TiO_2:Ta,N photoanode could be coupled to not only H_2 evolution but also CO_2 reduction using a molecular photocathode. Another important finding in heterogeneous photocatalysis using mixed anion materials is that the pyrochlore oxyfluoride $Pb_2Ti_2O_{5.4}F_{1.2}$ functions as a stable visible-light-responsive photocatalyst. Even though it is an oxyfluoride that had been believed to be unsuitable for visible-light photocatalysis, crystal engineering through the use of mixed anion would enable the development of oxyfluorides as new class of visible-light photocatalysts for solar-to-fuel energy conversion.

Approximately 20 years have passed since the research on mixed anion materials as water splitting photocatalysts started. While a satisfactory photocatalyst has not been developed to date, significant progress has been made in the recent 10 years, in collaboration with synthetic materials chemistry and spectroscopy. Recent research activities on mixed anion photocatalysts have also extended to construct a CO_2 reduction system with functional metal complexes, which is another important reaction in artificial photosynthesis [26–29]. This was a result of interdisciplinary interaction between different research fields (i.e. semiconductor photocatalysis and coordination chemistry).

A tentative goal of photocatalytic overall water splitting to produce H_2 is to develop a stable system that works with a solar-to-hydrogen energy conversion efficiency of ~10% [4]. The state of the art in water splitting photocatalyst is at most a few percent [24]. The major obstacle to achieve the goal is still the lack of an efficient photocatalytic material that is able to absorb longer wavelength photons and to show high quantum efficiency. Nevertheless, the material's constraint may be relaxed using a photovoltaic device and a photoelectrochemical cell [44], although such device combination appears unsuitable for large-scale application. Anyway, continuous efforts are still necessary to realizing artificial photosynthesis in the future.

References

1 Scaife, D.E. (1980). *Sol. Energy* 25: 41.
2 Kageyama, H., Hayashi, K., Maeda, K. et al. (2018). *Nat. Commun.* 9: 772.
3 Maeda, K. and Domen, K. (2007). *J. Phys. Chem. C* 111: 7851.
4 Maeda, K. and Domen, K. (2010). *J. Phys. Chem. Lett.* 1: 2655.
5 Maeda, K. and Domen, K. (2016). *Bull. Chem. Soc. Jpn.* 89: 627.
6 Chun, W., Ishikawa, A., Fujisawa, H. et al. (2003). *J. Phys. Chem. B* 107: 1798.
7 Hitoki, G., Takata, T., Kondo, J.N. et al. (2002). *Chem. Commun.*: 1698.
8 Hitoki, G., Ishikawa, A., Takata, T. et al. (2002). *Chem. Lett.* 7: 736.
9 Maeda, K., Takata, T., Hara, M. et al. (2005). *J. Am. Chem. Soc.* 127: 8286.
10 Lee, Y., Terashima, H., Shimodaira, Y. et al. (2007). *J. Phys. Chem. C* 111: 1042.
11 Maeda, K., Teramura, K., Lu, D. et al. (2006). *Nature* 440: 295.
12 Ohno, T., Bai, L., Hisatomi, T. et al. (2012). *J. Am. Chem. Soc.* 134: 8254.
13 Maeda, K., Lu, D., and Domen, K. (2013). *Chem. Eur. J.* 19: 4986.
14 Pan, C., Takata, T., Nakabayashi, M. et al. (2015). *Angew. Chem. Int. Ed.* 54: 2955.
15 Xu, J., Pan, C., Takata, T., and Domen, K. (2015). *Chem. Commun.* 51: 7191.
16 Abe, R., Takata, T., Sugihara, H., and Domen, K. (2005). *Chem. Commun.*: 3829.
17 Higashi, M., Abe, R., Teramura, K. et al. (2008). *Chem. Phys. Lett.* 452: 120.
18 Higashi, M., Abe, R., Ishikawa, A. et al. (2008). *Chem. Lett.* 37: 138.
19 Maeda, K., Higashi, M., Lu, D. et al. (2010). *J. Am. Chem. Soc.* 132: 5858.
20 Maeda, K., Abe, R., and Domen, K. (2011). *J. Phys. Chem. C* 115: 3057.
21 Maeda, K., Lu, D., and Domen, K. (2013). *ACS Catal.* 3: 1026.
22 Ma, G., Chen, S., Kuang, Y. et al. (2016). *J. Phys. Chem. Lett.* 7: 3892.
23 Fujito, H., Kunioku, H., Kato, D. et al. (2016). *J. Am. Chem. Soc.* 138: 2082.
24 Wang, Q., Hisatomi, T., Jia, Q. et al. (2016). *Nat. Mater.* 15: 611.
25 Pan, Z., Hisatomi, T., Wang, Q. et al. (2016). *ACS Catal.* 6: 7188.
26 Sekizawa, K., Maeda, K., Domen, K. et al. (2013). *J. Am. Chem. Soc.* 135: 4596.
27 Yoshitomi, F., Sekizawa, K., Maeda, K., and Ishitani, O. (2015). *ACS Appl. Mater. Interfaces* 7: 13092.
28 Muraoka, K., Kumagai, H., Eguchi, M. et al. (2016). *Chem. Commun.* 52: 7886.
29 Nakada, A., Kuriki, R., Sekizawa, K. et al. (2018). *ACS Catal.* 8: 9744.
30 Miyoshi, A., Nishioka, S., and Maeda, K. (2018). *Chem. Eur. J.* 24: 18204.
31 Asahi, R., Morikawa, T., Ohwaki, T. et al. (2001). *Science* 293: 269.
32 Kato, H. and Kudo, A. (2002). *J. Phys. Chem. B* 106: 5029.
33 Nukumizu, K., Nunoshige, J., Takata, T. et al. (2003). *Chem. Lett.* 32: 196.
34 Maeda, K., Shimodaira, Y., Lee, B. et al. (2007). *J. Phys. Chem. C* 111: 18264.

35 Nakada, A., Nishioka, S., Vequizo, J.J.M. et al. (2017). *J. Mater. Chem. A* 5: 11710.
36 Yamakata, A., Ham, Y., Kawaguchi, M. et al. (2015). *J. Photochem. Photobiol. A* 313: 168.
37 Yamakata, A., Vequizo, J.J.M., and Matsunaga, H. (2015). *J. Phys. Chem. C* 119: 24538.
38 Miyoshi, A., Vequizo, J.J.M., Nishioka, S. et al. (2018). *Sustainable Energy Fuels* 2: 2025.
39 Kuriki, R., Ichibha, T., Hongo, K. et al. (2018). *J. Am. Chem. Soc.* 140: 6648.
40 Oka, K., Hojo, H., Azuma, M., and Oh-ishi, K. (2016). *Chem. Mater.* 28: 5554.
41 Walsh, A., Payne, D.J., Egdell, R.G., and Watson, G.W. (2011). *Chem. Soc. Rev.* 40: 4455.
42 Wakayama, H., Utimula, K., Ichibha, T. et al. (2018). *J. Phys. Chem. C* 122: 26506.
43 Nakada, A., Uchiyama, T., Kawakami, N. et al. (2019). *ChemPhotoChem* 3: 37.
44 Grätzel, M. (2001). *Nature* 414: 338.

32

Electrocatalysts in Polymer Electrolyte Membrane Fuel Cells

Stephen M. Lyth[1,2] and Albert Mufundirwa[3]

[1] *Kyushu University, Platform of Inter/Transdisciplinary Energy Research (Q-PIT), 744 Motooka, Nishi-ku, Fukuoka 819-0395, Japan*
[2] *Kyushu University, International Institute for Carbon-Neutral Energy Research, 744 Motooka, Nishi-ku, Fukuoka 819-0395, Japan*
[3] *Kyushu University, Department of Hydrogen Energy Systems, Faculty of Engineering, 744 Motooka, Nishi-ku, Fukuoka 819-0395, Japan*

32.1 Introduction

Hydrogen reacts exothermically with oxygen resulting in explosive generation of water, heat, and light (Figure 32.1a). This looks impressive but is dangerous. In 1936 a hydrogen-filled zeppelin airship known as the Hindenburg ignited and exploded, resulting in 36 deaths (Figure 32.1b). In 2011 a tsunami in Japan triggered a failure of the cooling systems at Fukushima Daiichi nuclear power plant. Water reacted at hot zirconium surfaces in the reactor generating hydrogen, which exploded. This destroyed the containment buildings and helped spread radioactive materials into the environment.

Despite these isolated disasters, hydrogen can be relatively safe compared with liquid fuels like gasoline. Hydrogen fires generally burn quickly in a localized area and are over in a few seconds (Figure 32.1c). Meanwhile, gasoline fires burn for a long time and tend to be more destructive (Figure 32.1d) [1]. Another advantage of using hydrogen fuel is that it can be generated from electrolysis of water using renewable energy. This avoids CO_2 emissions and will help mitigate climate change. As such, hydrogen is being seriously considered as an alternative fuel for heating and powering homes, vehicles, and factories. But how can we control the potentially destructive force of hydrogen, and how can we utilize it safely? One of the simplest solutions is to physically separate the hydrogen and oxygen using a membrane and use electrochemistry to regulate the reaction.

Fuel cells are devices that generate electricity from electrochemical reactions between a fuel and an oxidant at the anode and cathode, respectively. These are physically separated by an electrolyte membrane. The principle is similar to a battery, but where batteries are self-contained, in fuel cells the reactants can be continuously supplied. Here we focus specifically on *polymer electrolyte fuel cells* (PEFCs) utilizing hydrogen as a fuel. The operation principle of a PEFC is

Heterogeneous Catalysts: Advanced Design, Characterization and Applications, First Edition.
Edited by Wey Yang Teoh, Atsushi Urakawa, Yun Hau Ng, and Patrick Sit.

Figure 32.1 (a) An exploding hydrogen-filled balloon. Source: Maxim Bilovitskiy (cropped) Licensed under CC BY SA-4.0. (b) The Hindenburg disaster. Sam Shere (cropped)/public domain. Fires in (c) a hydrogen fuel cell vehicle (FCV) and (d) a conventional gasoline vehicle. Source: US Department of Energy 2011 [1].

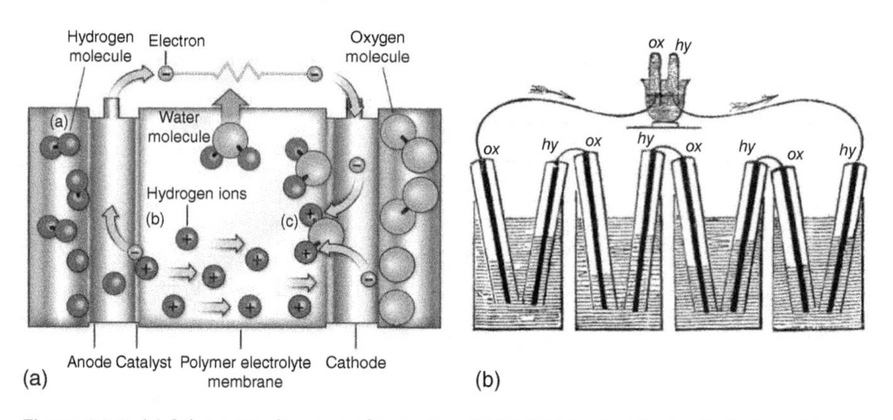

Figure 32.2 (a) Schematic diagram of a modern PEFC. (b) The first fuel cell, designed by William Grove in 1842. Source: (a) Reprinted with permission from Schlapbach and Züttel [2]. Copyright 2001, Springer Nature; (b) Grove 1842 [3]. Reprinted with permission of Taylor & Francis Ltd.

represented in Figure 32.2a [2]. At the anode, the *hydrogen oxidation reaction* (HOR) occurs on the surface of a platinum electrocatalyst, liberating protons and electrons (Eq. (32.1)). At the cathode, the *oxygen reduction reaction* (ORR) occurs: oxygen from air combines with protons and electrons and is reduced to form water (Eq. (32.2)). These two "*half-reactions*" generate a potential difference across the cell, driving protons through the electrolyte and electrons around the external circuit. Completing the circuit sustains the reaction indefinitely until the fuel or oxidant supply stops. The overall reaction is obtained by combining both half-reactions and canceling out the protons and electrons (Eq. (32.3)).

This balanced reaction is identical to the explosive combination of H_2 and O_2. However, in a PEFC most of the energy released can be captured as useful electricity and heat:

$$2H_2 \rightarrow 4H^+ + 4e^- \tag{32.1}$$

$$O_2 + 4H^+ + 4e^- \rightarrow 2H_2O \tag{32.2}$$

$$2H_2 + O_2 \rightarrow 2H_2O \tag{32.3}$$

Hydrogen fuel cells may seem like a futuristic concept, but they have been around for almost 180 years. In 1842, Grove discovered the fuel cell when he reversed a water electrolysis experiment (Figure 32.2b) [3]. In his pioneering experiment the electrodes were platinum foil, doubling as the electrocatalyst. Sulfuric acid was used as the electrolyte. Hydrogen and oxygen were trapped in tubes containing the anode and cathode, respectively. Grove linked several cells together, generating sufficient voltage to electrolyze water. He had built the world's first fuel cell, naming it the "gaseous voltaic battery."

However, Grove's fuel cell was inefficient and generated only a tiny current. One reason was the large distance between the electrodes, resulting in high ionic resistance (i.e. *ohmic losses*). Minimizing ohmic losses is key to improving performance and can be achieved by decreasing the distance between electrodes or by increasing the conductivity of the electrolyte. Another reason for the poor performance was *unoptimized electrode design.* For the HOR or ORR reactions to take place, three conditions must be met at the catalyst surface: (i) H_2 or O_2 molecules must be present, (ii) electron transport must be possible, and (iii) proton transport must be possible. The place where these three things can happen is the *triple phase boundary* [4, 5]. In Grove's cell, this is just the contact point between the gas, the acid, and the Pt foil, shown schematically by a dotted line in Figure 32.3a. In modern PEFC systems, a complex electrode design is employed to maximize the triple phase boundary. Pt nanoparticles are used in place of bulk platinum because of their larger surface area. These are supported on carbon black, which provides electronic transport and porosity for the gas supply. A very thin layer of polymer ionomer is coated on the surface to provide ionic transport (Figure 32.3b). The triple phase boundary of a modern electrocatalyst is shown

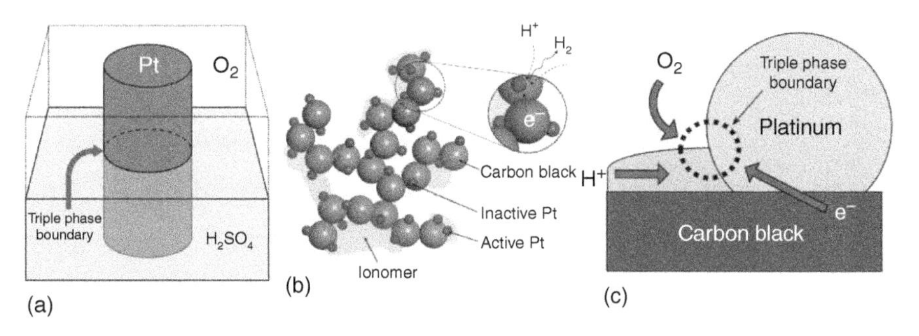

Figure 32.3 (a) Schematic of the triple phase boundary in Grove's fuel cell. (b) Structure of the electrocatalyst layer in a modern PEFC. Source: Image courtesy of David Rivera, Copyright 2018. (c) The triple phase boundary in a modern electrocatalyst layer.

in Figure 32.3c. In practice, the triple phase boundary is approximated using the *electrochemical surface area* (ECSA), normalized to the mass of catalyst. In modern systems, this is generally ~100 m^2/g, whereas in Grove's cell it would have been orders of magnitude lower. One more modern advance is the *solid polymer electrolyte*, replacing the aqueous electrolyte. This is usually Nafion, a sulfonated fluoropolymer with high proton conductivity (~0.1 S/cm). This is made into thin membranes (e.g. 20 μm thick) to decrease the ohmic losses. In addition, Nafion is mixed into the electrocatalyst layer to provide an interpenetrating electrolyte, increasing the triple phase boundary.

32.2 Platinum Electrocatalysts

Buying platinum jewelry is expensive, and using it in a fuel cell vehicle (FCV) is no different. In a typical PEFC it is estimated that the Pt electrocatalyst and its application make up 21–45% of the stack cost, depending on the production scale (Figure 32.4a) [6]. For a FCV such as the Toyota Mirai, this translates to ~US$3700 (estimated from Figure 32.4b) [7]. Intensive research over the decades has sought to improve the activity and thus reduce the cost through advances in nanoparticle synthesis, such as alloying with cheaper metals [8], or creation of complex nanoparticle geometries [9]. Modern commercial PEFCs use Pt loadings of $<0.2\ mg_{Pt}/cm^2$ [10].

Platinum nanoparticles undergo degradation during PEFC operation due to several different mechanisms (Figure 32.5) [11]. The binding energy between Pt

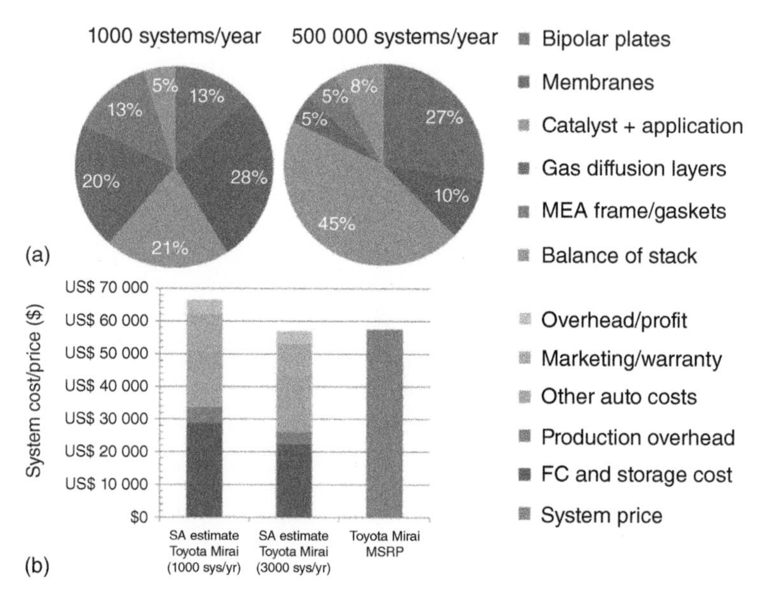

Figure 32.4 (a) Estimated cost of components of a PEFC stack at different production scales. (b) Estimated cost for components in a Toyota Mirai FCV, compared to the manufacturers recommended sale price (MRSP). GDLs, Gas diffusion layers. Source: (a) US Department of Energy 2015 [6]; (b) US Department of Energy 2017 [7]. (See online version for color figure).

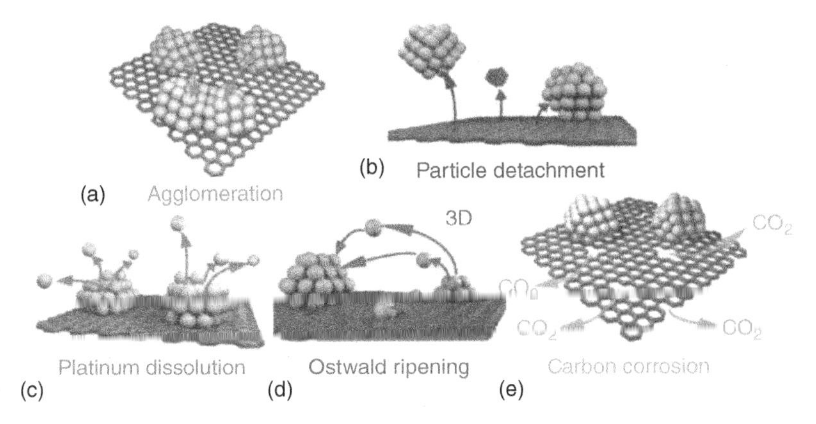

Figure 32.5 Summary of the different degradation mechanisms of platinum nanoparticle electrocatalysts. (a) agglomeration; (b) particle detachment; (c) platinum dissolution; (d) Ostwald ripening; and (e) carbon corrosion. Source: Meier et al. 2014 [11]. Licensed under CC BY 2.0.

and carbon is weak, so the nanoparticles move around on the surface of the support, like balls on a snooker table. When one Pt particle meets another, they stick together. This leads to gradual *agglomeration* into clumps (Figure 32.5a), decreasing the triple phase boundary by blocking the active surface of Pt. Similarly, the nanoparticles can fall off the carbon surface, known as *particle detachment*. This decreases the triple phase boundary by disrupting the electron pathway from the carbon to the platinum (Figure 32.5b). Another mode of catalyst degradation is *platinum dissolution*. Under ambient conditions solid platinum is stable in water, but at certain voltages and elevated temperatures, it can dissolve into solution (Figure 32.5c). A related process is *Ostwald ripening* (Figure 32.5d). As the voltage is cycled between low and high potential (e.g. by accelerating and decelerating in an FCV), Pt atoms from small nanoparticles dissolve and then reprecipitate onto larger and more thermodynamically stable nanoparticles [12]. This leads to a gradual increase in average particle size, decreasing the surface area. The final major mode of Pt degradation is *carbon corrosion*. During start-up and shutdown operation (e.g. turning an FCV on and off), the cell potential can reach 1.5 V. Carbon then reacts with oxygen or water to form surface oxides, CO, or CO_2 gas [13]. Carbon oxidation affects the electronic conductivity, increasing the ohmic losses, while gasification leads to particle detachment, reducing the triple phase boundary (Figure 32.5e).

There are many types of Pt-based electrocatalyst for PEFCs and different ways to prepare them. The support could be carbon black, graphene, carbon nanotubes, carbon foam, metal oxides, or a host of other nanomaterials [14–18]. Carbon black is the most common support, but even this is made in different ways by different suppliers. Platinum decoration can also be performed in different ways; herein we describe two common methods. One scheme uses thermal reduction to make nanoparticles from platinum acetylacetonate, known as the *Pt(acac) method* [17, 19]. For example, Pt(acac) is dissolved in dichloromethane, the carbon support is added, and then the dispersion is sonicated until the solvent completely evaporates, after which it is transferred to an infrared furnace and heated to decompose the precursor and form nanoparticles. The loading is controlled by varying the ratio of Pt(acac) to carbon. The *chloroplatinic acid method* uses chemical reduction to form nanoparticles [20]. For example, H_2PtCl_6 and $NaHSO_3$ are dissolved in distilled water; then H_2O_2 and NaOH solutions are added dropwise while maintaining a pH of 5.0. The resulting dispersion is filtered and dried in air and then heated at 200 °C in 5% H_2 gas.

32.3 Voltammetry

Voltammetry is used to analyze the electrochemical behavior of PEFC catalysts on an electrode, revealing information about the different species involved in redox reactions. This is usually performed in a *three-electrode cell* with (i) a *working electrode*, where the catalyst of interest is deposited; (ii) a *counter electrode*, which balances charge transfer; and (iii) a *reference electrode*, which monitors and helps control the potential (Figure 32.6a). Changes in current are recorded as the potential of the working electrode is varied, and the results are plotted in a *voltammogram*. This current corresponds to charge transfer to or from the electrode. Since they describe charge transfer at a single electrode rather than for the whole PEFC, they are known as *half-cell measurements*. Here, we will focus on the ORR at the cathode.

There are different protocols for preparing the working electrode, but here, we outline the method recommended by the Fuel Cell Commercialization Conference of Japan (FCCJ) [22]. The aqueous electrolyte is usually 0.1 M perchloric acid ($HClO_4$) or sulfuric acid (H_2SO_4). The working electrode is a disk with an area of 0.196 cm^2 and is made of glassy carbon (or gold). This is cleaned and polished with alumina paste to ensure that the surface is smooth and that the results are reproducible. Electrocatalyst ink is made by mixing 5.2 mg of electrocatalyst

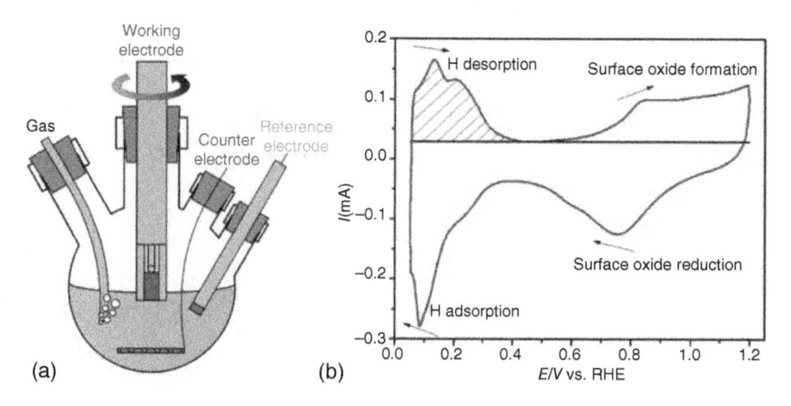

Figure 32.6 (a) Schematic of a three-electrode electrochemical cell. (b) Typical cyclic voltammogram for a Pt/C electrocatalyst in N_2-saturated solution. Source: (a) Hong et al. 2015 [21]. Reproduced with permission of Royal Society of Chemistry; (b) Courtesy of Jianfeng Liu, PhD thesis, Copyright 2014.

with 12 µl of 5 wt% Nafion dispersion, 0.72 ml of ethanol, and 2.28 ml of water, followed by sonication for 30 minutes. Then, 10 µl of this electrocatalyst ink is carefully deposited onto the electrode by micropipette, forming a hemispherical droplet that is then evaporated and dried at 60 °C for 15 minutes. The resulting Pt loading on the working electrode should be ~17.3 µg/cm^2. The role of the counter electrode is to balance the charge transfer at the working electrode. In PEFC studies a Pt wire or a graphite rod is generally used. Graphite is preferred, since Pt from the counter electrode may dissolve into solution and migrate to the working electrode, affecting the voltammograms in unpredictable ways. The reference electrode is a half-cell with known reduction potential with which to compare the working electrode. Current does not pass through this electrode – it is only used to control and measure the potential of the working electrode. Common choices include the reversible hydrogen electrode (RHE), the silver/silver chloride electrode (Ag/AgCl), or the saturated calomel electrode (SCE). It is possible to shift between these different reference potentials using a simplified version of the *Nernst equation* (Eq. (32.4), in saturated KCl solution at 25 °C):

$$E_{\text{RHE}} = E_{\text{Ag/AgCl}} + 0.197\ \text{V} = E_{\text{SCE}} + 0.244\ \text{V} \tag{32.4}$$

For example, the theoretical potential of the HOR is 0 V vs. RHE, −0.197 V vs. Ag/AgCl, or −0.244 V vs. SCE. Meanwhile, the theoretical potential of the ORR is 1.23 V vs. RHE, 1.033 V vs. Ag/AgCl, or 0.986 V vs. SCE. These values vary depending on the concentration and pH of the solution used, as well as the temperature.

32.4 Cyclic Voltammetry

In *cyclic voltammetry* the potential is scanned first in one direction and then the opposite direction. Cyclic voltammograms are static, meaning that the electrode is not rotated and the solution is not stirred. First, nitrogen gas is bubbled through the electrolyte solution for 30 minutes to displace dissolved oxygen. Before recording the data, the potential is cycled 50 times to remove surface contamination. A typical cyclic voltammogram for Pt/C is shown in Figure 32.6b. There are four characteristic peaks, corresponding to four distinct redox processes. Starting in the center of the potential range and sweeping in the positive direction, a positive current peak is observed at ~0.85 V vs. RHE. This corresponds to the formation of PtO_x at the platinum surface. Reversing the sweep direction reduces the PtO_x back to pristine platinum, resulting in a negative peak at ~0.75 V vs. RHE. The peak at ~0.10 V vs. RHE corresponds to adsorption of hydrogen ions onto the Pt from the electrolyte. Reversal of the sweep direction results in desorption of these protons back into solution. This hydrogen ion desorption step can be used to estimate the ECSA of the catalyst (with mass m_{Pt}). The number of hydrogen ions desorbed corresponds to a charge, Q_H, in coulombs. This charge can be obtained by integrating the area under the hydrogen desorption peak, as indicated by the shaded area in Figure 32.6. The ECSA is then estimated using Eq. (32.5), including a conversion factor of 2.1 C/m^2, which corresponds to one hydrogen atom per surface Pt atom:

$$\text{ECSA}\,[\text{m}^2/\text{g}] = \frac{Q_H}{2.1 m_{\text{Pt}}} \tag{32.5}$$

32.5 Linear Sweep Voltammetry

To probe the ORR, *linear sweep voltammetry* (LSV) is employed in a three-electrode cell. For this technique a *rotating disk electrode* (RDE) is required; the potential is only swept in one direction; and the electrolyte is saturated with oxygen gas. By rotating the working electrode, oxygen can be constantly replenished as it is consumed at the cathode (Figure 32.7a). First, nitrogen (or argon) gas is bubbled through the electrolyte for 30 minutes to remove dissolved oxygen for a reference scan. The electrode is set to rotate at 2500 rpm, the potential is scanned from 1.2 to 0.2 V vs. RHE, and the current is recorded. This is then repeated at 1600, 900, and 400 rpm. Next, oxygen is bubbled through the electrolyte for 30 minutes, and then the potential scans are repeated at the same electrode rotation speeds. Finally, the reference scan

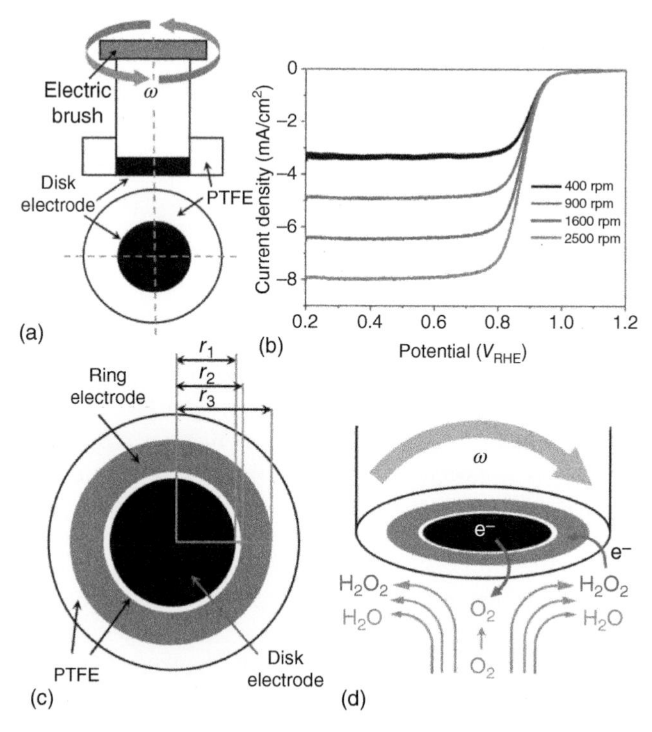

Figure 32.7 (a) Diagram of a rotating disk electrode (RDE). (b) Typical linear sweep voltammogram (LSV) for a Pt/C electrocatalyst at different electrode rotation speeds. (c) A rotating ring-disk electrode (RRDE) using (d) mass transport by centrifugal force. (See online version for color figure).

in N_2 is subtracted from the scan in O_2. A typical LSV for the ORR on Pt/C using an RDE is shown in Figure 32.7b. Several parameters are usually extracted from LSVs:

(i) *Onset potential*: The LSV scan starts at 1.2 V vs. RHE, where the net current density is zero because oxygen is not reduced. At ~1.0 V vs. RHE, the current density starts to increase. This is the *onset potential*, where the ORR starts to occur. The closer the onset potential is to the theoretical potential, the better the catalyst.

(ii) *Half-wave potential*: The current density then increases until ~0.6 V vs. RHE. The halfway point between the maximum and minimum current densities is the *half-wave potential*. This is an indicator of the performance of a catalyst under PEFC operating conditions and is affected by factors such as the catalytic activity, the microstructure, and the electronic conductivity of the electrocatalyst layer. The more positive the half-wave potential, the better the catalyst.

(iii) *Mass transport-limited current density*: The current density levels out below ~0.6 V vs. RHE, at −6.1 mA/cm². This is the *mass transport-limited current density*. Mass transport describes how fast O_2 can diffuse to the catalyst surface to replenish that consumed in the ORR. This depends on the diffusion coefficient of oxygen in the electrolyte and the rotation speed of the working electrode. If the rotation speed is high, the supply of oxygen is rapid, and the mass transport limited current will be large. If the rotation speed is low, supply of oxygen is slower, leading to a small mass transport-limited current.

(iv) *Specific activity (SA) and mass activity (MA)*: The *specific activity* (A/m^2_{Pt}) and *mass activity* (A/g_{Pt}) are used to measure the activity of a catalyst with respect to the ECSA and the mass of Pt, respectively. They are calculated using data from the LSV, with *Koutecky–Levich* (*KL*) *theory* [23]. The LSV comprises two different regions: the *kinetically controlled region* and the *mass transport-limited region*. The mass transport-limited current (i_M) is related to the concentration (C) and diffusion coefficient (D) of oxygen, the viscosity of the electrolyte (ν), the electron transfer number (n), the electrode area (A), the Faraday constant (F), and the angular rotation rate of the working electrode (ω) through the *Levich equation*:

$$i_M = 0.62nFAD^{\frac{2}{3}}\nu^{\frac{-1}{6}}C\omega^{\frac{1}{2}} \tag{32.6}$$

The kinetic current (i_k) reflects the fundamental kinetics of the reaction. The total working electrode current (i_D) can thus be written in terms of i_k and i_M in the *KL equation*:

$$\frac{1}{i_D} = \frac{1}{i_M} + \frac{1}{i_k} = \left(0.62nFAD^{\frac{2}{3}}\nu^{\frac{-1}{6}}C\omega^{\frac{1}{2}}\right)^{-1} + i_k^{-1} \tag{32.7}$$

Thus, i_k can be obtained from the intercept of a *KL plot* of i_D^{-1} vs. $\omega^{-½}$. This is dependent on the potential and is usually defined at 0.9 V vs. RHE for the ORR.

The specific activity and mass activity are then calculated by dividing i_k by the ECSA or the mass of Pt, respectively:

$$SA = i_k/ECSA\,(A/m^2) \tag{32.8}$$

$$MA = i_k/m_{Pt}\,(A/g) \tag{32.9}$$

32.6 Electron Transfer Number

The desired route for the ORR to proceed in a PEFC is reduction of oxygen to water in a four-electron transfer process (Eq. (32.2)). However, a competing reaction is hydrogen peroxide formation via two-electron transfer, with a theoretical potential of 0.695 V vs. RHE (Eq. (32.10)). Thus, the average *electron transfer number* (n) is usually lower than 4:

$$O_2 + 2e^- + 2H^+ \rightarrow H_2O_2 \tag{32.10}$$

This competing reaction reduces the PEFC efficiency, and H_2O_2 attacks the cell components, resulting in degradation. As such it is crucial to determine the electron transfer number and how much H_2O_2 is generated by any new electrocatalyst. One way to obtain this information is from the slope of the KL plot in which all the constants and variables except n are known.

A more advanced configuration of the three-electrode cell is the *rotating ring-disk electrode* (RRDE), which can be used to straightforwardly calculate the electron transfer number without varying the electrode rotation speed. In this, the working electrode is surrounded by a *ring electrode* with independently controlled voltage (Figure 32.7c). In ORR studies, the ring electrode acts as an electrochemical sensor to measure H_2O_2. Hydrogen peroxide formed at the working electrode is transported to the ring electrode via centrifugal force (Figure 32.7d). The ring electrode is held at 1.0 V vs. RHE, which reoxidizes the H_2O_2, generating a ring current (i_R):

$$H_2O_2 \rightarrow O_2 + 2H^+ + 2e^- \tag{32.11}$$

For lamellar flow, the ring current is related to the working (disk) electrode current through the collection efficiency (N). This can be estimated from the electrode geometry, i.e. the working electrode radius (r_1), the ring electrode inner radius (r_2), and the ring electrode outer radius (r_3) [24]. For example, in an RRDE system with $r_1 = 2.5$ mm, $r_2 = 2.75$ mm, and $r_3 = 4$ mm, the collection efficiency is 47.4%. Alternatively, N can be obtained empirically by measuring a known single-electron redox couple and comparing the ring and disk currents:

$$N = \frac{-I_R}{I_D} \tag{32.12}$$

Once the collection efficiency is known, the average electron transfer number can be calculated from the following equation:

$$n = \frac{4I_D}{I_D + \left(\frac{I_R}{N}\right)} \tag{32.13}$$

To calculate this accurately, it is important that lamellar flow is achieved, that the electrode is free from bubbles, and that the disk electrode is large enough. In addition, thick electrocatalyst layers can impede the diffusion of H_2O_2, affecting the collection efficiency and leading to artificially high estimates of *n*.

32.7 Durability Measurements in a Three-Electrode Cell

The durability of electrocatalysts in three-electrode cell configuration is commonly measured by two different cyclic voltammetry protocols. The *start–stop protocol* is used to simulate the conditions when a PEFC is turned on and off, promoting carbon oxidation and corrosion [25]. The *load cycling protocol* simulates normal operation (e.g. acceleration and deceleration in an FCV), promoting platinum dissolution and Ostwald ripening [26]. In Japan, specific protocols to measure these different degradation modes were developed by the FCCJ [27]. For the start–stop protocol, a two seconds triangular potential wave between 1.0 and 1.5 V vs. RHE is repeated for at least 60 000 cycles (Figure 32.8a). For the load cycling protocol, a six seconds square potential wave between −0.4 and 0 V vs. RHE is repeated for at least 60 000 cycles (Figure 32.8b). These protocols are performed in nitrogen-saturated electrolyte. At certain intervals (e.g. 500, 10 000, 30 000, and 60 000 cycles), an LSV is recorded in O_2 to probe the catalytic activity. Typical changes in the ECSA and the LSVs over the 60 000 cycles can be seen in Figure 32.8c,d.

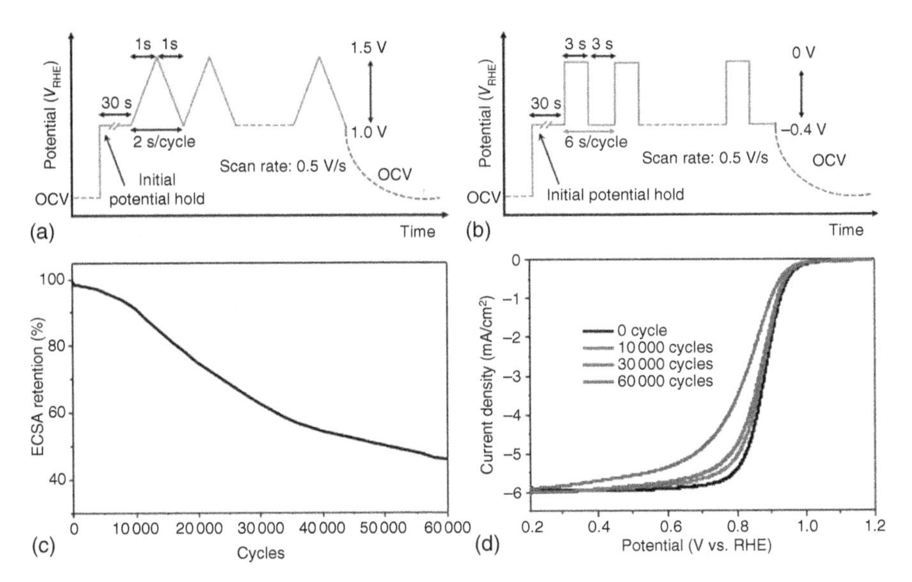

Figure 32.8 (a) Start–stop potential cycling protocol and (b) load potential cycling protocol for a three-electrode cell. Graphs showing typical experimental changes in (c) the ECSA and (d) the LSVs with cycle number during the start–stop protocol. (See online version for color figure).

32.8 Membrane Electrode Assembly (MEA) Fabrication

Three-electrode cell measurements are useful for characterizing electrocatalysts. However, to ensure good performance in a real PEFC, electrocatalysts should be tested in a *membrane electrode assembly* (MEA). This is essentially an ionomer membrane (e.g. Nafion), with electrocatalysts and a *gas diffusion electrode* (GDE) attached on each side. The *cell holder* approximates the rest of the PEFC system, providing electrical contacts and reactant gas supply. A schematic of the MEA within a cell holder is shown in Figure 32.9a. A generalized procedure for fabricating an MEA follows. First, cathode and anode *electrocatalyst inks* are prepared. The catalyst is sonicated for 30 minutes with water, ethanol, and commercially available 5 wt% Nafion dispersion. The solid content of the ink is ~3 wt%, and the Nafion-to-catalyst ratio is ~0.28 (but this should be optimized for each new tested catalyst). The resulting inks are then deposited onto both sides of a Nafion membrane using a mask to cover the desired area. This could be done by hand painting with a brush, by using an airbrush, or by using a purpose-built fully automated spray printing system. The final loading of the catalyst on the surface (e.g. 0.3 mg_{Pt}/cm^2) can be checked using a microbalance. This is then hot pressed at 132 °C and 0.6 kN for 180 seconds to improve the contact between the different layers by softening the Nafion. The catalyst area is then covered by sheets of hydrophobic carbon paper (e.g. Toray, EC-TP1-060T). These act as current collectors and GDEs.

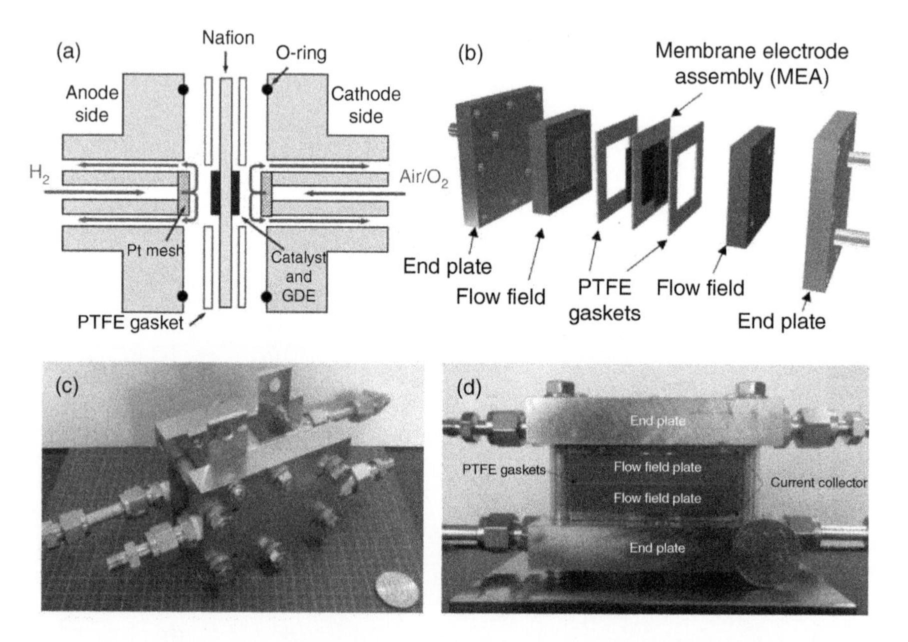

Figure 32.9 (a) Schematic diagram of an MEA in a simple cell holder. (b) Exploded diagram of an MEA in a NEDO cell holder. (c) Photo of a NEDO single cell holder and (d) cross section of the same cell. Source: (b) Courtesy of Thomas Bayer, Copyright 2016; (d) Courtesy of Thomas Bayer, Copyright 2015.

The MEA is then placed into the cell holder (e.g. a New Energy and Industrial Technology Development Organization (NEDO) cell holder, Japan; Figure 32.9b–d). Polytetrafluoroethylene (PTFE) *gaskets* are placed on either side of the MEA to seal the cell and avoid gas leakage. These are around the same size as the membrane and have windows corresponding to the position of the GDEs. The cell is then placed between two *flow-field plates*, made of graphite with gas flow channels on the side facing the GDEs. These plates conduct current to and from the GDEs, supply the gases, and help seal the cell. Finally, stainless steel *end plates* are clamped onto both sides of the cell and bolted together with a pressure of 2 N m using a torque wrench.

32.9 MEA Measurements

The current–voltage ($I-V$) characteristics of MEAs are evaluated using a *fuel cell test station*. This incorporates (i) an oven to control the cell temperature, (ii) a humidifier to control the humidity of the feed gases, (iii) a constant current pulse generator, and (iv) a gas flow control unit. The cell holder is connected to the gas supply and electrodes, placed into the oven, and heated to 80 °C. Experiments should be conducted under a hydrogen hood for safety. The gas channels are first purged with humidified N_2 gas, and then 100% humidified H_2 and air are supplied to the anode and cathode, respectively. Before $I-V$ measurements, a pretreatment step should be performed at a current density of 200 mA/cm^2 for four hours to "age" the cell and to ensure reproducible output. The current drawn from the cell is increased stepwise in intervals of 10 mA, and the cell voltage is measured. Once the cell voltage decreases to 0.2 V, the measurement is usually stopped. After the measurement, the gas lines should be purged with N_2 to remove residual H_2.

A typical MEA $I-V$ curve is shown in Figure 32.10. The shape is similar to the half-cell measurements (e.g. Figure 32.7) but rotated by 90° (since now the current rather than the voltage is being varied). The *theoretical cell voltage* is ~1.2 V, shown by the dotted line. This would be expected for a 100% efficient, perfect

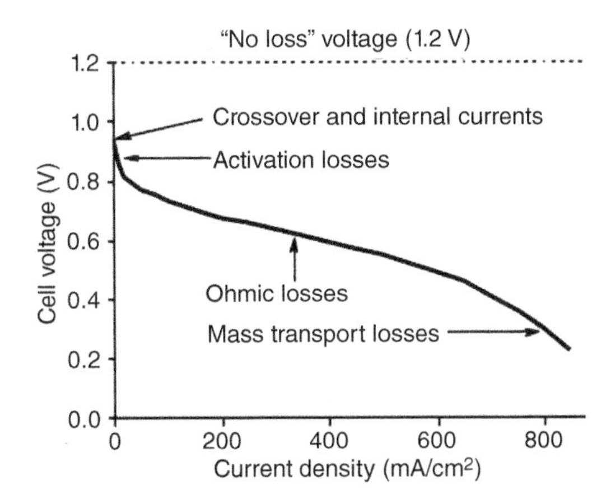

Figure 32.10 Typical current–voltage ($I-V$) characteristics of a membrane electrode assembly (MEA).

PEFC. The difference between this and the measured cell voltage is the *overpotential*, which describes efficiency losses. The *open-circuit voltage (OCV)* measured where no current flows is usually ~1.0 V. This initial overpotential of ~0.2 V is mainly attributed to *fuel crossover* (where H_2 passes through the membrane to react directly with O_2). This can be improved by using thicker membranes. As the current density increases, there is a small and rapid drop in cell voltage to ~0.8 V. This overpotential is due to *activation losses*, from the energy needed to break the bonds of the O_2 molecules. This can be minimized by using better electrocatalysts or increasing the temperature. Next a linear drop in voltage is observed as the current increases. This is due to *ohmic losses*, i.e. the ionic and electronic resistance of the PEFC. This can be minimized by increasing the conductivity and/or decreasing the thickness of the electrolyte. Finally, at high current density, a rapid drop in cell voltage occurs. This is due to *mass transport losses*, where the reaction occurs more quickly than the reactants can be supplied. This can be minimized by increasing the pressure or by optimizing the porosity of the electrocatalyst layer.

MEA durability can be tested using the same potential cycling protocols as for the half-cell measurements. For example, the anode is purged with H_2 and used as the reference/counter electrode, while the cathode working electrode is purged with N_2 and the potential is varied according to the protocols described earlier [28]. Alternatively, simple constant current density or constant voltage measurements can be performed.

32.10 Recent Electrocatalyst Research

The above discussion describes the general assembly and testing of PEFCs. As researchers, we can vary different factors in this setup to improve the performance or learn fundamental principles about the system. The electrocatalyst material, nanostructure, or loading, the type of support, the ionomer material, the membrane thickness, the humidity, and the temperature can all be varied. For example, we have carried out several studies in which the platinum support was changed. In one case, we investigated the effect of changing the support from carbon black to *graphene nanoplatelets* or *graphene-like carbon foam* [14, 16]. Switching to these supports resulted in enhanced mass activity, which was attributed to improved electronic conductivity, the open pore structure, the larger surface area, and the improved platinum particle size distribution. The durability was also improved, due to the strong binding energy between platinum and defects in the carbon foam, resulting in reduced Pt mobility, thus suppressing aggregation (Figure 32.11).

Carbon can also be replaced with *metal oxide supports*, such as tin oxide. Pt-decorated SnO_2 has been shown to have mass activity comparable with that of conventional Pt/C, but with much improved durability, since carbon oxidation during start–stop potential cycling is avoided [17, 29–31]. Atomic-resolution transmission electron microscopy (TEM) and density functional theory (DFT) simulations revealed that the binding energy between Pt nanoparticles and SnO_2

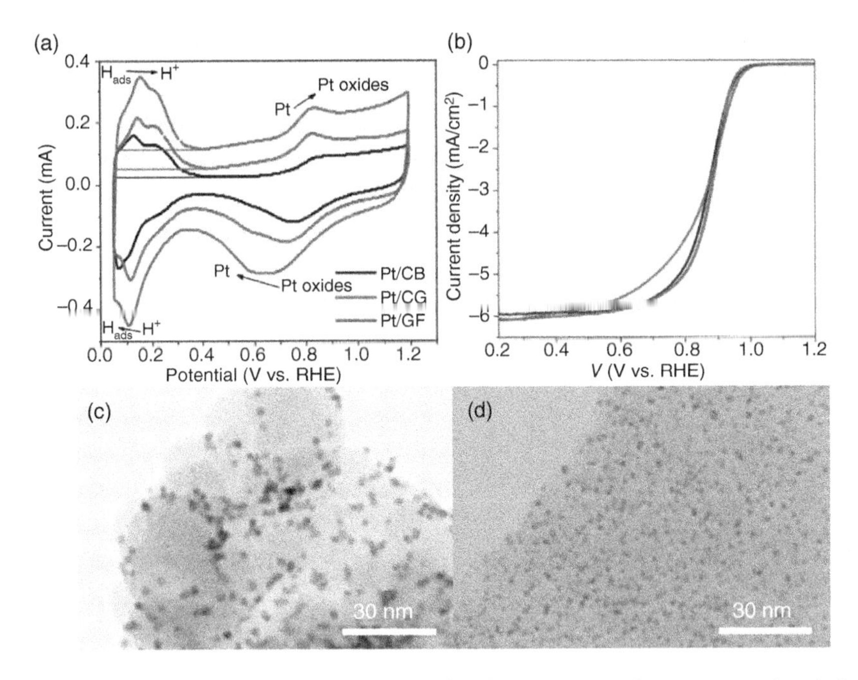

Figure 32.11 (a) Cyclic voltammograms and (b) linear sweep voltammograms for platinum decorated on different carbon supports. Source: Liu et al. 2014 [14]. Licensed under CC BY NC ND 4.0. TEM images of Pt-decorated (c) carbon black and (d) graphene-like carbon foam. (See online version for color figure).

is much stronger than between Pt and carbon, suppressing Pt agglomeration and dissolution [18] (Figure 32.12).

Another way to suppress carbon corrosion is to coat carbon supports with a thin layer of polybenzimidazole (PBI) before decoration with Pt. Platinum binds strongly to the nitrogen-containing azole group in PBI, helping to significantly enhance the durability by preventing agglomeration, without significantly sacrificing performance [28, 32]. Other strategies for suppressing degradation involve modifying the platinum electrocatalyst itself by alloying. For example, alloying platinum with cobalt to create structurally ordered core–shell nanoparticles has been shown to enhance the mass activity by 200% and the specific activity by 300% and result in negligible degradation over 5000 potential cycles. These improvements were attributed to the Pt-rich shell and the thermodynamically stable intermetallic Pt_3Co core atomic arrangement [33]. Bimetallic Pt_4Mo nanoparticles have been shown to be more tolerant to poisoning by carbon monoxide compared with more conventional PtRu electrocatalysts [34]. This was attributed to continuous oxidation of CO adsorbed on the Pt sites by oxygenated species activated at neighboring Mo atoms. Overall the shape, structure, porosity, and composition of platinum-based catalyst nanoparticles can be changed in a variety of different ways, resulting in complex changes in electrochemical efficiency, activity, durability, and resistance to poisoning [35].

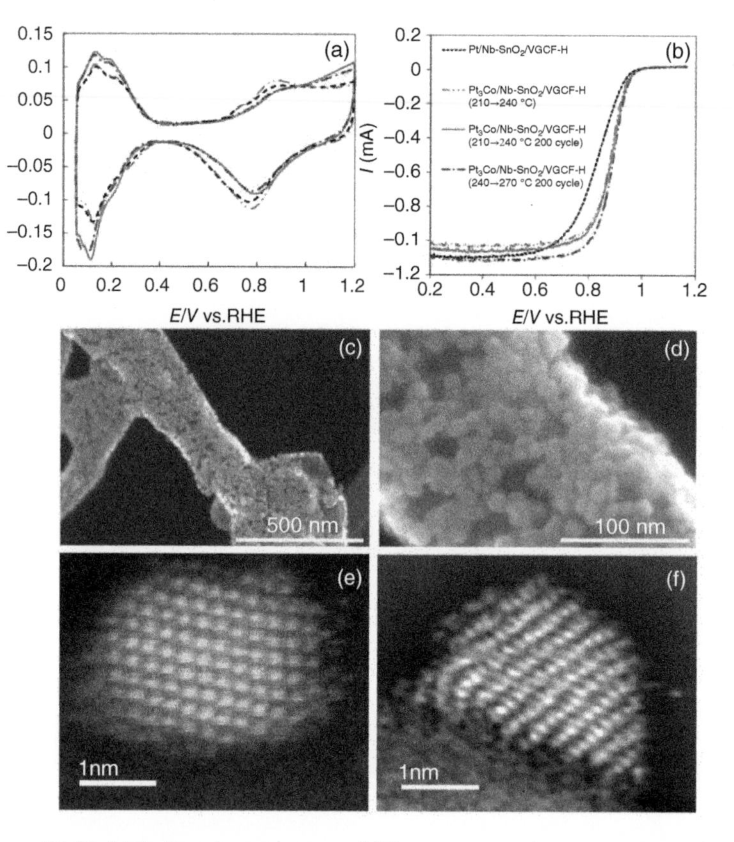

Figure 32.12 (a) Cyclic voltammograms, (b) linear sweep voltammograms and (c, d) scanning electron microscope (SEM) images of Pt nanoparticles decorated on tin oxide supports with carbon fiber fillers. Source: Republished with permission of The Electrochemical Society. Licensed under CC BY 4.0 [31]. TEM images of platinum nanoparticles decorated on (e) carbon black, and (f) tin oxide. Source: Daio, T. et al. 2015 [18]. Licensed under CC BY 4.0.

Finally, some of the issues with using Pt (e.g. high cost, degradation) can be avoided by using platinum-free electrocatalysts. For example, *nitrogen-doped carbon decorated with iron atoms (Fe–N–C)* is a potentially useful alternative (Figure 32.13a). Our lab developed an Fe–N–C foam (Figure 32.13b) with very large surface area (1600 m^2/g), which can be synthesized at low cost (~US$10/g) [37–40]. When the ORR half-cell reaction is measured in alkaline solution, the mass activity and half-wave potential are higher than conventional Pt/C and a commercially available Fe–N–C catalyst (Figure 32.13c). However, in load cycle durability tests, the half-wave potential decreases quickly after 10 000 cycles (Figure 32.13d). This was attributed to leaching of iron from the active sites and partial oxidation of the carbon. Further research is underway to improve the stability of this promising Pt-free electrocatalyst. While Fe–N–C catalysts receive the greatest attention and achieve the highest performances, other related Pt-free electrocatalyst systems under investigation include Co–N–C materials [41, 42] and metal-free nitrogen-doped carbons [43–45].

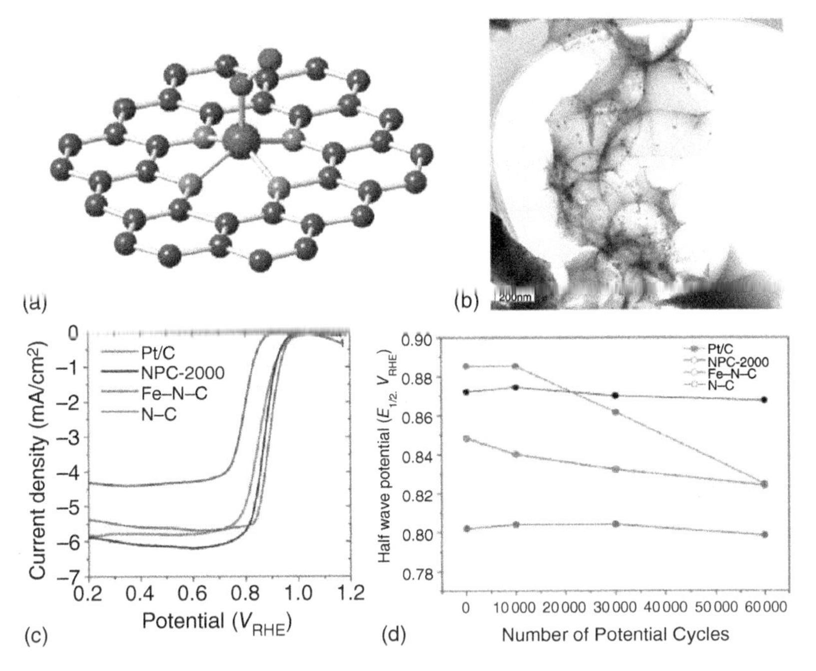

Figure 32.13 (a) Proposed active site in Fe–N–C electrocatalysts. (b) Transmission electron micrograph, (c) linear sweep voltammogram (LSV), and (d) change in the half-wave potential ($E_{1/2}$) with cycle number for Fe–N–C foam electrocatalysts developed in our lab compared with conventional catalysts, measured in 0.1 M KOH solution. Source: (a) Zitolo et al. 2017 [36]. Licensed under CC BY 4.0; (d) Mufundirwa et al. 2018 [37]. Reproduced with permission of Elsevier. (See online version for color figure).

32.11 Future Perspectives

The implementation of hydrogen fuel cell technologies is rapidly accelerating, and new emerging electrocatalysts with higher activity, lower cost, or improved durability have a good chance of contributing directly to this blossoming industry. Already the automotive sector has moved past conventional platinum nanoparticles decorated on carbon black and toward bimetallic PtCo nanoparticles decorated on different types of advanced carbon supports [46]. It is impossible to say which electrocatalyst technologies currently being developed by researchers across the globe will eventually be taken up by industry and incorporated into real fuel cell systems. However, students working on new catalyst systems today may well see their ideas applied in the real world in the near future.

One promising area for future applications is the use of alternative supports that are corrosion resistant, bind strongly to platinum, and enhance mass transfer, such as highly graphitized carbons, graphene-like materials, carbon nanotubes, fibrous carbons, nonporous carbons, carbon foams, metal oxides, inorganic materials, or multiphase composite materials [47, 48]. Another promising area is nanoscale engineering of electrocatalysts to enhance activity, including the creation of cage-like architectures, intermetallic structures, core–shell particles, nanoplatelets, and trimetallic catalysts. If the scalability, cost, and durability of such complex nanoparticle architectures can be improved, then these could be incorporated into fuel cells [9, 35]. Platinum-free electrocatalysts (i.e. Fe–N–C) are in the initial phase of industrialization in fuel cell systems and are on the verge of revolutionizing the industry, significantly reducing the cost of fuel cell systems. However, there are still problems to solve, such as the understanding the fundamental nature of the electrochemical active sites, improving the durability, and maximizing efficiency [36, 49]. A move toward fuel cells operating at higher temperature (e.g. 120–180 °C) is likely in the future, which would result in improved kinetics (i.e. reduced activation losses) and enhanced membrane conductivity. However, the degradation of platinum catalysts at higher temperature is still relatively fast [50, 51]. Finally, there has recently been a resurgence in the study of alkaline fuel cells due to the discovery of new high-performance anion exchange membranes, opening up the field to a range of new and radically different electrocatalysts [52, 53].

As such, this is an exciting time in the development of fuel cell electrocatalysts, and many new developments can be expected in the near future.

Acknowledgments

The authors gratefully acknowledge support from the Kyushu University Platform of Inter/Transdisciplinary Energy Research (Q-PIT), the Japan Science and Technology Agency (JST) through its Center of Innovation (COI) program, and the International Institute for Carbon-Neutral Energy Research (WPI-I2CNER), funded by the World Premier International Research Center Initiative (WPI), MEXT, Japan. This work was partially supported by the JSPS Bilateral Joint Research Projects fund. Albert Mufundirwa acknowledges support from a Ministry of Education, Culture, Sports, Science and Technology (MEXT) scholarship.

References

1 Michael R. Swain Fuel Leak Simulation, Proceedings of the 2001 U.S. DOE Hydrogen Program Review, Baltimore, Maryland April 17-19. (2001). https://www.nrel.gov/docs/fy01osti/30535.pdf.

2 Schlapbach, L. and Züttel, A. (2001). Hydrogen-storage materials for mobile applications. *Nature* 414: 353–358.

3 Grove, W.R. (1842). On a gaseous voltaic battery. *London, Edinburgh, Dublin Philos. Mag. J. Sci.* 21: 417–420.

4 Berg, P., Novruzi, A., and Volkov, O. (2008). Reaction kinetics at the triple-phase boundary in PEM fuel cells. *J. Fuel Cell Sci. Technol.* 5: 021007.

5 O'Hayre, R. and Prinz, F.B. (2004). The air/platinum/Nafion triple-phase boundary: characteristics, scaling, and implications for fuel cells. *J. Electrochem. Soc.* 151: A756.
6 Marcinkoski, J., Jacob Spendelow, Adria Wilson, Dimitrios Papageorgopoulos DOE Hydrogen and Fuel Cells Program Record: Fuel Cell System Cost – 2015. (2015). https://www.hydrogen.energy.gov/pdfs/15015_fuel_cell_system_cost_2015.pdf.
7 Adria Wilson, Gregory Kleen, and Dimitrios Papageorgopoulos Fuel Cell Vehicle Cost Analysis; DOE Hydrogen and Fuel Cells Program FY 2017 Annual Progress Report. (2017). https://www.hydrogen.energy.gov/pdfs/17007_fuel_cell_system_cost_2017.pdf.
8 Gasteiger, H.A., Kocha, S.S., Sompalli, B., and Wagner, F.T. (2005). Activity benchmarks and requirements for Pt, Pt-alloy, and non-Pt oxygen reduction catalysts for PEMFCs. *Appl. Catal., B* 56: 9–35.
9 Peng, Z. and Yang, H. (2009). Designer platinum nanoparticles: control of shape, composition in alloy, nanostructure and electrocatalytic property. *Nano Today* 4: 143–164.
10 Litster, S. and McLean, G. (2004). PEM fuel cell electrodes. *J. Power Sources* 130: 61–76.
11 Meier, J.C., Galeano, C., Katsounaros, I. et al. (2014). Design criteria for stable Pt/C fuel cell catalysts. *Beilstein J. Nanotechnol.* 5: 44–67.
12 Ostwald, W. (1886). *Lehrbuch der allgemeinen Chemie*. Leipzig: Wilhelm Engelmann.
13 Colmenares, L.C., Wurth, A., Jusys, Z., and Behm, R.J. (2009). Model study on the stability of carbon support materials under polymer electrolyte fuel cell cathode operation conditions. *J. Power Sources* 190: 14–24.
14 Liu, J., Takeshi, D., Sasaki, K., and Lyth, S.M. (2014). Defective graphene foam: a platinum catalyst support for PEMFCs. *J. Electrochem. Soc.* 161: F838–F844.
15 Fujigaya, T. and Nakashima, N. (2013). Fuel cell electrocatalyst using polybenzimidazole-modified carbon nanotubes as support materials. *Adv. Mater.* 25: 1666–1681.
16 Liu, J., Takeshi, D., Sasaki, K., and Lyth, S.M. (2014). Platinum-decorated nitrogen-doped graphene foam electrocatalysts. *Fuel Cells* 14: 728–734.
17 Tsukatsune, T., Takabatake, Y., Noda, Z. et al. (2014). Platinum-decorated tin oxide and niobium-doped tin oxide PEFC electrocatalysts: oxygen reduction reaction activity. *J. Electrochem. Soc.* 161: F1208–F1213.
18 Daio, T., Staykov, A., Guo, L. et al. (2015). Lattice strain mapping of platinum nanoparticles on carbon and SnO_2 supports. *Sci. Rep.* 5: 13126.
19 Kubota, T., Asakura, K., Ichikuni, N., and Iwasawa, Y. (1996). A new method for quantitative characterization of adsorbed hydrogen on Pt particles by means of Pt L-edge XANES. *Chem. Phys. Lett.* 256: 445–448.

20 Yoshitake, T., Shimakawa, Y., Kuroshima, S. et al. (2002). Preparation of fine platinum catalyst supported on single-wall carbon nanohorns for fuel cell application. *Physica B* 323: 124–126.

21 Hong, W.T., Risch, M., Stoerzinger, K.A. et al. (2015). Toward the rational design of non-precious transition metal oxides for oxygen electrocatalysis. *Energy Environ. Sci.* 8: 1404–1427.

22 Ohma, A., Shinohara, K., Iiyama, A. et al. (2011). Membrane and catalyst performance targets for automotive fuel cells by FCCJ membrane, catalyst, MEA WG. *ECS Trans.* 41: 775–784.

23 Koutecky, J.A. and Levich, V.G. (1958). The use of a rotating disk electrode in the studies of electrochemical kinetics and electrolytic processes. *Zh. Fiz. Khim.* 32: 1565–1575.

24 Allen, J. and Bard, L.R.F. (2001). *Electrochemical Methods: Fundamentals and Applications*, 2e. Wiley.

25 Maass, S., Finsterwalder, F., Frank, G. et al. (2008). Carbon support oxidation in PEM fuel cell cathodes. *J. Power Sources* 176: 444–451.

26 Ferreira, P.J., la O', G.J., Shao-Horn, Y. et al. (2005). Instability of Pt/C electrocatalysts in proton exchange membrane fuel cells. *J. Electrochem. Soc.* 152: A2256.

27 Fuel Cell Commercialization Conference of Japan (FCCJ). *Proposals of the Development Targets, Research and Development Challenges and Evaluation Methods Concerning PEFCs.* (2011).

28 Berber, M.R., Fujigaya, T., and Nakashima, N. (2014). High-temperature polymer electrolyte fuel cell using poly(vinylphosphonic acid) as an electrolyte shows a remarkable durability. *ChemCatChem* 6: 567–571.

29 Takabatake, Y., Noda, Z., Lyth, S.M. et al. (2014). Cycle durability of metal oxide supports for PEFC electrocatalysts. *Int. J. Hydrogen Energy* 39: 5074–5082.

30 Kanda, K., Noda, Z., Nagamatsu, Y. et al. (2014). Negligible start-stop-cycle degradation in a PEFC utilizing platinum-decorated tin oxide electrocatalyst layers with carbon fiber filler. *ECS Electrochem. Lett.* 3: F15–F18.

31 Matsumoto, S., Iwami, M., Noda, Z. et al. (2016). PEFC alloy electrocatalysts supported on SnO_2: a study on the preparation method. *ECS Trans.* 75: 851–858.

32 Fujigaya, T., Hirata, S., and Nakashima, N. (2014). A highly durable fuel cell electrocatalyst based on polybenzimidazole-coated stacked graphene. *J. Mater. Chem. A* 2: 3888.

33 Wang, D., Xin, H.L., Hovden, R. et al. (2013). Structurally ordered intermetallic platinum–cobalt core–shell nanoparticles with enhanced activity and stability as oxygen reduction electrocatalysts. *Nat. Mater.* 12: 81–87.

34 Mukerjee, S., Lee, S.J., Ticianelli, E.A. et al. (1999). Investigation of enhanced CO tolerance in proton exchange membrane fuel cells by carbon supported PtMo alloy catalyst. *Electrochem. Solid-State Lett.* 2: 12.

35 Wu, J. and Yang, H. (2013). Platinum-based oxygen reduction electrocatalysts. *Acc. Chem. Res.* 46: 1848–1857.

36 Zitolo, A., Ranjbar-Sahraie, N., Mineva, T. et al. (2017). Identification of catalytic sites in cobalt-nitrogen-carbon materials for the oxygen reduction reaction. *Nat. Commun.* https://doi.org/10.1038/s41467-017-01100-7.

37 Mufundirwa, A., Harrington, G.F., Smid, B. et al. (2018). Durability of template-free Fe–N–C foams for electrochemical oxygen reduction in alkaline solution. *J. Power Sources* 375: 244–254.

38 Liu, J., Takeshi, D., Orejon, D. et al. (2014). Defective nitrogen-doped graphene foam: a metal-free, non-precious electrocatalyst for the oxygen reduction reaction in acid. *J. Electrochem. Soc.* 161.

39 Liu, J., Yu, S., Daio, T. et al. (2016). Metal-free nitrogen-doped carbon foam electrocatalysts for the oxygen reduction reaction in acid solution. *J. Electrochem. Soc.* 163.

40 Lyth, S.M., Nabae, Y., Islam, N.M. et al. (2012). Solvothermal synthesis of nitrogen-containing graphene for electrochemical oxygen reduction in acid media. *e-J. Surf. Sci. Nanotechnol.* 10.

41 Dodelet, J.-P. (2013). The controversial role of the metal in Fe- or Co-based electrocatalysts for the oxygen reduction reaction in acid medium. In: *Electrocatalysis in Fuel Cells*, vol. 9 (eds. D. Shao, D. MinHiggins and Z. Chen), 271–338. London: Springer.

42 Kattel, S., Atanassov, P., and Kiefer, B. (2013). Catalytic activity of Co-N(x)/C electrocatalysts for oxygen reduction reaction: a density functional theory study. *Phys. Chem. Chem. Phys.* 15: 148–153.

43 Lin, Z., Song, M., Ding, Y. et al. (2012). Facile preparation of nitrogen-doped graphene as a metal-free catalyst for oxygen reduction reaction. *Phys. Chem. Chem. Phys.* 14: 3381–3387.

44 Jiang, G., Goledzinowski, M., Comeau, F.J.E. et al. (2016). Free-standing functionalized graphene oxide solid electrolytes in electrochemical gas sensors. *Adv. Funct. Mater.* https://doi.org/10.1002/adfm.201504604.

45 Lee, C.H., Jun, B., and Lee, S.U. (2018). Metal-free oxygen evolution and oxygen reduction reaction bifunctional electrocatalyst in alkaline media: from mechanisms to structure–catalytic activity relationship. *ACS Sustainable Chem. Eng.* 6: 4973–4980.

46 Mizutani, N. and Ishibashi, K. (2016). Enhancing PtCo electrode catalyst performance for fuel cell vehicle application. SAE Technical Paper 2016-01-1187, SAE 2016 World Congress and Exhibition.

47 Antolini, E. (2009). Carbon supports for low-temperature fuel cell catalysts. *Appl. Catal., B* 88: 1–24.

48 Park, Y.-C., Tokiwa, H., Kakinuma, K. et al. (2016). Effects of carbon supports on Pt distribution, ionomer coverage and cathode performance for polymer electrolyte fuel cells. *J. Power Sources* 315: 179–191.

49 Workman, M.J., Dzara, M., Ngo, C. et al. (2017). Platinum group metal-free electrocatalysts: effects of synthesis on structure and performance in proton-exchange membrane fuel cell cathodes. *J. Power Sources* 348: 30–39.

50 Matsumoto, K., Fujigaya, T., Sasaki, K., and Nakashima, N. (2011). Bottom-up design of carbon nanotube-based electrocatalysts and their application in high temperature operating polymer electrolyte fuel cells. *J. Mater. Chem.* 21: 1187–1190.

51 Roslia, R.E., Sulong, A.B., Daud, W.R.W. et al. (2017). A review of high-temperature proton exchange membrane fuel cell (HT-PEMFC) system. *Int. J. Hydrogen Energy* 42: 9293–9314.

52 Wang, L. and Varcoe, J.R. (2017). The first anion-exchange membrane fuel cell to exceed 1 W cm^{-2} at 70 °C with a non-Pt cathode. *Chem. Commun.* https://doi.org/10.1039/C7CC06392J.

53 Wang, L., Magliocca, E., Cunningham, E.L. et al. (2017). An optimised synthesis of high performance radiation-grafted anion-exchange membranes. *Green Chem.* 19: 831–843.

33

Conversion of Lignocellulosic Biomass to Biofuels

Cristina García-Sancho[1], *Juan A. Cecilia*[1], *and Rafael Luque*[2]

[1] *Universidad de Málaga, Departamento de Química Inorgánica, Cristalografía y Mineralogía, Campus de Teatinos, Facultad de Ciencias, 29071 Málaga, Spain*
[2] *Universidad de Córdoba, Campus de Rabanales, Departamento de Química Orgánica, Ctra. Nacional IV-A, Km. 396, 14014, Cordoba, Spain*

33.1 Introduction

Currently, fossil fuels account for the most of the energy worldwide consumption, and around 86% of fuels consumed are obtained from fossil resources such as petroleum, coal, and natural gas [1]. Moreover, world population is estimated to reach 8.5 billion in 2030 and 9.7 billion in 2050, which means an increase in the world energy consumption will be equal to approximately 48% in 2040 [2]. However, the depletion of fossil fuel reserves, together with the unsteady price of crude oil and the environmental concerns due to global warming and pollution, leads to the development of alternative and clean synthetic routes for producing chemicals and fuels from nonfossil carbon sources. Thus, new carbon-based fuels must be derived from photosynthetic products tackling problems such as greenhouse gases emission and energy shortage. Furthermore, these fuels must possess similar physicochemical properties to that of gasoline or diesel, which allow their use in internal combustion engines with or without modifications. They also must be produced economically and effectively [3]. In this sense, the use of biomass as a renewable source of carbon for the production of biofuels and valuable chemicals is a promising alternative and will lead to the development of biorefinery facilities. The biorefinery concept is analogous to the current petroleum refinery in that it integrates biomass conversion processes and equipment to produce fuels, power, heat, and value-added chemicals from biomass (Figure 33.1).

Lignocellulosic biomass is the most abundant source of biomass with an estimated annual production of about 2×10^{11} tons with an energy content of 3×10^{18} kJ/yr by photosynthesis, which is about 10 times the current annual energy consumption worldwide [4, 5]. Moreover, the cost of lignocellulosic biomass based on energy content is around 50% lower than other feedstocks [6]. Therefore, lignocellulosic biomass possesses a great potential to produce energy

Heterogeneous Catalysts: Advanced Design, Characterization and Applications, First Edition.
Edited by Wey Yang Teoh, Atsushi Urakawa, Yun Hau Ng, and Patrick Sit.

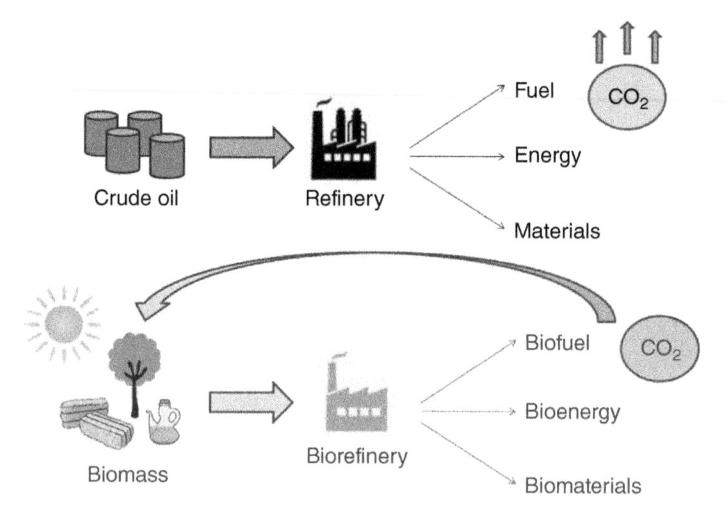

Figure 33.1 Comparison between the current refinery and biorefinery. (See online version for color figure).

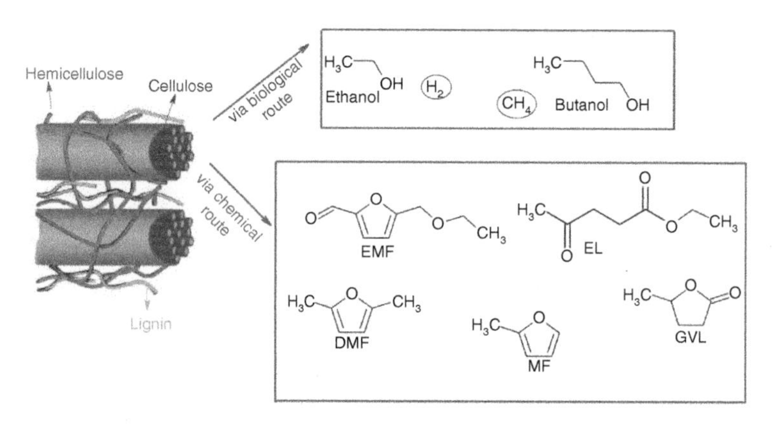

Figure 33.2 Biofuels produced from lignocellulosic biomass via biological and chemical pathways. (See online version for color figure).

owing to its inexpensive cost and wide availability, mainly from non-food lignocellulose such as agricultural and forest wastes.

The use of lignocellulosic biomass has been traditionally limited to burning for heating and cooking with subsequent negative environmental impacts such as land degradation and desertification [7]. However, lignocellulose can be transformed into different types of fuels by using both biological and chemical pathways (Figure 33.2) [8].

Bioethanol and biobutanol produced by carbohydrate fermentation have been extensively researched in the past 40 years [9–11]. Other high-quality fuels can be synthesized from carbohydrates in lignocellulosic materials via chemical route such as 5-ethoxymethylfurfural (EMF), ethyl levulinate (EL), 2,5-dimethylfuran (DMF), 2-methylfuran (MF), and γ-valerolactone (GVL) [8, 12, 13].

In this chapter, the main biofuels produced from lignocellulosic biomass are compiled. Their physicochemical properties and the catalysts employed in their production are discussed in detail.

33.2 Lignocellulosic Biomass: Composition and Resources

The lignocellulosic biomass is mainly composed of polymeric carbohydrates (cellulose and hemicellulose) and aromatic polymer lignin (Figure 33.2) with smaller amounts of pectin, proteins, and inorganic salts [14, 15]. The composition of hemicellulose, cellulose, and lignin in lignocellulosic biomass is 25–35%, 40–50%, and 15–20%, respectively [16]. The long chains of cellulose along with hemicellulose and lignin are bound via hydrogen bonds and van der Waals forces, forming a complex structure, which makes it necessary to break down or remove lignin for the exploitation of cellulose and hemicellulose components [17]. Therefore, different pretreatments of lignocellulosic biomass are required to release these polysaccharides [2].

The global potential of lignocellulosic biomass is significantly larger than other types of biomass since it is more independent on climatic and soil conditions – hence requiring fewer inputs of agrochemicals and competing less with food production [18]. Different resources of lignocellulosic biomass can be employed as feedstocks such as agricultural crop residues, food processing wastes, forest residues, mill wastes, and urban wood residues, among others [19]. Thus, a wide variety of lignocellulosic biomass can be transformed into biofuels. Biofuels can generally be classified by the biomass resources used for their production [20, 21]:

- *First-generation biofuels*: Produced from agricultural crops such as sugarcane, corn, wheat, rice, rapeseed, sunflower oil, soybean oil, palm oil. The first-generation biofuels, mainly bioethanol and biodiesel, are readily available, because the processes to obtain them are based on the use of highly mature technology. However, the use of these crops for biofuel production generates a conflict with their use as food, making it more recommendable to use other biomass resources.
- *Second-generation biofuels*: Produced from non-food crops or agricultural residues such as solid wheat straw, corn and wood wastes, used cooking oil, or industrial wastes. They are based on the biofuel production from the renewable resources including biofuels such as biodiesel, bioethanol, biobutanol, biohydrogen, DMF, EMF.
- *Third-generation biofuels*: Synthesized from algae feedstocks, such as biodiesel, bioethanol, biobutanol.
- *Fourth-generation biofuels*: Photobiological solar fuel and electrofuel based on the direct conversion of solar energy into fuel by the use of cheap and abundant feedstocks, such as biohydrogen, bioethanol.

The first-, second-, and third-generation biofuels can be obtained using lignocellulosic biomass as raw material, which makes the first-generation biofuels less advisable due to "food vs. energy" conflict although they can also reduce CO_2 emissions.

33.3 Biofuel Production from Lignocellulosic Biomass

A wide variety of biofuels can be obtained by using lignocellulosic biomass as feedstock. As mentioned earlier, these biofuels can be produced via biological or chemical routes. Biofuels such as bioethanol, biobutanol, biohydrogen, or biomethane have been largely studied in the literature, and they are produced via bioconversion processes from lignocellulosic biomass. For this reason, the production of these biofuels is not included in this chapter since this book is about heterogeneous catalysis. However, other biofuels can be produced by chemical transformation using lignocellulosic biomass as sustainable low-cost feedstock. The major component of plant-derived biomass is carbohydrates, being of great importance to developing efficient and green approaches to their valorization by conversion into high added-value products. Hexoses and pentoses, of which glucose, fructose, and xylose are the most common, are readily available by hydrolysis of biomass. Thus, carbohydrate-containing biomass can be converted into other important chemicals such as furfural or 5-hydroxymethylfurfural (HMF) [22]. These chemicals are considered building blocks and can be transformed to other products or fuels. In this context, furfural-derived fuels have attracted significant attention, which can be produced by etherification, hydrogenation, oxidation, and aldol condensation followed by hydrodeoxygenation (HDO) reaction [8]. Some of these routes to obtain biofuels are compiled in Figure 33.3.

These furfural-derived products can be directly blended with diesel or gasoline in suitable proportion or even could replace them [23, 24]. It is crucial that these biofuels are suitable for effective combustion and they exhibit similar physicochemical properties to gasoline or diesel so that they can be used in internal combustion engines with or without modifications to obtain high combustion efficiency with low emissions [25]. It is known that the presence of oxygen in biomass-derived molecules is high in comparison to fossil fuels. Hence, the removal of oxygen is a key step for biofuel production from biomass [26]. Table 33.1 compiles the physicochemical properties of the main biofuels produced from lignocellulosic biomass such as EMF, EL, DMF, MF, and GVL reported by several authors [3, 8, 27, 28], which are compared with gasoline, diesel, and bioethanol.

The world annual production of ethanol was higher than 85.6 billion liters in 2010 [21]. Balan reported that ethanol is mainly produced from corn (44%), sugarcane (34%), wheat or rice (3.7%), molasses (3.4%), and sugar beet (2%), while only 4.2% is produced from biomass, so it is considered the first-generation biofuel [29]. Moreover, despite the fact that ethanol is used currently as renewable fuel, it has several disadvantages such as high volatility, low energy density, high oxygen

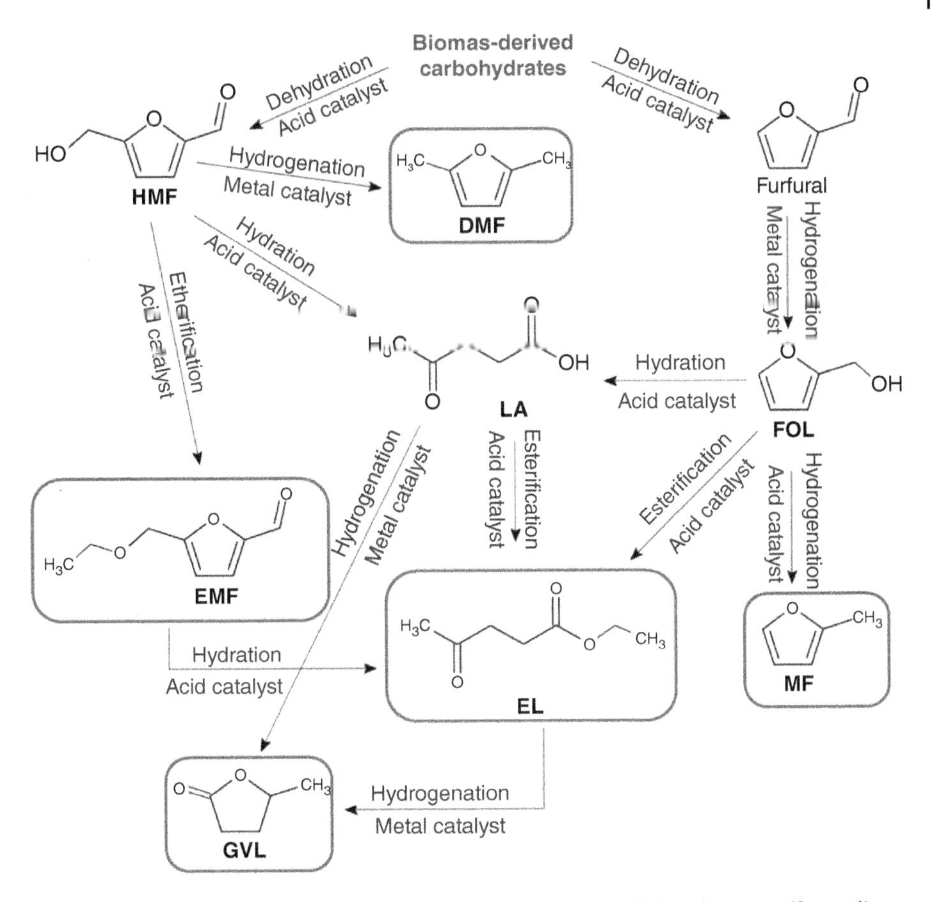

Figure 33.3 Main routes for biofuel production-derived lignocellulosic biomass. (See online version for color figure).

content, and the absorption of water from the atmosphere [3]. For these reasons, researchers have been trying to develop a new generation of renewable fuels in recent years. In this context, heterogeneous catalysis plays a key role to attaining high values of yield, which makes the recovery of catalyst for its reuse possible. Therefore, it is fundamental to design correctly this catalyst in such a way that its catalytic properties enhance the catalytic performance after optimizing reaction conditions with extended lifetime and minimum deactivation (Figure 33.4). These studies have been carried out by several researchers who have studied the production of different biofuels such as EMF, EL, DMF, MF, and GVL. Different heterogeneous catalysts have been evaluated and it will be discussed in the following sections.

33.3.1 Ethoxymethylfurfural (EMF)

EMF is considered a promising alternative fuel with an energy density significantly higher than that of ethanol and similar to that of gasoline and diesel (Table 33.1). It is one of the furanic-derived compounds with greater potential

Table 33.1 Physicochemical properties of gasoline, diesel, and biomass-derived fuels.

Properties	Gasoline	Diesel	Ethanol	EMF	EL	DMF	MF	GVL
Molecular formula	—	—	C_2H_6O	$C_8H_{10}O_3$	$C_7H_{12}O_3$	C_6H_8O	C_5H_6O	$C_5H_8O_2$
Molecular mass (g/mol)	100–105	—	46.07	154.16	144.17	96.13	82.1	100.12
H/C ratio	1.79	—	3	1.25	1.71	1.33	1.2	1.60
O/C ratio	0	—	0.5	0.37	0.43	0.17	0.2	0.4
Density (kg/m^3, 20 °C)	745	820	791	1100	1016	890	913	1050
Energy density (MJ/l)	31.9	33.6	21	30.3	26.4	29.3	28.5	27.7
Latent heat of vaporization (kJ/kg from 25 °C)	351	270–301	919.6	306	404	389.1	357	442.4
Research octane number (RON)	97	—	107	—	—	101	103	—
Boiling point (°C)	27–225	180–370	78	235	206	93	64	207–208
Flash point (°C)	−43	55–60	14	95.5	90.6	−1	−22	96

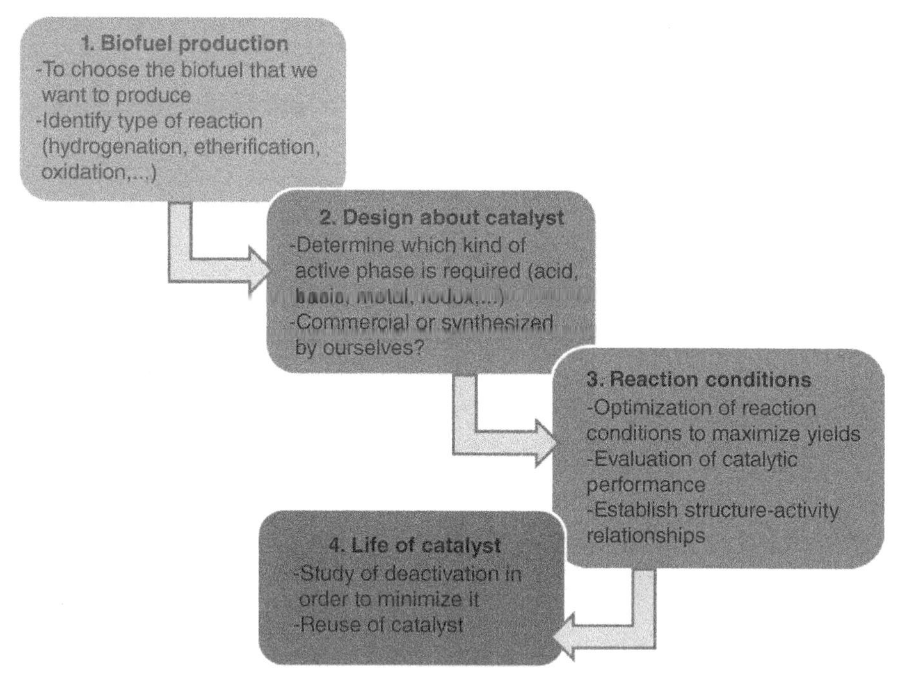

Figure 33.4 Design of steps for optimization of a heterogeneous catalytic process.

to be used as diesel additive or even as fuel on its own due to its high boiling point (235 °C) and low toxicity. The presence of EMF blended with diesel in engines also reduces the formation of SO_x, NO_x, and particles [30, 31]. It is synthesized by etherification of HMF derived from biomass and ethanol in the presence of acid catalysts, since etherification is the dehydration of alcohols to form ethers. The main by-product is generally EL, which will be studied below as a possible additive fuel or even biofuel. In addition to the higher stability of EMF in comparison with HMF molecule, which facilitates its isolation from reaction media, an advantage over other fuels (e.g. DMF) is that EMF does not need to be hydrogenated or to be blended with traditional diesel [30, 32]. Avantium employed EMF as additive for regular diesel blending up to 20% and demonstrated that EMF decreases SO_x emissions and solid particulate contamination; also, this blend worked in the engine without problems for several hours [33].

EMF has attracted the interest of many researchers over the last years. Sanborn patented a procedure for the synthesis and purification of EMF from HMF or fructose [34]. EMF synthesis has been carried out by direct HMF etherification in a temperature range of 70–140 °C by using several catalysts, including homogeneous mineral acids, inorganic salts, ion-exchange resins, and zeolites. Balakrishnan et al. compared different solid acid catalysts with H_2SO_4 and organic acids as catalysts for etherification of HMF to EMF [35]. An EMF yield of 55% was attained in the presence of Amberlyst-15, an ion-exchange resin, after 24 hours at 75 °C. Lanzafame et al. tested HY-zeolite for HMF etherification, but this material showed low activity (y_{EMF} = 8.5% after 24 hours at 70 °C) probably because its small pores led to diffusional problems [36]. They also evaluated mesoporous silica (MS) doped with heteroatoms such as Al-MCM-41 with different Si/Al ratios or Zr over SBA-15. With regard to Al content in Al-MCM-41, they checked that an increase of Al concentration increased the EL yield. The incorporation of Zr to SBA-15 enhanced EMF yield up to 76% after five hours at 140 °C. Nonetheless, the EL yield increased at the expense of EMF yield when Zr-SBA-15 was sulfated. These authors affirmed that the Lewis acid catalysts favored EMF production, whereas Brønsted acid sites promoted EL formation. Lewis acid sites accept an electron pair and Brønsted acid sites donate a hydrogen cation. Despite of this conclusion seemingly to facilitate the choice of catalyst as a function of the desired chemicals, this decision is not so simple. On the one hand, there are typical Lewis (Sn-zeolites, Al_2O_3, ZrO_2, etc.) and Brønsted (ion-exchange resins, heteropolyacids, sulfated zirconia, etc.) acid catalysts, but many catalysts show both types of acid sites, which are frequently required when several reactions in cascade are carried out. On the other hand, Lewis acid sites can be transformed into pseudo-Brønsted acid sites in the presence of water molecules evolved in these processes as by-product or sometimes as solvent depending on the catalyst and the reaction conditions. Therefore, it is complicated to establish a general rule that determines a more suitable catalyst to obtain one of these products.

Similarly, other mesoporous aluminosilicates, such as Al/TUD-1, were evaluated for HMF etherification by Neves et al. [37]. Al/TUD-1 with Si/Al = 21 displayed a promising catalytic performance, achieving an EMF yield of 70% after 24 hours at 140 °C. On the other hand, Che et al. supported a heteropolyacid over mesoporous silica catalyst, $H_4SiW_{12}O_{40}$/MCM-41, and attained high EMF yield (y_{EMF} = 77.4% after four hours at 90 °C), showing higher selectivity than *p*-toluenesulfonic acid (*p*-TSA), H_2SO_4, H_3PO_4, and Amberlyst-15, due to the acid strength of heteropolyanion [38]. Thus, other heteropolyacid-based catalysts have been evaluated for HMF etherification in order to obtain EMF. That includes HPW/MCM-41 [39], HPW/k-10 clay [40], Ag_1H_2PW [41], and Cs-exchanged silicotungstic acid [42].

However, the EMF production at larger scale is limited by the low stability of HMF that increases the cost of production. Therefore, it is required to use other cheaper feedstocks such as fructose, glucose, cellulose, or even directly from biomass for HMF production and subsequent etherification reaction so that the process is completed in only one pot (Figure 33.5). The presence of acid catalyst is also required for cellulose hydrolysis or dehydration of fructose or glucose.

In the past years, different solid acid catalysts have been reported for fructose conversion to EMF, but both temperature reaction and catalyst loading were generally increased in order to enhance EMF yield. Balakrishnan et al. also attained high EMF yields, y_{EMF} = 71% and 69%, by using fructose as feedstock in the presence of Amberlyst-15 resins and silica sulfuric catalysts, respectively, after 30 hours at 110 °C [35]. Similarly, high values of EMF yield (y_{EMF} = 72.8%) were achieved by using SO_3H-functionalized polymer catalysts for successive fructose dehydration and HMF etherification after 12.5 hours at 110 °C [43]. Ren et al. evaluated the Ag_1H_2PW catalyst for one-pot reaction from fructose,

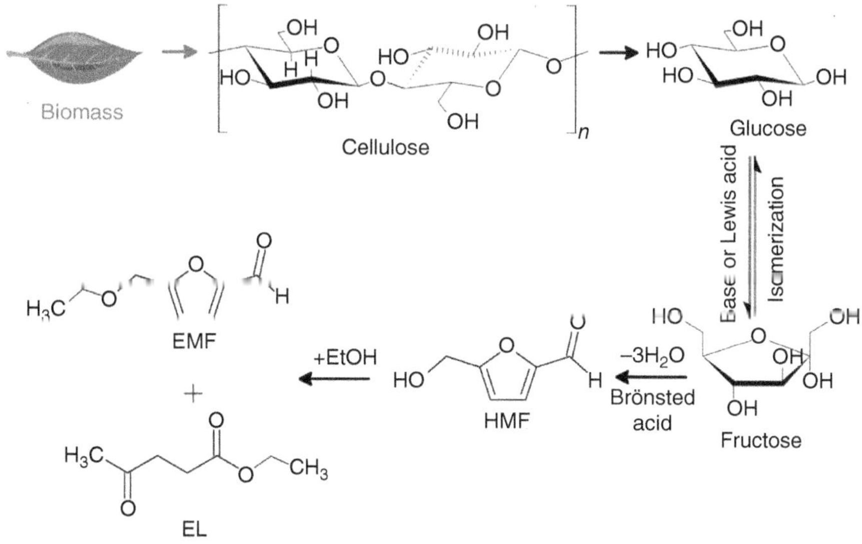

Figure 33.5 One-pot EMF production from different feedstocks. Source: Alipour et al. 1991 [30]. Reproduced with permission of Elsevier.

obtaining an EMF yield of 69.5% after 24 hours at 100 °C [41]. A relatively higher EMF yield (y_{EMF} = 63.1%) was achieved by Liu and Zhang by using SO_3H–silica catalysts using the same reaction conditions [44]. Similar values of EMF yields have been found for different solid acid catalysts supported over Fe_3O_4@SiO_2: Fe_3O_4@SiO_2–HPW (y_{EMF} = 54.8%) [45], Fe_3O_4@SiO_2–SO_3H (y_{EMF} = 72.3%) [46], and Fe_3O_4@SiO_2–SH–Im–SO_4H (y_{EMF} = 60.4%) [47]. Moreover, ethanol/tetrahydrofuran (THF) and ethanol/dimethyl sulfoxide (DMSO) mixtures have been evaluated as reaction medium to increase furan selectivity for one-pot fructose conversion into EMF [48, 49].

Although a lot of solid acid catalysts have been proposed for EMF production from fructose in the past years, the number of studies reported on the use of glucose as feedstock for EMF production is much lower in spite of glucose being the most abundant and the cheapest available monosaccharide [30, 50]. In this one-pot reaction, glucose isomerization to fructose, catalyzed by basic or Lewis acid catalysts, is frequently the limiting step [51, 52]. Lew et al. achieved an EMF yield of 31% from glucose by using the combination of Sn-BEA zeolite and Amberlyst-131 to catalyze glucose isomerization to fructose and fructose conversion into EMF, respectively [53]. Likewise, Li et al. carried out the EMF production from glucose employing different zeolites (H-USY, Sn-β, and dealuminated H-β) in combination with Amberlyt-15, obtaining values of EMF yields about 43–46% [54]. However, since the diffusion of reaction intermediates between the catalysts limits the efficient EMF production, the use of catalysts that allow for carrying out both steps is more recommended. Bai et al. synthesized a meso/microporous MFI-Sn/Al zeolite with Lewis acid sites attributed to Sn and Brønsted acid centers associated with Al–O(H)–Si sites, attaining an EMF yield of 44% [55]. The presence of both types of acid sites and meso/microporosity of zeolite allowed the three-step cascade reaction over a single catalyst in one pot.

Nevertheless, the most economically desirable way would be the EMF production from cellulose or even directly from biomass. Only a few studies have been reported on the cellulose conversion into EMF with low yields (6–22%) using homogeneous catalysis [12, 50].

33.3.2 Ethyl Levulinate (EL)

In general, levulinic acid (LA) esters can be used as fuel additive because they exhibit low toxicity, high lubricity, flash point stability, and moderate flow properties under low-temperature conditions in spite of their low cetane number [56]. One of the main problems with biodiesel is its high freezing point (3–7 °C), which leads to complications in low-temperature regions. Incorporation of additives such as EL, whose freezing point is −79 °C, improves diesel properties and reduces CO, SO_x, and NO_x emissions, providing higher efficiency and longer operating life to the engine [57–59]. Thus, EL is employed as a potential diesel fuel additive although mixtures with high concentrations of EL require the addition of a cetane-treating additive because its cetane number is less than 10. A blend based on 20% EL, 79% diesel, and 1% co-additive can be employed in regular diesel engines [58]. Therefore, it is a novel diesel miscible biofuel produced generally by esterification of LA in ethanol or even directly from C_6 carbohydrates or biomass in the presence of acid catalysts [25, 56]. In addition, it has been exposed that it is obtained as main by-product in the EMF production (Figure 33.5).

Different routes have been proposed for EL production. The most evident is the esterification of LA with ethanol attaining high selectivities, considering that esterification is the conversion of carboxylic acids to esters using acid and alcohols. Nandiwale et al. reported the use of desilicated zeolite H-ZSM-5 as support for heteropolyacids. The combination of dodecatungstophosphoric acid on desilicated H-ZSM-5 resulted in an efficient, robust catalyst, achieving a conversion of LA of 94% with 100% EL selectivity after four hours at 78 °C, being stable for four cycles [60]. These authors also evaluated the catalytic performance of desilicated H-ZSM-5 itself, obtaining an EL yield of 95% after seven hours at 120 °C [61]. Yan et al. synthesized mesoporous silica-supported Keggin heteropolyacid catalysts ($H_4SiW_{12}O_{40}$–SiO_2) for the esterification of LA in ethanol media, obtaining an EL yield of 67% after 66 at 75 °C [62]. Likewise, Pasquale et al. demonstrated that silica-supported Wells–Dawson heteropolyacid was a selective catalyst for the production of EL (70% yield at 78 °C after 10 hours) [63]. Due to the unique acid and textural properties of zeolites, Patil et al. evaluated the catalytic performance of H-BEA zeolite for the efficient conversion of EL [64]. They achieved LA conversion and EL selectivity of 40% and 98%, respectively, but the catalyst could be reused for four cycles. On the other hand, sulfonated materials have been studied as catalysts for LA esterification. Melero et al. prepared sulfonic mesostructured silica catalysts and evaluated their catalytic performance for this reaction in which SO_3H^- and SO_4^{2-} functional groups provided strong acidity to the catalysts, obtaining high EL yields [65]. Oliveira and da Silva attained a yield of

EL equal to 55% by using sulfonated carbon nanotubes [66]. In the same manner, different sulfated oxides have been studied such as SO_4^{2-}/SnO_2 and SO_4^{2-}/TiO_2, showing efficient conversion of LA into EL [67]. However, these authors found that Amberlyst-15 was more active compared to these sulfated oxides, although they did not provide data about EL selectivity.

EL may also be obtained directly from hexoses such as carbohydrates, fructose, or glucose in an anhydrous one-pot system by using alcohol as solvent in the presence of acid catalysts. Saravanamurugan et al. compared SO_3H-SBA-15 and zeolites for EL production. In the presence of zeolites, the conversion of fructose mainly stopped in the HMF ether, whereas the ester yield was low (10%). However, sulfonated SBA-15 materials provided high yields of EL from glucose (50%). Morales et al. reported the use of sulfated zirconia for the one-pot synthesis of EL from glucose. They affirmed that the EL formation depended on the reaction temperature, obtaining the highest yield (25%) at 140–150 °C after 24 hours of reaction since the humins (carbonaceous and heterogeneous compounds with unknown molecular structure that are insoluble in water) were produced at higher temperatures. Peng et al. also studied different sulfated oxides and confirmed that SO_4^{2-}/ZrO_2 was the most active catalyst, providing an EL yield above 30 mol% after three hours at 200 °C, and was reused for at least five times [68].

Another possibility is the direct conversion of biomass to EL. Although several studies have been reported on the synthesis of methyl levulinate from cellulose in the presence of solid acid catalysts [69–71], the El production from cellulose has almost not been studied. Deng et al. tested several solid acid catalysts (sulfated zirconia, sulfonated carbons, sulfonated resins, and heteropolyacids) for the treatment of cellulose in the presence of ethanol at 180 °C [72]. However, the main product obtained was ethyl glucoside, whereas EL was only achieved as by-product. Heteropolyacids were also reported as efficient catalysts for alkyl levulinates directly from cellulose.

On the other hand, another possible route for EL production could be based on furfuryl alcohol (FOL) as feedstock involving biomass-derived reactants. FOL obtained from furfural can be transformed to LA in water in the presence of an acid catalyst that can be converted into alkyl levulinates after treatment with alcohol [73]. Firstly, van de Graaf et al. evaluated the catalytic performance of different types of resins (Amberlyst and Dowex) and zeolites (H-ZSM, β, and mordenite) for ethanolysis of FOL [74, 75]. They affirmed that the accessibility of FOL to acid sites of catalyst is a key factor, being more important than acid strength. Thus, Dowex-type resins favored the diethyl ether formation because the acid sites were more accessible to ethanol molecules than FOL molecules, decreasing the EL production. Likewise, Neves et al. studied the relationship between the texture of zeolites or Amberlyst resins and the EL formation [76]. They corroborated the relevance of accessibility to acid sites, but they affirmed that a synergic effect between the acid site density, strength, and porosity was necessary. Finally, Chen et al. reported the direct conversion of furfural into levulinates in a one-pot reaction by using platinum supported on mesoporous phosphonated mixed oxides as catalyst [77]. Thus, high values of conversion and yield (92% and 76%, respectively) were attained, treating furfural in ethanol through hydrogenation to FOL and subsequent transformation to EL.

33.3.3 2,5-Dimethylfuran (DMF)

In the last years, DMF has been proposed as a potential alternative biofuel for internal combustion engines owing to its physicochemical properties such as high octane number and low oxygen content (Table 33.1). DMF possesses higher energy density, lower volatility, and lower solubility in water than ethanol, which make it more suitable to blend with gasoline [8, 78]. Moreover, DMF and gasoline exhibit similar viscosity (0.57 and 0.37–0.44 cSt, respectively), being beneficial for the establishment of the injection pressure of DMF in the fuel system [3]. Hu et al. studied the lubricity of DMF, concluding that it had better anti-wear effects compared to gasoline [79]. DMF has demonstrated comparable combustion and emission properties than commercial gasoline when Zhong et al. tested 99.8% DMF in a single-cylinder direction-injection research engine [80].

DMF is produced by selective hydrogenation of HMF. Thus, it is possible to produce from biomass-derived products by the transformation of C_6 monosaccharides to HMF followed by HDO of HMF to DMF. Román-Leshkov et al. reported DMF production by a two-step process: first, dehydration of fructose was carried out via catalysis by HCl in a biphasic system in the presence of inorganic salts. Then, HMF in 1-butanol obtained from the first step was employed for HDO reaction [81]. A biphasic system has been often used in the literature for dehydration of carbohydrates due to low water solubility of both furfural and HMF, favoring the products in the cosolvent and avoiding the side reactions. Also, the presence of inorganic salts increases the partition coefficient and enhances the catalytic performance. They studied the selective hydrogenation of HMF over $CuCrO_4$ using H_2 as hydrogen donor in 1-butanol, obtaining total conversion and a DMF yield of 61% at 220 °C for 10 hours and at 6.8 bar of H_2. However, the presence of chloride ions caused the deactivation of copper chromite. They proposed, instead of $CuCrO_4$, the use of carbon-supported copper–ruthenium catalyst (CuRu/C), which was affected, to a lesser extent, by the presence of chloride ions, achieving conversion and DMF yield of 100% and 71%, respectively.

Since then other catalytic systems, mainly based on Ru as active phase, have been developed for this process. Binder and Raines reported a DMF yield of 49% in a similar catalytic system to that of Román-Leshkov et al. by using Cu/Ru as catalyst for the selective hydrogenation of crude HMF from corn stover at 220 °C for 10 hours [82]. They also employed untreated corn straw as feedstock to obtain DMF in only one step, reaching a yield of 9%. Hu et al. found high selectivities to DMF (94.7%) by employing a commercial Ru/C catalyst using THF and H_2 [78]. A similar yield (93.4%) for total HMF conversion was obtained by Zu et al. with Ru supported over Co_3O_4 at 130 °C for 24 hours [83]. Likewise, Upare et al. achieved a DMF yield of 90% for hydrogenolysis of HMF in gas phase at 240 °C and atmospheric pressure (WHSV = 0.5 h^{-1}) by using a Ru–Sn/ZnO catalyst. They used HMF solutions obtained from fructose dehydration in 1-butanol catalyzed by Brønsted acid sites of Amberlyst-15 [84].

In addition to Ru, other metal phases such as palladium and nickel have been proposed in the literature. Saha et al. employed Pd/C catalyst in the presence of $ZnCl_2$ using THF as solvent and achieved a DMF yield of 85%, which decreased to 22% when the reaction was carried out from fructose [85]. Chidambaram and Bell found that phosphomolybdic acid helped to increase the efficiency of glucose conversion to HMF, which was subsequently converted into DMF in a two-step process in the presence of Pd/C catalyst and acetonitrile as additive [86]. They achieved a maximum of 17% DMF yield at 120 °C for one hour. It is notable that HMF was firstly transformed to MF and finally to DMF. Nishimura et al. synthesized Pd_xAu_y/C bimetallic catalysts and compared their catalytic performance with Pd/C and Au/C catalysts. The $Pd_{50}Au_{50}/C$ catalyst in the presence of HCl showed high activity for HMF hydrogenation, obtaining a DMF yield of 96% in THF at 60 °C for six hours [87]. Chatterjee et al. proposed a novel method for HMF hydrogenation by using supercritical CO_2 and H_2O as reaction media [88]. They attained DMF yield of 100% when 10 MPa CO_2 and 1 MPa H_2 were used for two hours at 80 °C in the presence of Pd/C catalyst. They also found that DMF selectivity depended on the CO_2/H_2O molar ratio. On the other hand, Huang et al. developed an active carbon-supported nickel–tungsten carbide (Ni–W_2C/AC) and attained 96% DMF yield after three hours at 180 °C [89], whereas Kong et al. achieved a DMF yield of 88.5% by using Raney Ni as catalyst [90]. Recently, Braun and Antonietti realized fructose conversion to a mixture of DMF and EL by flow process and achieved yields of 39% and 46%, respectively [91]. They converted fructose dissolved in ethanol and formic acid (FA) to HMF and EL by using Amberlyst-15 catalyst in a first column. Subsequently, the solution obtained in the first step was transformed to DMF by Ni-supported tungsten carbide catalyst.

Although it would be more interesting to produce DMF directly from carbohydrates or even biomass in only one pot, effective heterogeneous catalysts have not yet been found for these processes. Wei et al. developed an efficient catalytic system composed of a homogeneous Lewis–Brønsted acid mixture ($AlCl_3$, H_2SO_4, and H_3PO_4) and solid Ru/C catalyst using *N,N*-dimethylformamide as a solvent for the one-pot conversion of fructose to DMF via the dehydration/hydrogenolysis sequence [92]. They obtained a DMF yield of 66% at 200 °C with H_2 pressure of 1.5 MPa. Likewise, Saha and coworkers converted cellulosic biomass from algae to DMF by using a multicomponent system based on [*N,N*-dimethylacetamide]$^+$[CH_3SO_3]$^-$, Ru/C and FA as hydrogen donor and acid medium, attaining a maximum DMF yield of 32% in one hour [12].

With respect to the possible mechanism of DMF production from biomass, Hu et al. proposed the pathways as shown in Figure 33.6 [78].

Firstly, the HMF formation is carried out as shown in Figure 33.5. Then, C_2 aldehyde group and C_5 hydroxyl group in HMF are hydrogenated into 2,5-dihydroxymethylfuran and 5-methylfurfural, respectively (Figure 33.6). These products are subsequently hydrogenated into the same intermediate, 5-methylfurfuryl alcohol, which is further hydrogenated into DMF [78].

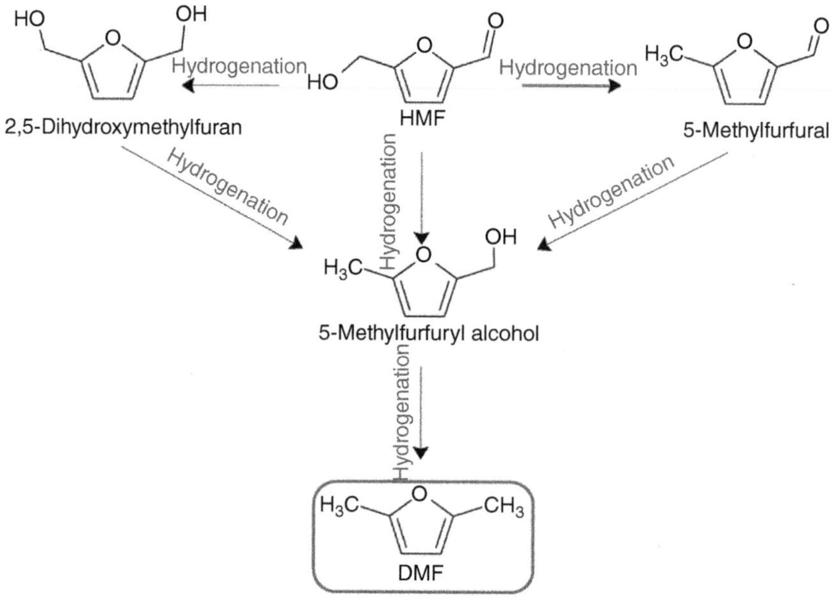

Figure 33.6 Mechanism of DMF formation from HMF. Source: Hu et al. 1991 [78]. Reproduced with permission of American Chemical Society.

33.3.4 2-Methylfuran (MF)

MF is a relevant furfural obtained from the hemicellulose fraction of biomass that is considered a promising biofuel. It can be used directly as fuel due to its high octane number, and its energy density is higher than that of ethanol and similar to that of DMF and gasoline (Table 33.1). It has been reported that MF possesses greater thermal efficiency than gasoline and DMF because of its fast burning rate and notably better knock suppression ability [93]. Similar to HMF, furfural can also be hydrogenated to MF, being the main by-product of FOL production. Thus, the MF formation requires hydrogenation of the C—O bond in the FOL molecules [56, 94]. Therefore, MF can be produced from biomass by a two-step process: first, the dehydration of xylose contained in the hemicellulose of biomass in the presence of acid catalysts and subsequent catalytic HDO of furfural [8].

However, most of the publications have focused on producing MF from furfural, by either gas-phase or liquid-phase hydrogenation. Different Cu-based catalysts were evaluated at high temperatures (200–300 °C), liquid hourly space velocity (LHSV) = 0.15–0.3 hours, 0.1 MPa, and a H_2/furfural molar ratio of 5–8 [95]. Raney Cu, Cu/Al_2O_3, and Cu-chromite displayed similar catalytic behavior, although the latter was more active and stable. Carbon-supported Cu-chromite showed the highest MF selectivity. Likewise, Burnett et al. found that copper chromite dispersed on activated charcoal was the most efficient catalyst in the vapor-phase hydrogenation of furfural to MF (y_{MF} = 90–95%) carried out at 1 atm and 200–230 °C, but the deactivation of catalyst took place [96]. Bimetallic Cu/Fe supported on SiO_2 catalysts was evaluated for vapor-phase hydrogenation of furfural at 1 bar of H_2, attaining a very high MF yield (y_{MF} = 98%) [97]. Zhu et al. obtained high MF yields (y_{MF} = 89%) for hydrogenation of furfural in a fixed-bed reactor at 212 °C by using Cu–Zn metallic as catalyst [98]. Recently, Jiménez-Gómez et al. synthesized a series of Cu-based catalyst (2.5–20 wt%) incorporating Cu species to mesoporous silica (MS) through complexation with the amine groups of dodecylamine employed as a structure-directing agent in the synthesis [99]. These catalysts were tested in the gas-phase hydrogenation of furfural and the catalyst with a Cu loading of 10 wt% showed the highest MF yield (y_{MF} = 95%) after five hours at 210 °C with a H_2/furfural molar ratio of 11.5 and a weight hourly space velocity (WHSV) of 1.5 h^{-1}. This catalyst still exhibited an MF yield of 80% after 14 hours of reaction.

In addition to copper, other metal phases have been used for MF production in the literature. Sitthisa et al. tested SiO_2-supported Ni and Ni–Fe bimetallic catalysts for the conversion of furfural under 1 bar H_2 in the temperature range of 210–250 °C [100]. They checked that monometallic Ni species favored the formation of FOL and furan as primary products and bimetallic species promoted the MF formation as major product via C–O hydrogenolysis. Likewise, Ni_2P-based catalysts supported on a commercial silica showed almost full conversion and high MF yields (y_{MF} = 73%) at 190 °C after five hours. The high selectivity toward MF was attributed to both the high hydrogenating capacity of the metallic sites leading to FOL formation and the presence of reduced P species that favored the hydrogenolysis process. Recently, Mo_2C catalysts provided MF as main product for vapor-phase HDO of furfural at 150 °C and atmospheric pressure has been reported [101]. It was affirmed that the high selectivity toward C—O bond could be due to the strong interaction between Mo_2C and C—O bond as supported by density functional theory (DFT) calculations and high-resolution electron energy loss spectroscopy (HREELS) experiments [102].

Due to the rapid deactivation of catalysts during the gas-phase production of MF, liquid-phase hydrogenation has been proposed to solve the problem. Early studies used Ni supported on a kieselguhr catalyst [103]. Linares and Nudelman reported the hydrogenolysis of furfural to MF by using Pd supported on carbon (Pd/C) at room temperature and 0.2 MPa H_2 [104]. Yan and Chen used Cu/Fe catalysts for furfural hydrogenation to obtain MF and achieved conversion and yield of 99% and 51%, respectively, after 14 hours at 220 °C after 20 bar H_2 [105].

33.3.5 γ-Valerolactone (GVL)

GVL is a potential intermediate for the production of chemicals and fuels derived from renewable feedstocks and can be employed as fuel additive due to its excellent properties (Table 33.1). GVL possesses an energy density slightly higher than that of ethanol and a high boiling point. GVL is stable in water and air and has low toxicity. The main risk is flammability. The risk is, however, low due to the low volatility of GVL [106]. It has also been reported that the presence of GVL in blends with diesel reduced CO and smoke emissions in automobile exhaust [107]. A comparative study of GVL and ethanol as fuel additives demonstrated that a mixture of 90 v/v% gasoline and 10 v/v% GVL or ethanol exhibited similar fuel properties such as octane numbers. These improved combustion properties are attributed to its lower vapor pressure [27, 108].

Different routes to produce GVL have been proposed. On the one hand, the LA hydrogenation produces an unstable intermediate, γ-hydroxyvaleric acid, which ring closes by intramolecular esterification, losing a water molecule to form GVL [106]. Another possible route for GVL formation is the LA dehydration to angelica lactone and subsequent hydrogenation to GVL in the presence of acid catalysts [109]. In the case of LA esters, the ester hydrogenation takes place to produce hydroxy levulinic ester, which ring closes by intramolecular transesterification, producing GVL [110]. GVL can be also produced from 4-pentenoic acid [106]. The LA hydrogenation to produce GVL has been widely studied in the literature. Quaker Oats described a continuous process for the vapor-phase LA hydrogenation to GVL using $CuO–Cr_2O_3$ mixed oxides as catalyst [111]. Since then, La hydrogenation into GVL has been studied by using Ru, Pd, Pt, Ni, and Cu catalysts under relatively moderate H_2 pressures and temperatures [56, 94]. Hengne et al. evaluated Ru-based catalysts for the GVL production and found that Ru/C catalyst provided the highest values of conversion and selectivity (95% and 91%, respectively), being stable for five catalytic runs [112]. Upare et al. compared different carbon-supported Ru, Pt, and Pd catalysts for the selective vapor-phase hydrogenation of LA and Ru/C provided high GVL selectivity [113]. In the same manner, Manzer tested several metals supported on carbon for GVL production, and Ru/C exhibited the best catalytic performance, attaining a GVL yield of 97% at 150 °C under 34.5 bar H_2 [114]. Likewise, a new process for GVL production combining the use of water as cosolvent and the use of supercritical CO_2 was proposed by Bourne et al. [115]. They carried out the reactions at 200 °C and 10 MPa H_2 in the presence of Ru/SiO_2 as catalysts, attaining high GVL yields (>99%). Considering the CO_2 advantages, Yan et al. synthesized Pd particles supported on SiO_2 and Al_2O_3 by using liquid CO_2, showing high activity and durability for LA hydrogenation [116, 117]. On the other hand, Du et al. studied several metals (e.g. Pd, Pt, Ru, Au) supported on TiO_2, SiO_2, ZrO_2, and C, showing the Au/ZrO_2 catalyst to have the best performance in terms of high conversion with high yields of ~97% [118, 119]. Likewise, Son et al. confirmed that the supported Au nanoparticle catalyst was efficient for the synthesis of GVL ($y_{GVL} = 90\%$) after screening a series of Ru/C, Ru/SBA, Au/ZrC, and Au/ZrO_2 catalysts [120].

Due to the high cost of noble metal catalysts, other inexpensive metal-based catalysts have been proposed in the literature. Upare et al. attained high GVL yields ($y_{GVL} = 96\%$) produced by LA hydrogenation on Al_2O_3-supported Ni–Cu bimetallic catalysts at 265 °C and 25 bar H_2 [121]. Haan et al. patented a Ni catalyst that provided a 71% yield of GVL [122]. On the other hand, bimetallic nanoparticle catalysts tend to display better performances for liquid-phase hydrogenation of LA to GVL. Shimizu et al. evaluated a series of base metal (Ni, Co, Cu, and Fe) and metal oxides (Mo, V, and W oxides) co-loaded carbon (C) and Ni-loaded metal oxides for the hydrogenation of LA to GVL. The best catalytic behavior was found in $Ni–MoO_x/C$ (99% yield) [123]. Obregón et al. attained high GVL yields for the hydrogenolysis of LA by using the Al_2O_3-supported Ni and Ni–Cu catalysts after two hours at 50 °C and 6.5 MPa [124]. Recently, Hengne and Rode reported the use of nanocomposites of

$Cu–ZrO_2$ and $Cu–Al_2O_3$ as catalysts for the LA hydrogenation to GVL, obtaining GVL selectivity of 90% [125].

Another alternative route for GVL production is the catalytic transfer of hydrogen (CTH) of LA using alcohol as hydrogen donor through the Meerwein–Ponndorf–Verley (MPV) reaction. Chia and Dumesic reported the use of different metal oxides such as ZrO_2, γ-Al_2O_3, MgO/Al_2O_3, MgO/ZrO_2, and $CeZrO_x$ to produce GVL by CTH [126]. The highest GVL yield ($y_{GVL} = 92\%$) was found in the presence of ZrO_2 as catalyst by using 2-butanol at 150 °C.

A large number of studies about GVL production from LA have been reported in the literature. However, the transformation from carbohydrates or even biomass would be more interesting economically, which makes the use of bifunctional catalysts able to hydrogenate and containing acid or basic properties necessary. The GVL production from fructose is considerably easier than from other carbohydrates. Zhou et al. reported the one-pot conversion of fructose into GVL, with EL being the intermediate [127]. A polymer of divinylbenzene highly cross-linked with an acidic ionic liquid was employed as heterogeneous acid catalyst for the alcoholysis of fructose into EL in a mixed solvent of ethanol and DMSO, attaining full conversion and EL yield of 65% after 12 hours at 150 °C. Then, EL was extracted by ethyl acetate, and direct hydrogenation was carried out at 130 °C for six hours under 3 MPa H_2 over TiO_2-supported Co nanoparticles, achieving a GVL yield of 55%. However, the GVL yield decreased from third cycle due to Co leaching. Son et al. also carried out the one-pot conversion of fructose into GVL, but they employed FA as the hydrogen source [120]. It has been previously mentioned that first they found Au-supported catalysts were the most active for LA hydrogenation to GVL. Then, they realized the one-pot reaction from fructose in the presence of FA in which FA acted as both acid catalyst for fructose dehydration to LA and hydrogen source for LA hydrogenation to GVL, attaining a final GVL yield of 48%. Yang et al. carried out GVL production from fructose [128]. Firstly, they realized alcoholysis of fructose in ethanol by using Amberlyst-15 obtaining EL after 20 hours at 120 °C, which was used for GVL production ($y_{GVL} = 50\%$) using 2-propanol as hydrogen source and Raney Ni as catalyst.

The use of glucose as feedstock is preferred to fructose because it is more abundant and inexpensive. One of the main disadvantages for this process is the deactivation of the heterogeneous catalysts by poisoning due to the presence of acids and humins. To solve this problem, alternatives such as the use of an organic cosolvent have been proposed in the literature. Alonso et al. carried out the GVL synthesis from corn stover using 0.5 M H_2SO_4 aqueous solution and alkylphenol to extract LA [129]. Then, the LA hydrogenation to GVL took place over Ru–Sn catalyst. Molinari et al. used 2-methyltetrahydrofuran (MTHF) to extract LA from aqueous feeds obtained by the H_2SO_4-catalyzed hydrolysis of lignocellulosic materials [130]. The LA extracted by MTHF was successfully converted into GVL in the presence of Raney Ni, with an LA conversion >99% and a GVL yield of 96%. The use of MTHF is preferred because of its low cost, sustainability, and availability. Wettstein et al. designed a biphasic reaction system for GVL production from cellulose using an aqueous-phase solution containing HCl saturated with NaCl and GVL as solvent to extract LA (Figure 33.7) [131]. This biphasic system provided high yields of LA and FA (e.g. 70%). The LA recovered in

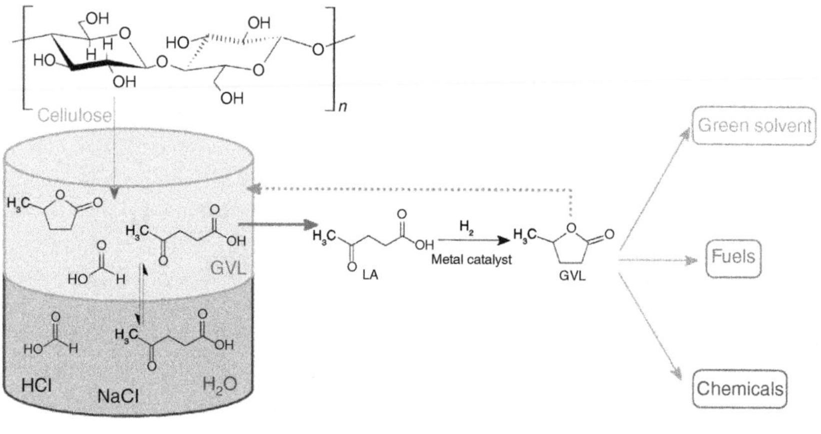

Figure 33.7 Schematic representation of biphasic reaction system for GVL production from cellulose and main GVL applications. Source: Wettstein et al. 1991 [131]. Reproduced with permission of Royal Society of Chemistry. (See online version for color figure).

GVL was subsequently converted into GVL over a carbon-supported Ru–Sn catalyst. These research groups improved this catalytic system using Amberlyst-70 as solid acid catalyst instead of HCl for conversion of cellulose into LA [132]. They affirmed that the solvent had a positive effect on LA production since unconverted cellulose and degradation products were detected after reaction using pure water. These products were not observed for GVL/water systems. High LA yields were obtained from cellulose and real biomass (69% and 54%, respectively) after 16 hours and 160 °C in the presence of Amberlyst-70. Finally, LA was converted into GVL by using $RuSn_4/C$ as catalyst, which was stable for 60 hours of time onstream.

Although homogeneous solutions of mineral acids were very active for the hydrolysis of lignocelluloses or carbohydrates into LA and FA, they are difficult to recycle and lead to environmental problems, which makes the use of heterogeneous catalysts advisable. Ding et al. reported a one-pot two-step method for the synthesis of GVL from cellulose [133]: They used mesoporous Al–$NbOPO_4$ and commercial 5% Ru/C as solid catalysts for LA and GVL production, respectively. Without LA separation, it attained a GVL yield of 57%. Al–$NbOPO_4$ accelerated LA formation due to its high acid concentration and the coexistence of Lewis and Brønsted acid sites. Putro et al. reported a similar procedure for the production of GVL from sugarcane bagasse over Pt/TiO_2 and acid-activated bentonite as catalysts, obtaining an interesting performance with respect to LA conversion (100%) and GVL selectivity (95%) [134].

33.4 Outlook and Conclusions

Lignocellulosic biomass conversion to biofuels is a promising alternative to traditional fossil fuels since lignocellulose is a low-cost and sustainable feedstock and that is widely available. It has been demonstrated that carbohydrate-containing biomass can be converted to platform molecules such as furfural and HMF, in which further transformation to other chemicals and fuels is feasible. In this chapter, recent advances in procedures to obtain different furanic-derived fuels including EMF, EL, DMF, MF, and GVL have been discussed. These biofuels can be directly blended with diesel or gasoline with or without modifications, obtaining high combustion efficiency and reducing the negative environmental impact of traditional fuels.

Currently, the conversion of lignocellulosic biomass into biofuels involves multistep processes, but one pot reaction directly from biomass would be more economical and sustainable since lignocellulose is an abundant and inexpensive feedstock. The design of multifunctional catalysts plays a key role in the development of these processes as they are required to carry out the production of these biofuels directly from biomass in one-pot synthesis. This fact saves cost, time, and energy for an economic industrial-scale production. These multifunctional catalysts must possess a variety of catalytic functions, including acid sites, basic sites, and metal hydrogenation sites. Moreover, these catalysts must be recoverable and must maintain their catalytic performance for multiple catalytic runs. The main bottleneck of the development of these catalysts is the catalytic deactivation found in most of these processes. The side reactions involved in these processes due to the high reactivity of HMF and furfural molecules are also well known. This deactivation is still more pronounced if biomass is directly employed as feedstock since more number of secondary reactions can take place. The formation of soluble and insoluble by-products such as humins accounts for 10–50% carbon loss of the feed, causing unfavorable process economics. These species are frequently deposited on the catalytic surface, resulting in deactivation of catalyst. Therefore, a deeper analysis on the deactivation processes is required in order to understand and to assist in the design of effective catalysts specifically tailored to produce biofuels from lignocellulosic biomass. By identifying, optimizing, and tuning important catalytic parameters to prevent deactivation, the design of multifunctional catalysts for cascade reactions will be more effective in decreasing the coke formation, improving the selectivity to desirable chemicals, and increasing the rate of reaction. Therefore, future attention may also focus more on the deactivation and regeneration of catalysts to enhance their stability and subsequently their economic viability.

Finally, the development of efficient method of separation and purification of these biofuels is also required. With these technological advances, the synthesis of these biofuels will make a more promising progress and play an important role in the transportation sector.

References

1 Hu, L., Lin, L., Wu, Z. et al. (2017). *Renewable Sustainable Energy Rev.* 74: 230–257.

2 De Bhowmick, G., Sarmah, A., and Sen, R. (2018). *Bioresour. Technol.* 247: 1144–1154.

3 Qian, Y., Zhu, L., Wang, Y., and Lu, X. (2015). *Renewable Sustainable Energy Rev.* 41: 633–646.

4 Wang, T., Nolte, M., and Shanks, B. (2014). *Green Chem.* 16: 548–572.

5 Demirbas, A. (2001). *Energy Convers. Manage.* 42: 1357–1378.

6 FitzPatrick, M., Champagne, P., Cunningham, M.F., and Whitney, R.A. (2010). *Bioresour. Technol.* 101: 8915–8922.

7 Lynd, L., Sow, M., Chimphango, A. et al. (2015). *Biotechnol. Biofuels* 8 (1): 18.

8 Saha, B., Bohre, A., Dutta, S., and Abu-Omar, M. (2015). *ACS Sustainable Chem. Eng.* 3: 1263–1277.

9 Agarwal, A. (2007). *Prog. Energy Combust. Sci.* 33: 233–271.

10 Balat, M. (2011). *Energy Convers. Manage.* 52: 858–875.

11 Demirbas, A. (2007). *Prog. Energy Combust. Sci.* 33: 1–18.

12 De, S., Dutta, S., and Saha, B. (2012). *ChemSusChem* 5: 1826–1833.

13 Dutta, S., De, S., Alam, M. et al. (2012). *J. Catal.* 288: 8–15.

14 Taherzadeh, M. and Karimi, K. (2008). *Int. J. Mol. Sci.* 9: 1621–1651.

15 Brandt, A., Grasvik, J., Hallett, J., and Welton, T. (2013). *Green Chem.* 15: 550–583.

16 Alonso, D., Bond, J., and Dumesic, J. (2010). *Green Chem.* 12: 1493–1513.

17 Avanthi, A., Kumar, S., Sherpa, K., and Banerjee, R. (2017). *Biofuels-UK* 8: 431–444.

18 Kubicka, D., Kubickova, I., and Cejka, J. (2013). *Catal. Rev.* 55: 1–78.

19 Chandra, R., Takeuchi, H., and Hasegawa, T. (2012). *Appl. Energy* 94: 129–140.

20 Demirbas, A. (2011). *Appl. Energy* 88: 17–28.

21 Kumari, D. and Singh, R. (2018). *Renewable Sustainable Energy Rev.* 90: 877–891.

22 van Putten, R., van der Waal, J., de Jong, E. et al. (2013). *Chem. Rev.* 113: 1499–1597.

23 Zeitsch, K.J. (2000). *The Chemistry and Technology of Furfural and Its Many By-Products*. Elsevier Science B.V.

24 Lange, J.P., Van Der Heide, E., Van Buijtenen, J., and Price, R. (2012). *ChemSusChem* 5: 150–166.

25 Mascal, M. and Nikitin, E. (2008). *Angew. Chem. Int. Ed.* 47: 7924–7926.

26 Tang, X., Wei, J., Ding, N. et al. (2017). *Renewable Sustainable Energy Rev.* 77: 287–296.

27 Horvath, I., Mehdi, H., Fabos, V. et al. (2008). *Green Chem.* 10: 238–242.

28 Yan, K., Wu, G., Lafleur, T., and Jarvis, C. (2014). *Renewable Sustainable Energy Rev.* 38: 663–676.

29 Balan, V. (2014). Current challenges in commercially producing biofuels from lignocellulosic biomass. *ISRN Biotechnol.* 2014: 463074 (31). Hindawi Publishing Corporation ISRN Biotechnology. https://doi.org/10.1155/2014/463074.

30 Alipour, S., Omidvarborna, H., and Kim, D. (2017). *Renewable Sustainable Energy Rev.* 71: 908–926.

31 Gruter, G.J.M. and Dautzenberg, F. Method for the synthesis of 5-hydroxymethylfurfural ethers and their use. US Patent 2011/0082304 A1, filed 15 December 2010 and issued 7 April 2011.

32 Ras, E., Maisuls, S., Haesakkers, P. et al. (2009). *Adv. Synth. Catal.* 351: 3175–3185.

33 Imhof, P., Dias, A.S., de Jong, E., and Gruter, G.J. (2009). *OA02 Furanics: Versatile Molecules for Biofuels and Bulk Chemicals Applications.* Avantium Technologies BV.

34 Sanborn, A.J. (2008). Processes for the preparation and purification of hydroxymethylfuraldehyde and derivatives. US Patent 2006/0142599, filed 09 December 2005 and issued 29 June 2006.

35 Balakrishnan, M., Sacia, E., and Bell, A. (2012). *Green Chem.* 14: 1626–1634.

36 Lanzafame, P., Temi, D., Perathoner, S. et al. (2011). *Catal. Today* 175: 435–441.

37 Neves, P., Antunes, M., Russo, P. et al. (2013). *Green Chem.* 15: 3367–3376.

38 Che, P., Lu, F., Zhang, J. et al. (2012). *Bioresour. Technol.* 119: 433–436.

39 Liu, A., Zhang, Z., Fang, Z. et al. (2014). *J. Ind. Eng. Chem.* 20: 1977–1984.

40 Liu, A., Liu, B., Wang, Y. et al. (2014). *Fuel* 117: 68–73.

41 Ren, Y., Liu, B., Zhang, Z., and Lin, J. (2015). *J. Ind. Eng. Chem.* 21: 1127–1131.

42 Raveendra, G., Rajasekhar, A., Srinivas, M. et al. (2016). *Appl. Catal., A* 520: 105–113.

43 Li, H., Zhang, Q., and Yang, S. (2014). *Int. J. Chem. Eng.*: 481627. https://doi.org/10.1155/2014/481627.

44 Liu, B. and Zhang, Z. (2013). *RSC Adv.* 3: 12313–12319.

45 Wang, S., Zhang, Z., Liu, B., and Li, J. (2013). *Catal. Sci. Technol.* 3: 2104–2112.

46 Zhang, Z., Wang, Y., Fang, Z., and Liu, B. (2014). *ChemPlusChem* 79: 233–240.

47 Yin, S., Sun, J., Liu, B., and Zhang, Z. (2015). *J. Mater. Chem. A* 3: 4992–4999.

48 Wang, H., Deng, T., Wang, Y. et al. (2013). *Green Chem.* 15: 2379–2383.

49 Morales, G., Paniagua, M., Melero, J., and Iglesias, J. (2017). *Catal. Today* 279: 305–316.

50 Kazi, F., Patel, A., Serrano-Ruiz, J. et al. (2011). *Chem. Eng. J.* 169: 329–338.

51 Zhao, H., Holladay, J., Brown, H., and Zhang, Z. (2007). *Science* 316: 1597–1600.

52 Kunkes, E., Simonetti, D., West, R. et al. (2008). *Science* 322: 417–421.

53 Lew, C., Rajabbeigi, N., and Tsapatsis, M. (2012). *Ind. Eng. Chem. Res.* 51: 5364–5366.

54 Li, H., Saravanamurugan, S., Yang, S., and Riisager, A. (2016). *Green Chem.* 18: 726–734.

55 Bai, Y., Wei, L., Yang, M. et al. (2018). *J. Mater. Chem. A* 6: 7693–7705.

56 Mariscal, R., Maireles-Torres, P., Ojeda, M. et al. (2016). *Energy Environ. Sci.* 9: 1144–1189.

57 Joshi, H., Moser, B., Toler, J. et al. (2011). *Biomass Bioenergy* 35: 3262–3266.

58 Silva, J., Grekin, R., Mariano, A., and Maciel, R. (2018). *Energy Technol.* 6: 613–639.
59 Unlu, D., Ilgen, O., and Hilmioglu, N. (2016). *Chem. Eng. J.* 302: 260–268.
60 Nandiwale, K., Sonar, S., Niphadkar, P. et al. (2013). *Appl. Catal., A* 460: 90–98.
61 Nandiwale, K., Niphadkar, P., Deshpande, S., and Bokade, V. (2014). *J. Chem. Technol. Biotechnol.* 89: 1507–1515.
62 Yan, K., Wu, G., Wen, J., and Chen, A. (2013). *Catal. Commun.* 34: 58–63.
63 Pasquale, G., Vazquez, P., Romanelli, G., and Baronetti, G. (2012). *Catal. Commun.* 18: 115–120.
64 Patil, C., Niphadkar, P., Bokade, V., and Joshi, P. (2014). *Catal. Commun.* 43: 188–191.
65 Melero, J., Morales, G., Iglesias, J. et al. (2013). *Appl. Catal., A* 466: 116–122.
66 Oliveira, B. and da Silva, V. (2014). *Catal. Today* 234: 257–263.
67 Fernandes, D., Rocha, A., Mai, E. et al. (2012). *Appl. Catal., A* 425: 199–204.
68 Peng, L., Lin, L., Zhang, J. et al. (2011). *Appl. Catal., A* 397: 259–265.
69 Peng, L., Lin, L., Li, H., and Yang, Q. (2011). *Appl. Energy* 88: 4590–4596.
70 Liu, Y., Liu, C., Wu, H., and Dong, W. (2013). *Catal. Lett.* 143: 1346–1353.
71 Kuo, C., Poyraz, A., Jin, L. et al. (2014). *Green Chem.* 16: 785–791.
72 Deng, W., Liu, M., Zhang, Q., and Wang, Y. (2011). *Catal. Today* 164: 461–466.
73 Demolis, A., Essayem, N., and Rataboul, F. (2014). *ACS Sustainable Chem. Eng.* 2: 1338–1352.
74 Lange, J., van de Graaf, W., and Haan, R. (2009). *ChemSusChem* 2: 437–441.
75 van de Graaf, W.D. and Lange, J.P. (2007). A catalytic process for the convers ion of furfuryl alcohol into levulinic acid or alkyl levulinate. US Patent WO2007023173, filed 24 August 2006 and issued 01 March 2007.
76 Neves, P., Lima, S., Pillinger, M. et al. (2013). *Catal. Today* 218: 76–84.
77 Chen, B., Li, F., Huang, Z. et al. (2014). *ChemSusChem* 7: 202–209.
78 Hu, L., Lin, L., and Liu, S. (2014). *Ind. Eng. Chem. Res.* 53: 9969–9978.
79 Hu, E., Hu, X., Wang, X. et al. (2012). *Tribol. Int.* 55: 119–125.
80 Zhong, S., Daniel, R., Xu, H. et al. (2010). *Energy Fuels* 24: 2891–2899.
81 Román-Leshkov, Y., Barrett, C., Liu, Z., and Dumesic, J. (2007). *Nature* 447: 982–986.
82 Binder, J. and Raines, R. (2009). *J. Am. Chem. Soc.* 131: 1979–1985.
83 Zu, Y., Yang, P., Wang, J. et al. (2014). *Appl. Catal., B* 146: 244–248.
84 Upare, P., Hwang, D., Hwang, Y. et al. (2015). *Green Chem.* 17: 3310–3313.
85 Saha, B., Bohn, C., and Abu-Omar, M. (2014). *ChemSusChem* 7: 3095–3101.
86 Chidambaram, M. and Bell, A. (2010). *Green Chem.* 12: 1253–1262.
87 Nishimura, S., Ikeda, N., and Ebitani, K. (2014). *Catal. Today* 232: 89–98.
88 Chatterjee, M., Ishizaka, T., and Kawanami, H. (2014). *Green Chem.* 16: 1543–1551.
89 Huang, Y., Chen, M., Yan, L. et al. (2014). *ChemSusChem* 7: 1068–1072.
90 Kong, X., Zhu, Y., Zheng, H. et al. (2014). *RSC Adv.* 4: 60467–60472.
91 Braun, M. and Antonietti, M. (2017). *Green Chem.* 19: 3813–3819.
92 Wei, Z., Lou, J., Li, Z., and Liu, Y. (2016). *Catal. Sci. Technol.* 6: 6217–6225.
93 Wang, C., Xu, H., Daniel, R. et al. (2013). *Fuel* 103: 200–211.

94 De, S., Saha, B., and Luque, R. (2015). *Bioresour. Technol.* 178: 108–118.
95 Bremner, J.G.M. and Keeys, R.K.F. (1947). *J. Chem. Soc.*: 1068–1080.
96 Burnett, L.W., Johns, I.B., Holdren, R.F., and Hixon, R.M. (1948). *Ind. Eng. Chem.* 40: 502–505.
97 Lessard, J., Morin, J., Wehrung, J. et al. (2010). *Top. Catal.* 53: 1231–1234.
98 Zhu, Y., Xiang, H., Li, Y. et al. (2003). *New J. Chem.* 27: 208–210.
99 Jiménez-Gómez, C., Cecilia, J., Moreno-Tost, R., and Maireles-Torres, P. (2017). *ChemSusChem* 10: 1448–1459.
100 Sitthisa, S., An, W., and Resasco, D. (2011). *J. Catal.* 284: 90–101.
101 Lee, W., Wang, Z., Zheng, W. et al. (2014). *Catal. Sci. Technol.* 4: 2340–2352.
102 Xiong, K., Lee, W., Bhan, A., and Chen, J. (2014). *ChemSusChem* 7: 2146–2149.
103 Peters, F.N. (1933). Method for the reduction of furfural and furan derivates US Patent 1906873, filed 26 April 1928 and issued 02 May 1933.
104 Linares, G. and Nudelman, N. (2003). *J. Phys. Org. Chem.* 16: 569–576.
105 Yan, K. and Chen, A. (2014). *Fuel* 115: 101–108.
106 Alonso, D., Wettstein, S., and Dumesic, J. (2013). *Green Chem.* 15: 584–595.
107 Bereczky, A., Lukács, K., Farkas, M., and Dóbé, S. (2014). *Nat. Resour.* 5: 177–191.
108 Fábos, V., Koczó, G., Mehdi, H. et al. (2009). *Energy Environ. Sci.* 2: 767–769.
109 Serrano-Ruiz, J.C., West, R.M., and Dumesic, J.A. (2010). *Annu. Rev. Chem. Biomol. Eng.* 1: 79–100.
110 Gurbuz, E., Alonso, D., Bond, J., and Dumesic, J. (2011). *ChemSusChem* 4: 357–361.
111 Dunlop, A.P. and Madden, J.W. (1957). Process of preparing gamma-valerolactone. US Patent 2786852, filed 19 August 1953 and issued 26 March 1957.
112 Hengne, A., Biradar, N., and Rode, C. (2012). *Catal. Lett.* 142: 779–787.
113 Upare, P., Lee, J., Hwang, D. et al. (2011). *J. Ind. Eng. Chem.* 17: 287–292.
114 Manzer, L. (2004). *Appl. Catal., A* 272: 249–256.
115 Bourne, R., Stevens, J., Ke, J., and Poliakoff, M. (2007). *Chem. Commun.*: 4632–4634.
116 Yan, K., Lafleur, T., Wu, G. et al. (2013). *Appl. Catal., A* 468: 52–58.
117 Yan, K., Jarvis, C., Lafleur, T. et al. (2013). *RSC Adv.* 3: 25865–25871.
118 Du, X., Bi, Q., Liu, Y. et al. (2011). *ChemSusChem* 4: 1838–1843.
119 Du, X., He, L., Zhao, S. et al. (2011). *Angew. Chem. Int. Ed.* 50: 7815–7819.
120 Son, P., Nishimura, S., and Ebitani, K. (2014). *RSC Adv.* 4: 10525–10530.
121 Upare, P., Lee, J., Hwang, Y. et al. (2011). *ChemSusChem* 4: 1749–1752.
122 Haan, R., Lange, J.P., Petrus, L., and Petrus-Hoogenbosch, C. (2007). Hydrogenation process for the conversion of a carboxylic acid or an ester having a carbonyl group. US Patent 20070208183 A1, filed 28 February 2007 and issued 06 September 2007.
123 Shimizu, K., Kanno, S., and Kon, K. (2014). *Green Chem.* 16: 3899–3903.
124 Obregón, I., Corro, E., Izquierdo, U. et al. (2014). *Chin. J. Catal.* 35: 656–662.
125 Hengne, A. and Rode, C. (2012). *Green Chem.* 14: 1064–1072.

126 Chia, M. and Dumesic, J. (2011). *Chem. Commun.* 47: 12233–12235.
127 Zhou, H., Song, J., Kang, X. et al. (2015). *RSC Adv.* 5: 15267–15273.
128 Yang, Z., Huang, Y., Guo, Q., and Fu, Y. (2013). *Chem. Commun.* 49: 5328–5330.
129 Alonso, D., Wettstein, S., Bond, J. et al. (2011). *ChemSusChem* 4: 1078–1081.
130 Molinari, V., Antonietti, M., and Esposito, D. (2014). *Catal. Sci. Technol.* 4: 3626–3630.
131 Wettstein, S., Alonso, D., Chong, Y., and Dumesic, J. (2012). *Energy Environ. Sci.* 5: 8199–8203.
132 Alonso, D., Gallo, J., Mellmer, M. et al. (2013). *Catal. Sci. Technol.* 3: 927–931.
133 Ding, D., Wang, J., Xi, J. et al. (2014). *Green Chem.* 16: 3846–3853.
134 Putro, J., Kurniawan, A., Soetaredjo, F. et al. (2015). *RSC Adv.* 5: 41285–41299.

34

Conversion of Carbohydrates to High Value Products

Isao Ogino

Hokkaido University, Faculty of Engineering, , N13W8 Kita-ku, Sapporo, 060-8628, Japan

34.1 Introduction

Biomass is an organic matter typically formed from CO_2 and H_2O through photosynthesis:

$$x\mathrm{CO_2} + y\mathrm{H_2O} \xrightarrow{h\nu} \mathrm{C}_x(\mathrm{H_2O})_y + x\mathrm{O_2}$$

Lignocellulose consists of cellulose (40–50%), hemicellulose (15–30%), and lignin (Figure 34.1) [1] and has been considered as the most renewable source of carbon to produce value-added chemicals. Because the chemical formula of cellulose and hemicellulose is represented by $C_x(H_2O)_y$ and most carbon atoms in them bear a molecule of water attached in the form of an H and an OH, they are also referred to as in the group of carbohydrates ("hydrates of carbon") [2]. Carbohydrates are usually defined as polyhydroxy aldehydes and ketones or substances that hydrolyze to yield polyhydroxy aldehydes and ketones. Carbohydrates that cannot be hydrolyzed into simpler carbohydrates are called monosaccharides. Glucose, fructose, and xylose are examples of monosaccharides. Carbohydrates that undergo hydrolysis to form two or three molecules are called disaccharides and trisaccharides, respectively. Cellulose is an example of polysaccharides that give a large number of molecules of monosaccharides (>10) upon hydrolysis. Cellulose is a crystalline polymer of glucose in which β-D-glucopyranose units are linked through β-glycosidic bonds (Figure 34.1) [3]. It exhibits high resistance to chemical attack because of its crystalline structure with extensive hydrogen bonding through OH groups.

Hemicelluloses are other examples of polysaccharides that have β-(1→4)-linked backbones with an equatorial configuration (Figure 34.1) [4]. They are amorphous with low degrees of polymerization (~200) and branched polymers. The structure of hemicelluloses depends on the source and may consist of C_5 carbohydrates (pentoses) such as xylose and arabinose and C_6 sugars (hexoses) such as glucose, fructose, galactose and uronic acids.

A general strategy to convert carbohydrates to value-added chemicals starts with deconstruction of polysaccharides into smaller molecules (Figure 34.2) [6].

Heterogeneous Catalysts: Advanced Design, Characterization and Applications, First Edition.
Edited by Wey Yang Teoh, Atsushi Urakawa, Yun Hau Ng, and Patrick Sit.

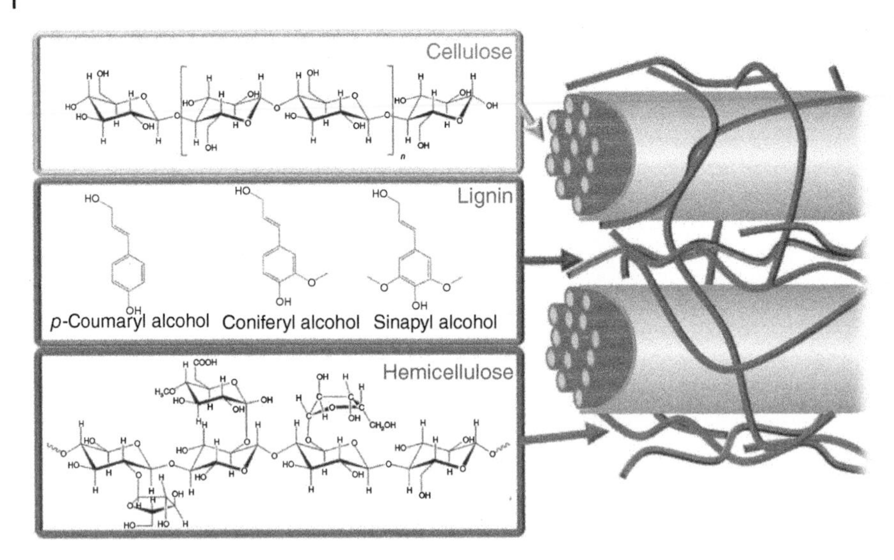

Figure 34.1 Structure of lignocellulosic biomass with cellulose, hemicellulose, and lignin represented. The building blocks of lignin, *p*-coumaryl, coniferyl, and sinapyl alcohol are also shown. Source: Adapted from Alonso et al. [1]. Copyright 2012, Royal Society of Chemistry. (See online version for color figure).

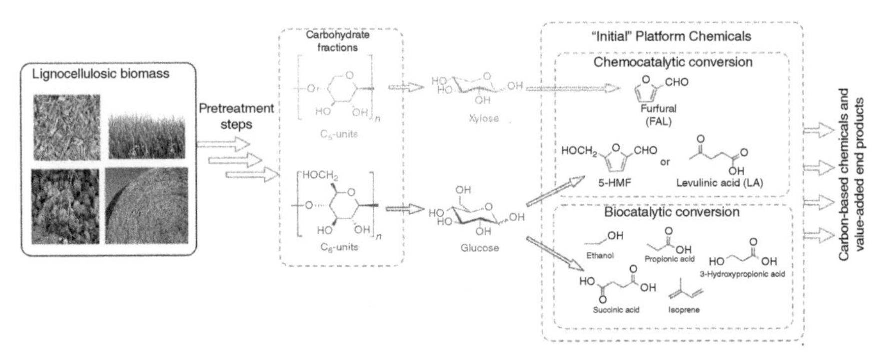

Figure 34.2 Synthesis of platform chemicals from biomass. Source: Adapted from Mika et al. [5]. Copyright 2018, American Chemical Society. (See online version for color figure).

The molecules derived from polysaccharides are often rich with functionality such as –OH, –C=O, and –COOH, which are reactive toward by-products. Thus, it is difficult to control selectivity to produce target chemicals in a single processing step. Hence, these molecules are converted into other molecules that possess lower oxygen content and reduced reactivity of functional groups [5]. The derived molecules serve as building blocks for further transformations [7]. The US Department of Energy has selected such building molecules and reported them as platform chemicals (Figure 34.2) [8]. Some of the platform chemicals are derived by biochemical methods, whereas others by chemical methods.

In this chapter, some examples of the recent and emerging technologies related to heterogeneous catalysts are described for the conversion of carbohydrates to value-added chemicals.

34.2 Overview of Strategy for Catalyst Development and Routes for Conversion of Carbohydrates

The current trends include the syntheses of nontoxic green solvents and polymer precursors that potentially substitute petroleum-derived compounds. Heterogeneous catalysts offer technological advantages over homogeneous catalysts because they allow easier separation of reaction products and enable continuous flow reaction processes [9]. Continuous flow processes generally facilitate mass production of chemicals, which is critical to providing biomass derived chemicals at reasonable prices. General strategy for development of catalysts starts by identifying the type of reactions to convert certain functional groups or configurations of a molecule to those of a target molecule. Types of reactions often used in the conversion of carbohydrates and molecules derived from carbohydrates are dehydration, hydrolysis, isomerization, hydrogenation, hydrogenolysis, and C–C coupling. Some of these reactions (e.g. dehydration, hydrolysis, and isomerization) are often catalyzed by acid or base sites, whereas others (e.g. hydrogenation) are catalyzed by metal sites or by a combination of acid sites and metal sites. Then, a catalyst is synthesized by aiming to incorporate active sites (acid or basic sites or metal sites) possibly in a desirable environment (nonpolar or polar) in nanometer-sized pores of a catalyst. When conversion of carbohydrate-derived molecules requires sequential reactions at acid and metal sites, those sites are coupled as a bifunctional catalyst or as a packed-bed reactor consisting of a dual bed of acid catalysts and supported metal catalysts. Some of these catalysts may be similar to those used in petroleum industry. However, the conversion of carbohydrate-derived molecules is often conducted in the presence of water, sometimes under hydrothermal conditions at elevated temperatures, and possibly in an acidic or alkaline solution. Thus, new catalysts that are robust and stable under these conditions are required. Carbon materials possess hydrophobic surfaces and exhibit high resistance against acidic or alkaline media. Some metal oxides such as titania, niobia, and zirconia exhibit higher stability than other metal oxides and thus may be desirable for the conversion of carbohydrates. Alumina and aluminosilicates, which have been used widely in petroleum industry, undergo structural collapse in liquid-phase reactions for biomass conversion [10]. Coating of surface of alumina particles with graphitic carbon is known to improve their hydrothermal stability [11]. An ultimate goal in catalyst development is to synthesize a robust solid catalyst that enables high selectivity under mild reaction conditions like enzyme catalysts. Various efforts have been paid to synthesize solid catalysts that mimic enzyme-like performance. Representative examples are crystalline silicate-based catalysts that possess Lewis acid sites in the molecular-sized channels. These materials possess the same topology as some zeolites. Because they are crystalline and possess molecular-sized pores and may allow entrance and conversion of specific molecules, they are usually referred to as molecular sieves. A representative example is a molecular sieve with β topology that contains Sn (Sn-β) as the Lewis acidic center (Figure 34.3) [12]. Zeolites are crystalline aluminosilicates consisting of interconnected SiO_4 and AlO_4^- tetrahedra. The negative charge

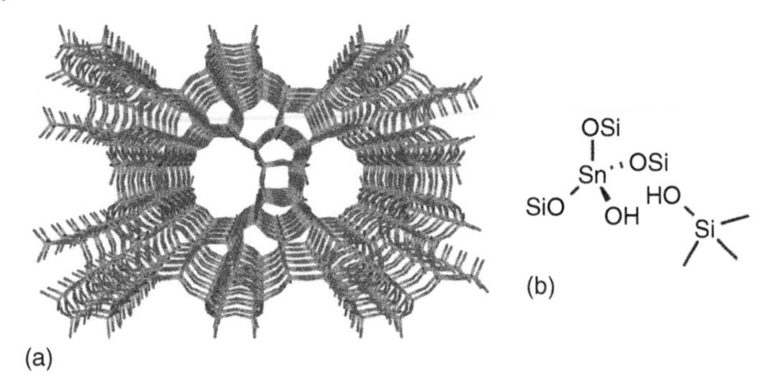

Figure 34.3 Schematic drawings of zeolite β framework (a) and an open site (b). (See online version for color figure).

on the AlO_4^- tetrahedra is balanced by extra-framework cations such as H^+ and Na^+, rendering ion-exchange properties. When protons are present as extra-framework cations, they act as strong Brønsted acids. The framework Si in pure silica molecular sieves may be substituted with other heteroatoms such as Fe and Sn. Because Sn has the same number of valence electron as Si, it requires no extra-framework cations like H^+, and hence Sn sites act as Lewis acid sites. Many metal cations bind water strongly and thus are poisoned by it and not desirable for the conversion of carbohydrates. However, Sn sites in Sn-β exhibit weaker interaction with water and still activate functional group such as carbonyl groups. In addition, when Sn-containing molecular sieves are highly crystalline, they offer highly hydrophobic environments that mimic those in active sites of enzymes [13]. Furthermore, the pore apertures of Sn-β molecular sieves (6–7 Å) are large enough to accommodate glucose. The environment inside the pores is highly hydrophobic and seems to be unfavorable for the adsorption of hydrophilic glucose. However, a favorable entropy change upon adsorption of glucose from an aqueous solution into the pore is known to drive its facile transfer [14]. Various works indicate that the location of Sn sites may not be random [15] and that a so-called open site consisting of $(-Si-O)_3-Sn-OH$ with a neighboring SiOH (Figure 34.3) enables high selectivity in isomerization of glucose to fructose [13].

Representative routes to derive platform chemicals and their derivatives with examples of catalysts used in the reactions are shown in Figure 34.4, and details are described in Section 34.3.

A sequence of selective conversions of cellulose yields dihydrolevoglucosenone, a nontoxic green solvent, dihydrolevoglucosenone, via levoglucosenone (LGO) as will be described in Section 34.3.1 [16]. Dehydration of C_6 carbohydrates (glucose and fructose) forms a platform chemical, 5-hydroxymethylfurfural (HMF) [17], which can be converted further into various chemicals as will be described in Sections 34.3.2–34.3.4 [18]. Glucose is more abundant than fructose as a feedstock. However, dehydration of fructose to HMF is more facile than that of glucose to HMF. Thus, isomerization of glucose to fructose is desirable [19]. A Sn-β can isomerize glucose to fructose efficiently [12, 13, 20] and allows a one-pot synthesis of HMF from glucose when it is coupled with dehydration reaction catalyzed by a Brønsted acid catalyst [21].

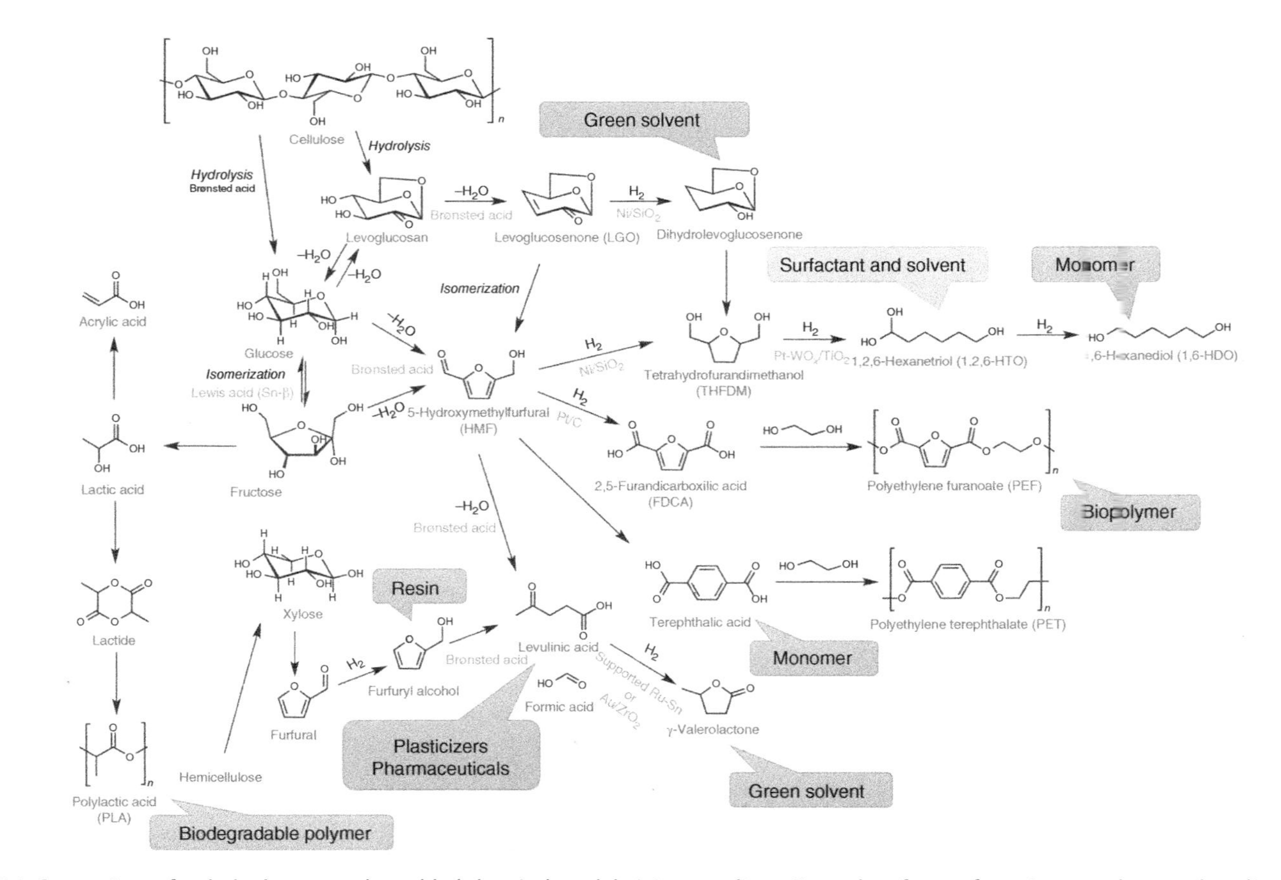

Figure 34.4 Conversions of carbohydrates to value-added chemicals and their intermediates. Examples of type of reactions, catalysts, and applications are presented. (See online version for color figure).

Hydrogenation of HMF forms tetrahydrofurandimethanol (THFDM). THFDM may also be obtained from LGO and can be converted further into 1,2,6-hexanetriol (1,2,6-HTO) that serves as surfactant and solvent or 1,6-hexanediol (1,6-HDO) that serves as polymer precursor [22]. Oxidation of HMF yields furandicarboxylic acid (FDCA), which has been considered as a substitute of petroleum-based terephthalic acid for the production of polyethylene terephthalate (PET) as will be described in Section 34.3.3 [23]. Diels–Alder cycloaddition of reduced or oxidized HMF with ethylene forms terephthalic acid as shown in Section 34.3.4 [24], which can be used to synthesize PET.

Acid-catalyzed dehydration of xylose or hemicellulose produces furfural, a five-carbon building block [25]. Solid acid catalysts combined with a biphasic reaction system allow the continuous flow synthesis of furfural from xylan hemicellulose and xylose [25]. Partial or total hydrogenation of furfural or HMF produces furfuryl alcohol or tetrahydrofurfuryl alcohol, respectively [26]. Furfuryl alcohol has been used in the synthesis of foundry resin. Tetrahydrofurfuryl alcohol produces 1,5-pentanediol, which can be used as a polymer precursor. Acid-catalyzed conversion of hydration of furfuryl alcohol yields another platform chemical, levulinic acid [27], which serves as a building block for pharmaceuticals, plasticizers, and cosmetics [28]. Alkyl esters formed from levulinic acid via esterification reaction with alcohol using acid catalysts [29] may be used for flavors, fragrances, and plasticizers. Hydrogenation of levulinic acid forms γ-valerolactone (GLV) [30], which has been demonstrated as an effective solvent for various biomass-related conversions [31]. For example, HMF readily undergoes condensation reactions at moderate temperatures like 373 K in the presence of an acid or a base. Thus, dehydration of fructose to HMF in water results in low yields of HMF (<20%). Organic solvents such as GLV limit the degradation of HMF by condensation reactions and allow high yields of HMF (>70%) from fructose.

Fermentation of pentoses or hexoses forms lactic acid (2-hydroxypropanoic acid), which is another platform molecule that can be converted into various useful molecules [32]. Because the fermentation pathway to obtain lactic acid is costly, heterogeneous catalytic systems have been developed as will be described in Section 34.3.5 [33]. Lactic acid also serves as the building block for polylactic acid (PLA), which is a biodegradable polymer. A heterogeneous catalytic system based on molecular sieving catalysts has been developed to enable an economical route to produce an intermediate molecule for PLA as will be described in Section 34.3.6 [34].

Some of the platform chemicals have been produced on a large scale in industry. For example, AVA Biochem has produced HMF on a commercial scale by hydrothermal processing. GFBiochemicals has produced levulinic acid from C_6 carbohydrates, which are derived from biomass through an acid-catalyzed hydrolysis treatment under high temperature and pressure. In addition to the platform chemicals, other useful chemicals have been produced on a large scale from biomass. For example, Circa has developed a Furacell™ technology, which converts biomass to LGO through a continuous reaction system.

34.3 Synthesis of Value-Added Chemicals from Carbohydrates

34.3.1 Dihydrolevoglucosenone

When dehydration of cellulose is conducted in aprotic solvents, it forms LGO. Partial hydrogenation of LGO over a supported Pd catalyst forms dihydrolevoglucosenone (Cyrene™), which has been produced on a large scale by Circa. Like this reaction, hydrogenation reactions are generally carried out using supported metals such as Pd, Pt, Ni, or Ru to saturate C—C bonds (C=C and C≡C bonds) and C=O bonds [35]. Hydrogen atoms formed on supported Pd metals from H_2 add C=C bond in LGO. Cyrene has been certified as a nontoxic solvent by the EU recently. However, Cyrene exhibits solvent properties similar to conventional aprotic solvents such as dimethylformamide (DMF) and *N*-methylpyrrolidone (NMP), which are toxic. Beneficial effects of Cyrene as a solvent have been demonstrated in several chemical reactions such as fluorination of an aromatic ring, which is of significant interest in pharmaceutical industry. LGO and Cyrene can be converted to THFDM via hydrogenation of C=O bond to form threo- and erythro-levoglucosanol (Lgol) and a subsequent hydrogenolysis reaction [36].

34.3.2 1,6-Hexanediol

1,6-HDO serves as a monomer for polyesters and polyurethanes [37]. It has been produced on a large scale via a petrochemical route using catalytic hydrogenation of adipic acid or its esters. Rennovia has developed a pilot-scale plant for an alternative biomass route and has reported that it enables a better profit margin than the petrochemical route. 1,6-HDO may be obtained by hydrogenation of HMF or LGO to THFDM, followed by hydrogenolysis [22]. Hydrogenolysis is a reaction that cleaves C—C and C—O bonds by hydrogen and is often referred to as hydrodeoxygenation (HDO). HDO is typically carried out on a bifunctional solid catalyst that performs both hydrogenation at metal sites and dehydration reactions at acid sites. The conversion of HMF to 1,6-HDO has been accomplished at high yields by using Pd/SiO_2 and $Ir\text{-}ReO_x/SiO_2$ catalysts in a H_2O/tetrahydrofuran (THF) mixture [38]. Other researchers have demonstrated a high-yield synthesis of 1,6-HDO by hydrogenation of LGO by a Ni/SiO_2 or $Pt/SiO_2\text{–}Al_2O_3$ catalyst and hydrogenolysis of THFDM over a $Pt\text{–}WO_x/TiO_2$ or $Pt\text{–}WO_x/ZrO_2$ catalyst (Figure 34.4) [22].

34.3.3 Furandicarboxylic Acid (FDCA)

FDCA has been considered as a substitute of petroleum-based terephthalic acid for the production of polyesters with ethylene glycol (EG). Polymerization of FDCA with EG yields polyethylene 2,5-furandicarboxylate (PEF) (Scheme 34.1). PEF is considered to reduce energy usage and greenhouse gas emissions by about 50% and an improved oxygen permeability relative to PET. PET is generally synthesized by polymerization of EG and dimethyl terephthalate or terephthalic acid.

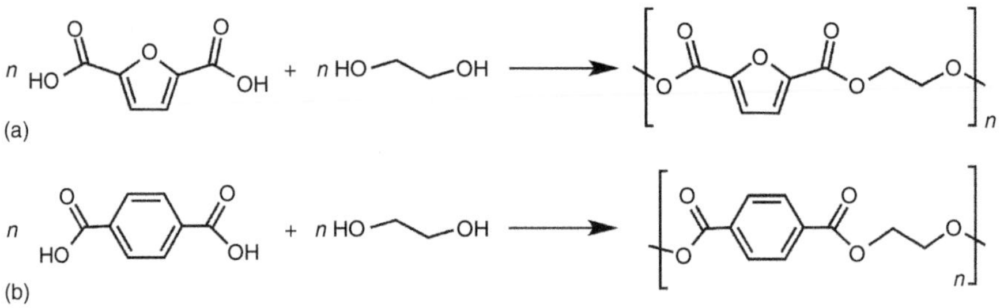

Scheme 34.1 Synthesis of polyethylene 2,5-furandicarboxylate (a) and polyethylene terephthalate (b).

Terephthalic acid has been produced by selective oxidation of *p*-xylene. The manufacture of PET is the largest and about one-fifth in the total production of polymer in the world. It is used as bottles for water and soft drinks as well as food packaging.

Avantium has produced a homogeneous catalyst system that converts methoxymethylfurfural to FDCA. FDCA can also be obtained by the oxidation of HMF using solid catalysts. Selective oxidation of HMF to FDCA may be achieved by the chemistry used in the synthesis of terephthalic acid from *p*-xylene with some modifications. However, FDCA exhibits low solubility in many solvents, and two to four equivalent bases are often required to prevent deactivation of solid catalysts by precipitation of FDCA on their surfaces. Consequently, a subsequent neutralization step, which produces a large amount of wastes, is required to obtain FDCA. To overcome this challenge, solid base materials such as MgO and hydrotalcites [39] were used as catalyst supports for metals that enable the oxidation reaction. Hydrotalcites are layered double hydroxides consisting of octahedrally coordinated Mg^{2+} and Al^{3+} (Figure 34.5). A partial substitution of Mg sites in brucite $Mg(OH)_2$ with Al forms positive charges in the layer, which are compensated by interlayer anions. Hydrotalcites show anion-exchange ability [40] and exhibit strong Brønsted basicity when it incorporates OH^- anions [41]. Calcination of hydrotalcites forms Mg–Al mixed oxides that exhibit strong Lewis basicity, depending on the Mg–Al ratio.

The approach described in the preceding paragraph often needs a highly diluted HMF feed to prevent precipitation of products. A more recent work demonstrated that fructose can be converted to HMF at 70% yield in a

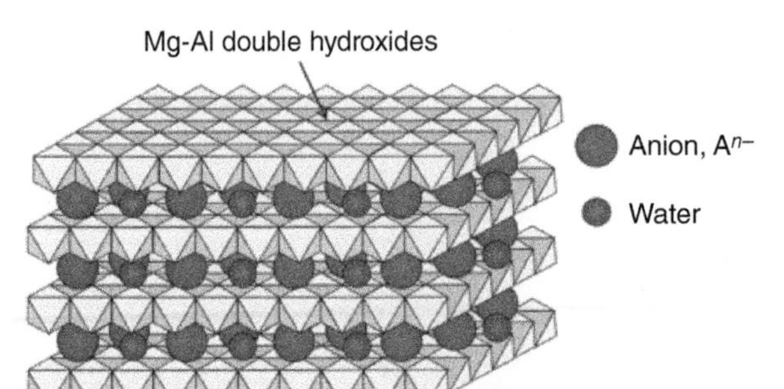

Figure 34.5 Schematic drawing of Mg–Al layered double hydroxides. (See online version for color figure).

γ-valerolactone (GVL)/H_2O mixed solvent and subsequently oxidized to FDCA at 93% yield over a carbon-supported platinum nanoparticle catalyst (Pt/C catalyst) [23]. The high solubility of FDCA in GVL/H_2O solution enables the oxidation at high concentrations. Thus, the Pt/C catalyst can be used to yield FDCA at a high yield without a homogeneous base catalyst. In addition, FDCA can be separated from the solvent by crystallization to obtain >99% pure FDCA. Additionally, FDCA acts as an effective catalyst to dehydrate fructose to HMF. Therefore, corrosive mineral acid catalysts such as HCl and H_2SO_4 are not required. By-products formed by the dehydration of carbohydrates can be separated and converted into activated carbon.

Another work has demonstrated the synthesis of FDCA by oxidation of an acetal derivative of HMF using a CeO_2-supported gold nanoparticle catalyst and Na_2CO_3 in water [23]. The acetal derivative of HMF is formed by the reaction of HMF with 1,3-propanediol (Scheme 34.2) and exhibits high stability against humin formation. The work has suggested two crucial steps in the reaction mechanism through first principles calculations (Figure 34.6). Partial hydrolysis of the acetal into 5-formyl-2-furan carboxylic acid is catalyzed by OH^- and Lewis acid sites on CeO_2, and subsequent oxidative dehydrogenation of the in situ generated hemiacetal involves Au nanoparticles.

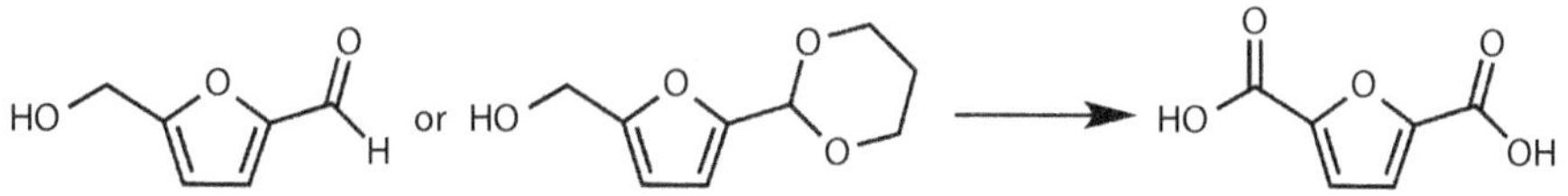

HO—(HMF) or HO—(acetal) ⟶ HO—(FDCA)—OH

Scheme 34.2 Synthesis of FDCA from HMF or an acetal derivative of HMF.

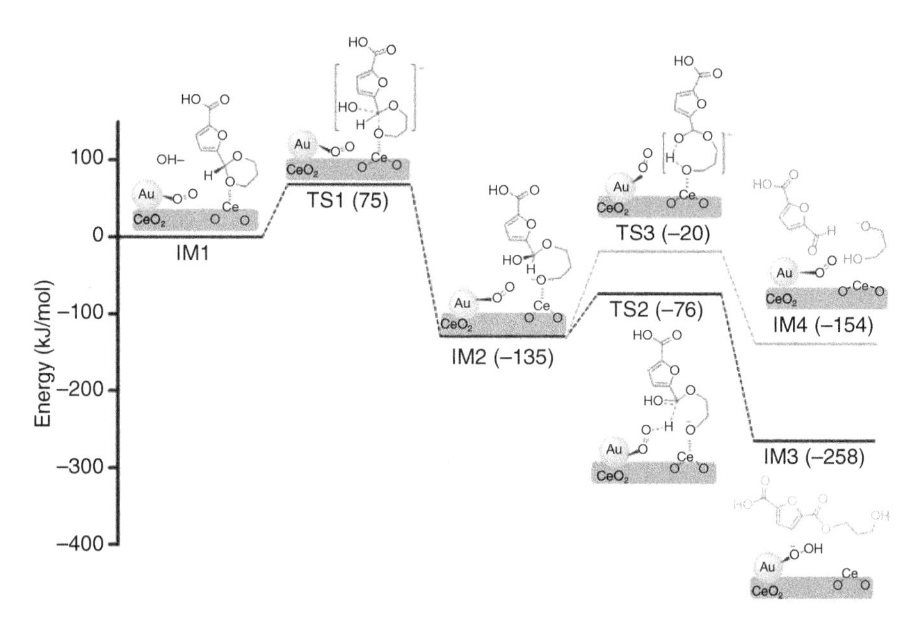

Figure 34.6 Computed reaction energy diagram for the rate-determining step during the oxidation of acetal-derivative of HMF by 1,3-propanediol (PD-HMF) into FDCA over the CeO_2-supported Au catalyst. Source: Adapted from Kim et al. [23]. Copyright 2018, Wiley-VCH. (See online version for color figure).

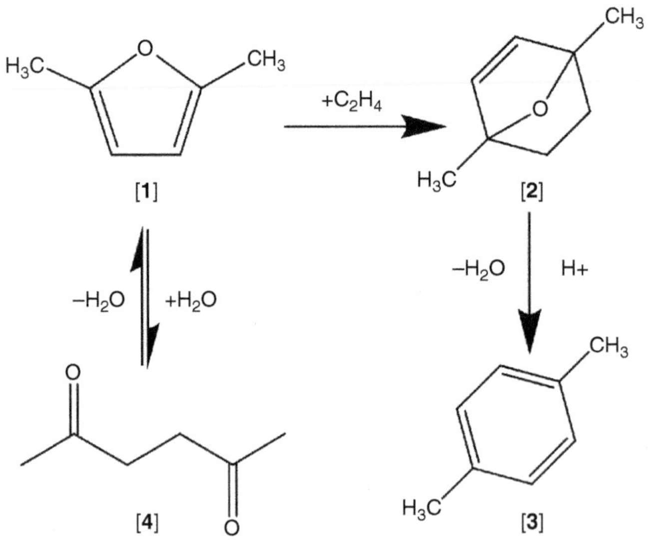

Figure 34.7 Diels–Alder cycloaddition of dimethylfuran [1] and ethylene produces an oxanorbornene cycloadduct [2], which dehydrates to *p*-xylene [3]. Water hydrolyzes dimethylfuran [1] to 2,5-hexadione [4] in equilibrium. Source: Adapted from Chang et al. [24]. Copyright 2014, Royal Society of Chemistry.

34.3.4 Terephthalic Acid

Recently, a new catalytic route from HMF to terephthalic acid has been developed using Diels–Alder reaction (Figure 34.7) [24]. A Diels–Alder reaction proceeds between a conjugated diene and a compound containing a double bond. The reaction forms two σ bonds at the expense of two π bonds. HMF may be converted to dimethylfuran by hydrogenolysis using a carbon-supported CuRu bimetallic catalyst. Then, the dimethylfuran can be reacted with ethylene in heptane solvent at 523 K using various solid acid catalysts, which results in the symmetry-allowed [4 + 2] Diels–Alder cycloaddition of ethylene to dimethylfuran to form an oxanorbornene and subsequent Brønsted-acid-catalyzed dehydration to yield *p*-xylene. A β zeolite that possesses Brønsted acid sites exhibited the best mass-specific activity and a high selectivity to *p*-xylene of ~90% at conversion of 99%. When the conversion of dimethylfuran is low, the reaction forms a side product of 2,5-hexanedione by hydration. However, as the conversion was increased, it reverted back to dimethylfuran and was converted to *p*-xylene.

Because hydrogenation of HMF requires a supported precious metal catalyst, a more economical route, which uses oxidized derivatives of HMF, has been reported (Figure 34.8) [24, 42]. In the work, a partially oxidized HMF, 5-(hydroxymethyl)furoic acid (HMFA), and each of the ether and ester derivatives of HMFA were reacted with ethylene to yield aromatic products. The fully oxidized HMF, which is FDCA, can be reacted with ethylene to form *p*-xylene. However, the strong electron-withdrawing effects of two –COOH groups in FDCA deactivate the diene and result in a prohibitory slow rate of reaction.

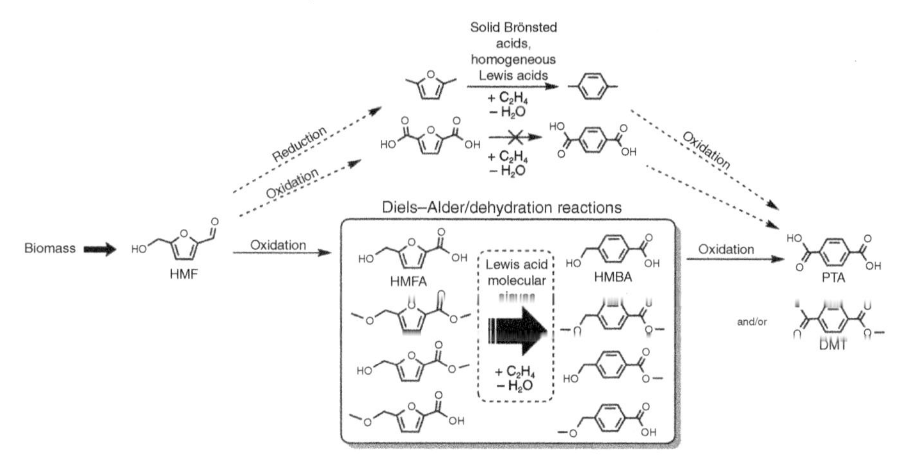

Figure 34.8 Synthesis of *p*-xylene through Diels–Alder cycloaddition of reduced or oxidized HMF. Source: Adapted from Pacheco and Davis [24]. Copyright 2014, American Chemical Society.

The partially oxidized derivatives of HMF allows the reaction of ethylene at reasonable rates over molecular sieves with zeolite β topology containing Lewis acid sites (Sn^{4+} and Zr^{4+}). The molecular-sized pores present in the catalysts facilitate the cycloaddition reaction by confinement effects.

34.3.5 Lactic Acid

PLA has been currently produced on a large scale through an anaerobic fermentation route. The sequential synthesis route consists of fermentation of carbohydrates (pentoses or hexoses) to lactic acid, conversion of lactic acid to a cyclic dimer called as lactide, and ring-opening polymerization of lactide to form PLA [43].

The first fermentation step dominates the total cost of the production of PLA. Thus, to achieve a more economical production of PLA, various catalytic routes that replace the fermentation route have been investigated over a decade. Initially, a homogeneous Lewis acid catalyst based on Sn halide was used to convert trioses to alkyl lactates [44]. Later, heterogeneous acid catalysts based on molecular sieves have been investigated as potential catalysts to convert trioses to lactic acid or esters [45].

Conversion of hexoses to lactic acid is more challenging because it involves scissions of C—C bonds, which requires more harsh conditions and consequently facilitates the formation of side products. Direct conversion of fructose or sucrose to methyl lactate was demonstrated by using solid Lewis acid catalysts such as Sn-β molecular sieves [33]. The reaction was performed in methanol solvent at 433 K, forming methyl lactate in carbon-based yields of up to 44% for fructose and 68% for sucrose. The reaction proceeds via a retro-aldol reaction that forms two trioses from fructose, which undergo sequential dehydration and addition of methanol (Figure 34.9).

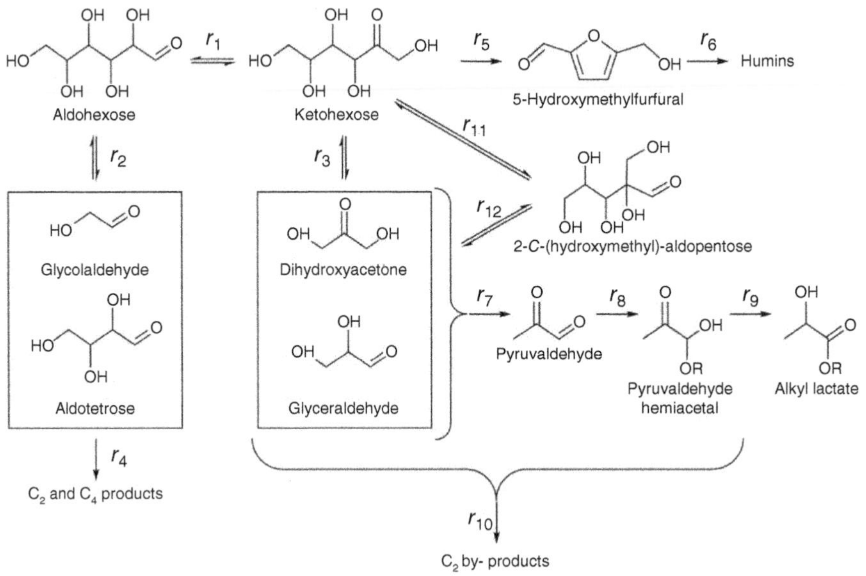

Figure 34.9 Schematic representation of reaction network in which ketohexoses can isomerize to aldohexoses via 1,2-hydride shift (r_1) and to 2-C-(hydroxymethyl)-aldopentoses via 1,2-carbon shift (r_{11}) reactions. Retro-aldol reactions of hexose species (r_2, r_3, and r_{12}) lead to the formation of C_2, C_3, and C_4 carbohydrate fragments. Lewis acids can then catalyze the formation of α-hydroxy carboxylic acids from these smaller fragments (e.g. r_7, r_8, and r_9 in the formation of alkyl lactate from trioses). Side reactions, involving dehydration reactions of fructose to 5-HMF (r_5), redox and fragmentation reactions of unstable intermediates, and various humin-forming condensation reactions, lead to loss yield of desired products. Source: Adapted from Orazov and Davis [33]. Copyright 2015, American Chemical Society.

The high-temperature conditions used in the preceding work results in the formation of HMF and subsequent degradation of HMF to insoluble humins when sugars are converted. Moreover, the Sn-β facilitates aldose–ketose isomerization via 1,2-hydride shift, which results in the formation of C_2 and C_4 products from aldoses in addition to desirable C_3 products. To minimize the undesirable reactions, a tandem catalyst system has been developed recently [33]. The catalyst system combined heterogeneous catalysts such as molybdenum oxide, alkali-exchanged stannosilicate molecular sieves, or amorphous TiO_2–SiO_2 coprecipitates, which are coupled with solid Lewis acid catalysts such as Sn-containing molecular sieves with MFI topology (Sn-MFI). The former catalysts catalyze retro-aldol reactions but do not readily catalyze the aldose–ketose isomerization reaction. Because Sn-MFI possesses smaller pore apertures (5–6 Å) than Sn-β, it prevents access of hexoses to Lewis acid sites and hence does not cause undesirable side reactions of aldose–ketose isomerization.

34.3.6 Lactide

The current process converts lactic acid to lactide via polycondensation of lactic acid with removal of water to a prepolymer and a subsequent reaction to form

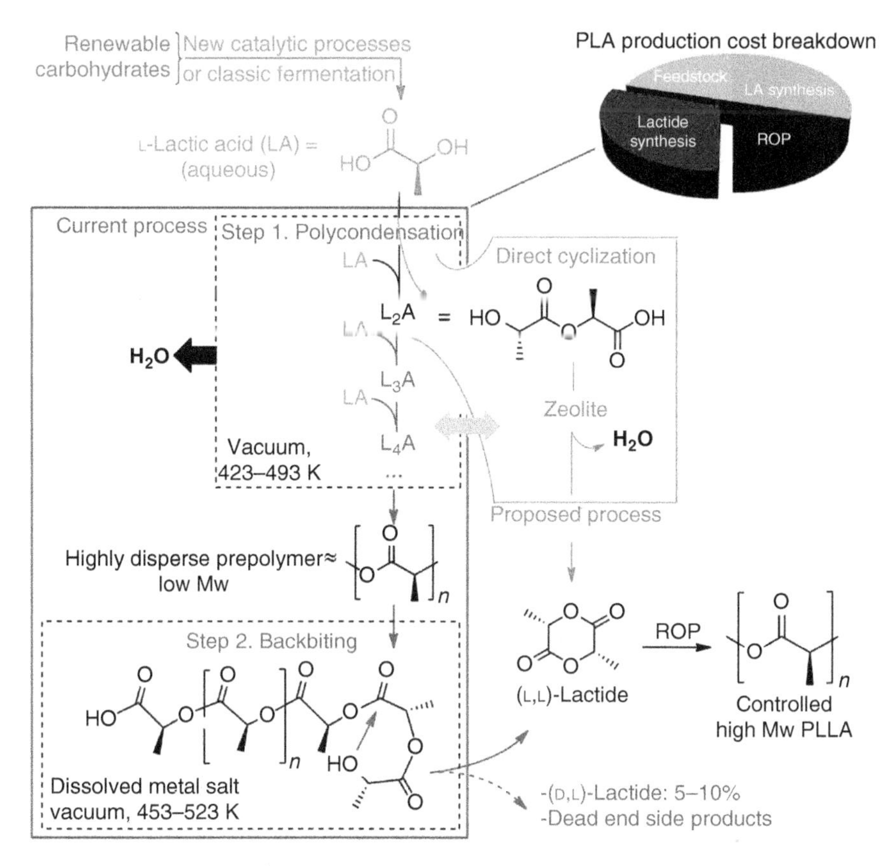

Figure 34.10 Current and new chemical process for making lactide and polylactic acid from lactic acid. Red: current industrial two-step process; blue: proposed direct lactide synthesis by selectivity control during condensation. L_nA, linear oligomer of *n* lactyl units; ROP, ring-opening polymerization. Pie chart shows a cost estimate for current PLA production from sugar to pellet. Source: Adapted from Dusselier et al. [34]. Copyright 2015, AAAS. (See online version for color figure).

lactide from the prepolymer (Figure 34.10), both of which are energy-intensive conversions. In particular, the second synthesis step requires continuous removal of lactide by an energy-intensive separation process like distillation to shift the equilibrium transesterification reaction. A more energy-efficient route has been demonstrated using zeolites as shape-selective catalysts [46]. In general, Brønsted-acid-catalyzed conversion of lactic acid forms lactide as well as undesired lactyl oligomers. However, molecular-sized pores present in zeolites restrict the formation of oligomers and facilitate the cyclization to form lactide. A β zeolite forms lactide with nearly 79% yield at full conversion of lactic acid. In addition, an integrated process has been proposed. In this process, a stirred tank reactor was used to perform the shape-selective conversion of lactic acid under reflux, water formed in the reaction is continuously removed from the reaction mixture, and a dry solvent is recycled to the reactor. A more work from the same group demonstrated the synthesis of lactide from alkyl lactates in the gas-phase reaction over a supported TiO_2/SiO_2 catalyst [34].

34.4 Perspective

As shown by examples shown in this chapter, promising solid catalysts and catalytic processes have been developed to convert carbohydrates into value-added products. The ultimate goal in catalyst development in twenty-first century is to design and synthesize a solid catalyst that achieves 100% selectivity and high activity under mild reaction conditions like enzyme catalysts. To achieve this goal, the research work in twenty-first century should aim to deepen the fundamental understanding of structure–performance relationships to design catalyst architectures to enable such performance. Systematic investigations of solid catalysts with a series of well-defined structures using state-of-the-art analytical techniques that track their dynamic structural changes at the atomic scale will facilitate such efforts as described in other sections. In addition, theoretical calculations that guide interpretation of experimental results and predict desirable structures will facilitate the rational design of new catalysts further. In addition, a synthetic method that creates active sites with atomic precision [47] is required to enable the synthesis of solid catalysts with structures predicted through rational design. Achieving such goal may be still far away. However, as demonstrated in the examples of Sn-β catalysts and other various catalysts, efforts to achieve the ultimate success have continued.

References

1 (a) Gibson, L.J. (2012). *J. R. Soc. Interface* 9: 2749–2766. (b) Alonso, D.M., Wettstein, S.G., and Dumesic, J.A. (2012). *Chem. Soc. Rev.* 41: 8075–8098.
2 Solomons, T.W.G. and Fryhle, C.B. (2011). *Organic Chemistry*, 10e. Hoboken, NJ: Wiley.
3 O'Sullivan, A.C. (1997). *Cellulose* 4: 173–207.
4 Scheller, H.V. and Ulvskov, P. (2010). *Annu. Rev. Plant Biol.* 61: 263–289.
5 (a) Corma, A., Iborra, S., and Velty, A. (2007). *Chem. Rev.* 107: 2411–2502. (b) Mika, L.T., Csefalvay, E., and Nemeth, A. (2018). *Chem. Rev.* 118: 505–613.
6 Onda, A., Ochi, T., and Yanagisawa, K. (2008). *Green Chem.* 10: 1033–1037.
7 (a) Sheldon, R.A. (2014). *Green Chem.* 16: 950–963. (b) Zhang, X., Wilson, K., and Lee, A.F. (2016). *Chem. Rev.* 116: 12328–12368.
8 Bozell, J.J. and Petersen, G.R. (2010). *Green Chem.* 12: 539–554.
9 (a) Ertl, G., Knözinger, H., Schüth, F., and Weitkamp, J. (2008). *Handbook of Heterogeneous Catalysis*, vol. 8. Wiley. (b) Gates, B.C. (1992). *Catalytic Chemistry*. New York, NY: Wiley.
10 Ravenelle, R.M., Copeland, J.R., Kim, W.-G. et al. (2011). *ACS Catal.* 1: 552–561.
11 Pham, H.N., Anderson, A.E., Johnson, R.L. et al. (2012). *Angew. Chem. Int. Ed.* 51: 13163–13167.

12 (a) Corma, A., Nemeth, L.T., Renz, M., and Valencia, S. (2001). *Nature* 412: 423–425. (b) Moliner, M., Román-Leshkov, Y., and Davis, M.E. (2010). *Proc. Natl. Acad. Sci. U.S.A.* 107: 6164–6168. (c) Davis, M.E. (2015). *Top. Catal.* 58: 405–409.

13 (a) Bermejo-Deval, R., Assary, R.S., Nikolla, E. et al. (2012). *Proc. Natl. Acad. Sci. U.S.A.* 109: 9727–9732. (b) Gounder, R. and Davis, M.E. (2013). *AlChE J.* 59: 3349–3358.

14 Bai, P., Siepmann, J.I., and Deem Michael, W. (2013). *AlChE J.* 59: 3523–3529.

15 (a) Bare, S.R., Kelly, S.D., Sinkler, W. et al. (2005). *J. Am. Chem. Soc.* 127: 12924–12932. (b) Yang, G., Pidko, E.A., and Hensen, E.J.M. (2013). *J. Phys. Chem. C* 117: 3976–3986. (c) Wolf, P., Valla, M., Núñez-Zarur, F. et al. (2016). *ACS Catal.* 6: 4047–4063.

16 Cao, F., Schwartz, T.J., McClelland, D.J. et al. (2015). *Energy Environ. Sci.* 8: 1808–1815.

17 (a) Román-Leshkov, Y., Chheda, J.N., and Dumesic, J.A. (2006). *Science* 312: 1933. (b) Chen, S.S., Maneerung, T., Tsang, D.C.W. et al. (2017). *Chem. Eng. J.* 328: 246–273. (c) Simeonov, S.P., Coelho, J.A., and Afonso, C.A. (2012). *ChemSusChem* 5: 1388–1391.

18 (a) van Putten, R.J., van der Waal, J.C., de Jong, E. et al. (2013). *Chem. Rev.* 113: 1499–1597. (b) Rosatella, A.A., Simeonov, S.P., Frade, R.F.M., and Afonso, C.A.M. (2011). *Green Chem.* 13: 754–793.

19 (a) Takagaki, A., Ohara, M., Nishimura, S., and Ebitani, K. (2009). *Chem. Commun. (Cambridge, UK)*: 6276–6278. (b) Osatiashtiani, A., Lee, A.F., Granollers, M. et al. (2015). *ACS Catal.* 5: 4345–4352.

20 Ren, L., Guo, Q., Kumar, P. et al. (2015). *Angew. Chem. Int. Ed.* 54: 10848–10851.

21 Nikolla, E., Román-Leshkov, Y., Moliner, M., and Davis, M.E. (2011). *ACS Catal.* 1: 408–410.

22 He, J., Burt, S.P., Ball, M. et al. (2018). *ACS Catal.* 8: 1427–1439.

23 (a) Motagamwala, A.H., Won, W.Y., Sener, C. et al. (2018). *Sci. Adv.*: 4. (b) Kim, M., Su, Y., Fukuoka, A. et al. (2018). *Angew. Chem. Int. Ed.* 57: 8235–8239.

24 (a) Chang, C.-C., Green, S.K., Williams, C.L. et al. (2014). *Green Chem.* 16: 585–588. (b) Pacheco, J.J. and Davis, M.E. (2014). *Proc. Natl. Acad. Sci. U.S.A.* 111: 8363–8367.

25 Aellig, C., Scholz, D., Dapsens, P.Y. et al. (2015). *Catal. Sci. Technol.* 5: 142–149.

26 Nakagawa, Y., Tamura, M., and Tomishige, K. (2013). *ACS Catal.* 3: 2655–2668.

27 (a) González Maldonado, G.M., Assary, R.S., Dumesic, J., and Curtiss, L.A. (2012). *Energy Environ. Sci.* 5: 6981–6989. (b) Weingarten, R., Cho, J., Xing, R. et al. (2012). *ChemSusChem* 5: 1280–1290.

28 (a) Démolis, A., Essayem, N., and Rataboul, F. (2014). *ACS Sustainable Chem. Eng.* 2: 1338–1352. (b) Pileidis, F.D. and Titirici, M.M. (2016). *ChemSusChem* 9: 562–582.

29 Ogino, I., Suzuki, Y., and Mukai, S.R. (2018). *Catal. Today* 314: 62–69.

30 Du, X.L., He, L., Zhao, S. et al. (2011). *Angew. Chem. Int. Ed.* 50: 7815–7819.

31 (a) Delidovich, I., Leonhard, K., and Palkovits, R. (2014). *Energy Environ. Sci.* 7: 2803–2830. (b) Qi, L., Mui, Y.F., Lo, S.W. et al. (2014). *ACS Catal.* 4: 1470–1477.

32 Dusselier, M., Van Wouwe, P., Dewaele, A. et al. (2013). *Energy Environ. Sci.* 6: 1415.

33 Orazov, M. and Davis, M.E. (2015). *Proc. Natl. Acad. Sci. U.S.A.* 112: 11777–11782.

34 (a) Dusselier, M., Van Wouwe, P., Dewaele, A. et al. (2015). *Science* 349: 78. (b) De Clercq, R., Dusselier, M., Makshina, E., and Sels, B.F. (2018). *Angew. Chem. Int. Ed.* 57: 3074–3078.

35 Ogino, I., Serna, P., and Gates Bruce, C. (2017). Supported catalysts. In: *Handbook of Solid State Chemistry*, vol. 6 (eds. R. Dronskowski, S. Kikkawa and A. Stein), 313–337. Wiley-VCH.

36 Krishna, S.H., Assary, R.S., Rashke, Q.A. et al. (2018). *ACS Catal.* 8: 3743–3753.

37 Krishna, S.H., Huang, K., Barnett, K.J. et al. (2018). *AlChE J.* 64: 1910–1922.

38 Xiao, B., Zheng, M., Li, X. et al. (2016). *Green Chem.* 18: 2175–2184.

39 Ardemani, L., Cibin, G., Dent, A.J. et al. (2015). *Chem. Sci.* 6: 4940–4945.

40 Ogino, I., Tanaka, R., Kudo, S., and Mukai, S.R. (2018). *Microporous Mesoporous Mater.* 263: 181–189.

41 Ebitani, K., Motokura, K., Mori, K. et al. (2006). *J. Organomet. Chem.* 71: 5440–5447.

42 Pacheco, J.J., Labinger, J.A., Sessions, A.L., and Davis, M.E. (2015). *ACS Catal.* 5: 5904–5913.

43 De Clercq, R., Dusselier, M., and Sels, B.F. (2017). *Green Chem.* 19: 5012–5040.

44 Hayashi, Y. and Sasaki, Y. (2005). *Chem. Commun. (Cambridge, UK)*: 2716–2718.

45 Dapsens, P.Y., Menart, M.J., Mondelli, C., and Pérez-Ramírez, J. (2014). *Green Chem.* 16: 589–593.

46 Holm, M.S., Saravanamurugan, S., and Taarning, E. (2010). *Science* 328: 602.

47 Hermans, S. and Visat de Bocarmé, T. (2015). *Atomically-Precise Methods for Synthesis of Solid Catalysts.* Cambridge: Royal Society of Chemistry.

35

Enhancing Sustainability Through Heterogeneous Catalytic Conversions at High Pressure

Nat Phongprueksathat and Atsushi Urakawa

Delft University of Technology, Department of Chemical Engineering, Catalysis Engineering, Van der Maasweg 9, 2629 HZ Delft, The Netherlands

35.1 Importance of High-Pressure Reaction Condition

The most successful example of the high-pressure chemistry in heterogeneous catalysis and chemical processing is the commercialization of ammonia synthesis process (300–700 bar) developed by Haber and Bosch in 1913 [1]. This breakthrough had a significant impact on large-scale production of fertilizers supporting almost half of the world's population through enhanced food production [2]. Following this innovation, high-pressure methanol synthesis process (300 bar) was commercialized by BASF in 1923 [3]. This marked the emergence of the mass production of organic chemicals since methanol is considered one of the most promising building blocks to obtain more complex chemical structures in petrochemical industries [4]. Another success is the development of a still up-to-date polymerization process of ethylene at an ultrahigh pressure (3000 bar) in 1935 [5]. Since then, polyethylene has evolved into a material critical to our life. The major characteristics of these high-pressure processes are still similar to those of the state-of-the-art processes existing today.

Nowadays, the global trend of those high-pressure catalytic processes tends to shift toward a milder condition at lower pressure and temperature or even at ambient conditions [6–8]. This movement was partially triggered by the famous "12 principles of green chemistry." At a glance, the high-pressure reaction condition does not comply with the "design for energy efficiency" and "safety" principles. Still, the high-pressure condition offers distinctive advantages for many catalytic reactions by significantly increasing the product yield and improving the incorporation of materials used in the process into the final product, thus "maximizing atom economy." The smaller but more productive process enabled by the high-pressure condition can render such processes safer and more energy efficient. Moreover, high pressure allows the use of supercritical states of various media such as carbon dioxide and water as a "safer solvent or auxiliary" instead of often hazardous organic solvents. The unique advantages of the high-pressure condition can be so prominent that the overall catalytic processes can be designed

Heterogeneous Catalysts: Advanced Design, Characterization and Applications, First Edition.
Edited by Wey Yang Teoh, Atsushi Urakawa, Yun Hau Ng, and Patrick Sit.

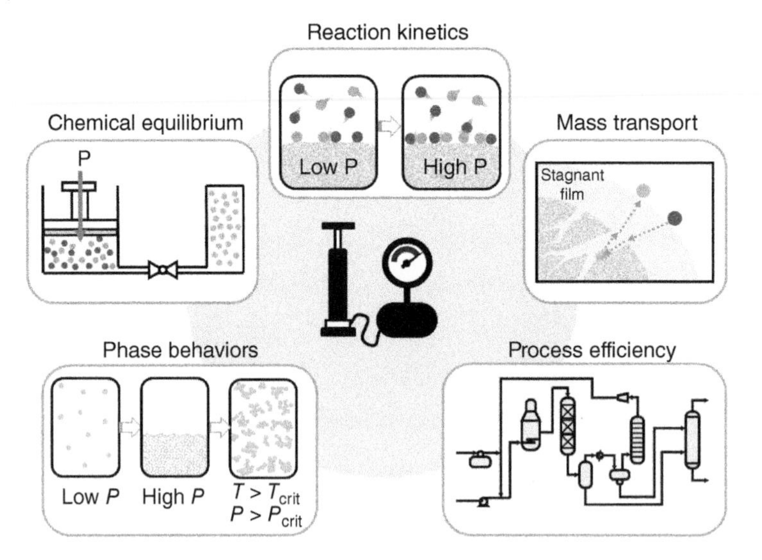

Figure 35.1 The main motivations to use high-pressure conditions for heterogeneously catalyzed reactions.

in accordance with the green chemistry principles despite the apparent imaginal discrepancy.

In the industrial chemical and catalytic processes, the motivation to use high-pressure reaction condition varies. Figure 35.1 summarizes common advantageous characteristics offered by such reaction conditions, and each point is described below.

35.1.1 Chemical Equilibrium (One Phase)

The most common motivation to employ high-pressure condition is to induce thermodynamic advantages by shifting chemical equilibrium when the number of molecules (thus volume for gas-phase reaction) reduces toward product formation, according to Le Châtelier's principle. The shift in chemical equilibrium ultimately affects the maximum value of attainable conversion and product selectivity. The syntheses of ammonia, methanol, and low-density polyethylene are obvious examples of exploiting high-pressure reaction conditions for such advantages. The role of catalyst in this respect is to accelerate the reaction so that thermodynamics play more dominant roles in defining reaction performance. Generally, high-pressure conditions offer fewer advantages for liquid-phase reaction due to negligible or the less compressive nature of liquids.

35.1.2 Phase Behavior (Multiphase)

35.1.2.1 Phase Separation

High-pressure conditions are prevalently used for phase separation in the industrial processes, mainly to liquefy some condensable components out of the gas mixture based on an increased vapor density with pressure. Some liquid-phase reactors take advantage of high-pressure conditions to prevent vaporization of the light component such as light hydrocarbons, thus retaining reactive components with a catalyst in the liquid phase. Relatively novel utilization of high-pressure condition for phase separation is an *in situ* condensation of product from the gas phase during the reaction. In this case, the reaction performance is ruled by multiphase equilibrium (chemical and vapor–liquid equilibrium) and can override the limit defined by the one-phase thermodynamic equilibrium of the reaction. This concept resembles multiphase equilibrium exploitation in reactive distillation that integrates chemical reaction and product separation in a single unit. An example is the methyl tert-butyl ether (MTBE) production where MTBE is continuously removed from a catalytic distillation column to enhance equilibrium-limited isobutene conversion [4].

35.1.2.2 Supercritical State

Increasing pressure of a medium can force two separate phases to collapse into one. When pressure and temperature are above the critical point, the phase boundary is indistinguishable, and the fluid density is somewhere between that of gas and liquid at standard conditions. The diffusion rate of molecules in their supercritical state is comparable with that of the gas phase. Dense medium and enhanced molecular diffusion are favorable for chemical reactions since molecular collisions and consequently the reaction rate can be enhanced. Also, high fluid density promotes heat transfer, which is helpful for heat removal. These characteristics enable the use of supercritical fluids as a unique solvent or even as reactant (e.g. CO_2 conversion under a supercritical state of CO_2) as frequently applied in hydrogenation and water oxidation under supercritical conditions.

35.1.3 Mass Transfer and Kinetics

In a heterogeneous reaction involving a solid catalyst, the chemical reaction is merely one of several steps. The required interaction of reactants with catalyst surface highlights the importance of mass transport, which can affect the progress of reaction apart from reaction equilibrium and kinetics. For instance, convective and diffusive transport of reactants from the bulk fluid into catalyst pore and adsorption, desorption, and surface diffusion of reactants on catalyst surface can play decisive roles in defining the reaction rate. Also, high-pressure conditions are known to induce higher coverage of reactants and consequently the concentration of surface intermediates at elevated pressures. This generally leads to enhanced surface reaction rate, favorable for overall reaction kinetics [9, 10].

35.1.3.1 Molecular Diffusion

The aforementioned enhanced molecular diffusion under pressurized conditions, e.g. supercritical states, can be advantageous for reaction kinetics. In reality, the effect of pressure on diffusion rate in heterogeneous catalysis is complex since reaction pressure affects more than one parameter at the same time (e.g. viscosity of fluids). Another important nature of diffusion is that the driving force for molecular diffusion is not only intrinsic but also extrinsic. For example, the concentration gradient, the main driving force for molecular diffusion as illustrated by Fick's law, is related to partial pressure in the gas phase. Thus, the partial pressure driving force can be conveniently generated and tuned in case of gas-phase reaction by creating pressure difference in a reaction system. A convenient approach that applies such pressure gradient is membrane reactor, where a product could be continuously separated by a membrane, and thus reaction equilibrium is not limited by one-phase thermodynamics.

35.1.3.2 Multiphase Reaction

In case of a multiphase reaction taking place in a trickle-bed or slurry reactor, dissolution and mass transport of reactant from the gas phase into the liquid phase are often enhanced by increasing reaction pressure (Henry's law). This was applied for high-pressure hydrotreating processes where hydrogen transfer into liquid heavy oil and the reaction occurs on the solid surface of the catalyst after the diffusion of molecular hydrogen through the liquid medium.

35.1.4 Process Efficiency and Economy

High-pressure operation effectively decreases reactor volume and thus the amount of catalyst required, which could offer economic advantages despite obvious higher capital cost for thicker-wall reactor and compression. Another note is that an industrial catalytic process hardly consists of only one reaction unit. Usually, a product from the unit will be treated and/or processed by the preceding or the subsequent units for further reaction and separation operated at different temperature and pressure. For high-pressure reactions, prior compression of reactants is required, and it is one of the most expensive operations in a catalytic process and consumes a large amount of energy. Therefore, it is imperative to evaluate the process efficiency and economy not only on the basis of the catalytic reactor but also as the whole process. For example, in the steam reforming of methane to produce syngas, the steam reformer is operated at a relatively high pressure (30 bar), although the reaction is not thermodynamically favorable at higher pressures and it requires higher reaction temperature and more excess steam to counterbalance the negative effect on the reaction equilibrium [4]. Still, most applications of syngas such as methanol and ammonia synthesis require syngas at high pressure (50–150 bar) or even higher [1, 3], and operation of the steam reforming at high pressure makes sense as the whole process.

Furthermore, by taking advantage of the thermodynamic stability of the reaction products, typical compression could be avoided. An example is the generation of hydrogen from formic acid. Using a nano-Pd heterogeneous catalyst, the system can generate high-pressure gases containing only hydrogen and carbon dioxide at over 360 bar under mild temperature (80 °C) [11]. This system is suitable for the practical application of fuel cell vehicles (FCVs), which requires hydrogen at high pressure over 350 bar [12].

35.2 State-of-the-Art Application of High Pressure in Heterogeneous Catalysis

35.2.1 Boosting CO_2 Conversion and Surpassing One-Phase Chemical Equilibrium by *In situ* Phase Separation at High-Pressure Reaction Condition

Hydrogenation of CO_2 allows the production of hydrocarbons and oxygenates such as methane, methanol, dimethyl ether, and formic acid. Production of these chemicals will play pivotal roles in energy storage technology and utilization of CO_2 known as the most emitted greenhouse gas. Particularly, methanol has gained increasing attention due to the ease in its transport and safety compared with hydrogen as a chemical energy carrier [13]. Methanol can be conveniently handled by the existent liquid fuels infrastructure and serve as a starting material (C1 feedstock) for producing various chemicals (olefin and gasoline, to name a few) [14]. Methanol production utilizing waste CO_2 (e.g. contained in flue gases) and H_2 produced by water electrolysis sourced from natural and renewable energy, such as wind and solar power, is a sustainable path for carbon recycling urged to be implemented and spread in near future.

Methanol synthesis by hydrogenation of CO_2 (i) is more complex compared with that from synthesis gas (ii) due to the fact that reverse water–gas shift (RWGS) reaction (35.3) takes place in parallel with methanol synthesis reaction (35.1):

$$CO_2 + 3H_2 \rightleftharpoons CH_3OH + H_2O \tag{35.1}$$

$$CO + 2H_2 \rightleftharpoons CH_3OH \tag{35.2}$$

$$CO_2 + H_2 \rightleftharpoons CO + H_2O \tag{35.3}$$

According to the stoichiometry of these reactions, the number of molecules reduces for the methanol synthesis reactions (35.1) and (35.2), whereas the number remains the same for the RWGS reaction. This means that increasing reaction pressure can shift the chemical equilibrium toward methanol formation since the pressure dependency of the RWGS reaction is negligible [15]. In addition, the methanol synthesis reactions (35.1) and (35.2) are exothermic, while RWGS reaction is endothermic; thus lower temperature should be favored for high-yield methanol synthesis. This thermodynamic nature of the methanol synthesis reaction from CO_2 and H_2 is illustrated by the equilibrium CO_2 conversion and methanol selectivity at the stoichiometric ratio ($CO_2 : H_2 = 1:3$) in the temperature range where generally reported catalysts show activity [16]. Obviously, the reaction pressure has a drastic influence on the yield of methanol (molar basis – not time yield), showing a clear advantage of operating the reaction at higher pressures. At 100 and 200 bar, there is a steep increase in CO_2 conversion toward lower reaction temperature (Figure 35.2, gray line, at c. 170 and 240 °C, respectively), and this is due to the condensation of the liquid products (methanol and water) below the transition temperatures. This condensation removes the products from the gas-phase reaction and shifts

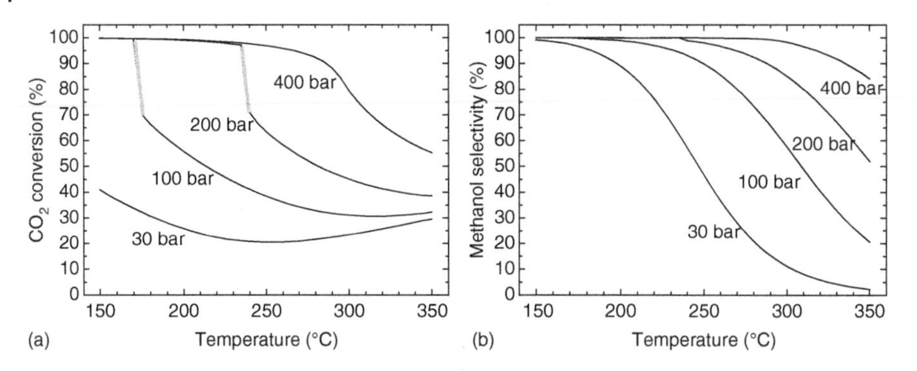

Figure 35.2 Equilibrium CO_2 conversion (a) and methanol selectivity (b) at different temperatures in stoichiometric CO_2 hydrogenation at 30, 100, 200, and 400 bar. Only CO was considered as another carbon-containing chemical produced by the reaction. The thick gray line in CO_2 conversion at 100 and 200 bar shows the phase transition due to liquid-phase condensation at lower temperatures. Source: Álvarez et al. 2017 [16]. Adapted with permission of American Chemistry Society.

forward the equilibrium of Reaction (35.1); thus >95% methanol yield is within reach in theory.

The phase separation of product stream was demonstrated at 150 bar and 206 °C using a high-pressure view cell by Kommoß et al. [17]. In this cell, a mixture of H_2, CO_2, CH_3OH, and H_2O was prepared to simulate a product stream outlet from a reactor at 50% CO_2 conversion and 67% CH_3OH selectivity. This suggested a possible *in situ* phase separation of the products within the reactor. At the same time, the results imply the necessity to perform the reaction at a kinetically unfavorable temperature lower than 206 °C to allow the condensation of methanol and water from the gas phase and eventually shift the equilibrium toward higher methanol yield. It should be noted that the phase separation from methanol synthesis from synthesis gas feed has also been demonstrated for the first time under reaction conditions by van Bennekom et al. [18]. The liquid condensation was observed at 200 °C and 200 bar from a view cell, as shown in Figure 35.3a. When the phase condensation takes place *in situ* in CO_2 hydrogenation, one can expect gradual phase separation along the axial direction of the catalytic reactor (Figure 35.3b).

The optimal temperature for this reaction is reported to be at 260–280 °C to maximize CO_2 conversion, CH_3OH selectivity, and thus methanol yield. However, as evident from Figure 35.2, it is not possible to take advantage of both phase separation and optimal reaction rate at 100 bar, since the phase condensation temperature (170 °C) is lower than optimal reaction temperature (260–280 °C). Increasing liquid condensation temperature can be achieved by increasing reaction pressure or rather such a pressure increase induces the creation of one dense phase as indicated by the smoothened CO_2 conversion curve at 400 bar (Figure 35.2a). Under such very high-pressure conditions, it is possible to achieve c. 90% methanol yield under continuous operation with unprecedentedly high weight time yield (gram of methanol produced per gram of catalyst per hour) when mass transfer limitations are minimized [19].

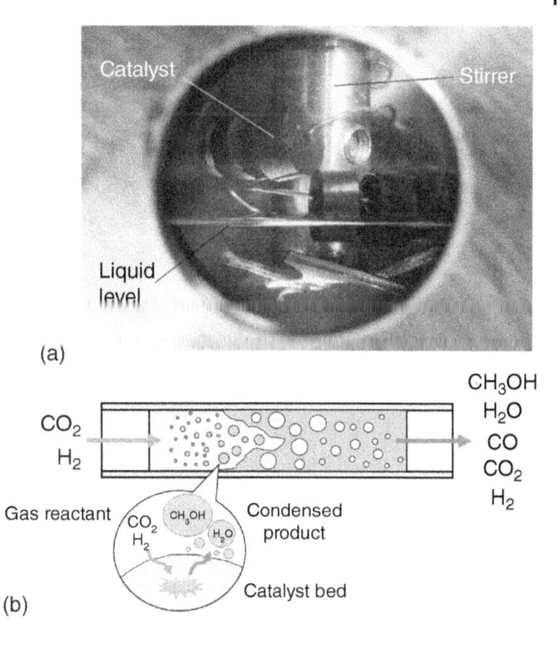

Figure 35.3 (a) Liquid product formation during methanol synthesis from syngas over a commercial $Cu/ZnO/Al_2O_3$ catalyst. $P = 200$ bar, $T = 200\,°C$, syngas: $H_2/CO/CO_2 = 0.70/0.28/0.02$. (b) The proposed *in situ* phase separation within the catalyst bed inside the reactor. Source: van Bennekom et al. 2013 [18]. Reproduced with permission of Elsevier.

35.2.2 Exploitation of Supercritical Fluid Properties for Catalytic Reactions

In industrial processes, supercritical fluids have been used as solvent, reactant, and catalyst itself in catalytic reactions to play vital roles. The unique properties of supercritical fluids can offer advantages for heterogeneously catalyzed reactions. As discussed above, supercritical fluids exhibit a high diffusion rate similar to the gas phase while providing a high collision rate due to its high density. The mixed and distinct characteristics are due to the molecular heterogeneity of supercritical fluids resulting from the formation of local clusters, which can be observed and simulated for supercritical CO_2 and H_2O [20, 21].

The polarity of fluids often changes under supercritical state compared with that under subcritical state due to a considerable decrease in dielectric constant. A highly polar water exhibits almost nonpolarity at supercritical condition (above 374 °C and 221 bar), which makes it suitable for dissolving organic molecules. For this reason, many supercritical reactions dealing with organic matters are considered a green process since water can be used instead of organic solvents. Complete miscibility between water and organic compounds also facilitates a single-phase reaction without mass transport limitation at phase boundaries. This advantage is now commercially utilized for the treatment of water containing hazardous waste [22]. Moreover, the acidity (and basicity) of water is greatly increased due to an increase in the dissociation constant, which can be exploited for acid- and base-catalyzed reactions. The combination of the increased acidity and organic solubility of water can be applied for acid-catalyzed reactions of organic compounds such as hydrolysis of esters for biodiesel production, Friedel–Crafts alkylation for aromatic substitution, and alcohol dehydration to olefin [23]. Catalysts are also employed to circumvent the requirement of harsh reaction conditions and improve the economics of the supercritical process, although most of them are catalyzed homogenously. The state-of-the-art process in this category is Hydrofaction™ developed by Steeper Energy that combines supercritical water chemistry (390–420 °C and 300–350 bar) and homogeneous catalysis to produce renewable crude oil from woody biomass [24]. On the other hand, supercritical water oxidation and metal-catalyzed heterogeneous reaction are considered the last option because of deactivation and metal leaching problem [23, 25]. More efforts are needed to improve the stability of catalysts by preventing the active site leaching and by improving sulfur tolerance as one of the major applications of supercritical water, biomass conversion, typically deals with sulfur-containing organic substances that often poison catalysts.

A relatively new heterogeneous catalytic process employs supercritical properties of water for biomass conversion to produce renewable fuels. A notable example is supercritical water gasification (SCWG) that produces high-quality biofuel such as hydrogen and methane from any organic compound. This technology provided a high thermal efficiency (70–77%) with a short residence time (<30 minutes) [26]. Supercritical water becomes weakly polar solvent allowing inorganic salt precipitation prior to gasification [27]. On the other hand, tar and coke formation is suppressed because of their high solubility in supercritical water [28]. However, the SCWG is usually performed at a high temperature (600–700 °C) and pressure (280 bar) to achieve high carbon conversion efficiency, and the main product is hydrogen [29]. By using supported metal catalysts (e.g. Ru, Rh, Ni, Cr, Co, Zn, Pd, Pt, Ti, and Mo), methane (synthetic natural gas [SNG]) instead of hydrogen can be produced, and this is called catalytic supercritical water gasification (CSCWG). It can be operated at a moderate temperature (374–500 °C) with a high gasification rate and high methane selectivity [30].

In recent years, CSCWG has shown a great potential for the conversion of wet biomass (e.g. microalgae, manure, and sewage sludge), since no drying is required unlike pyrolysis process [31]. Biofuel production from microalgae has attracted more than 150 companies (e.g. ExxonMobil, Dow Chemical, and Shell) [32] because it can be easily cultivated and does not affect the food supply. However, CSCWG process is still not commercially available because of technological challenges. The main challenge to overcome is, once again, fast deactivation of the catalyst by minerals and sulfur-containing in biomass. A new development of a continuous CSCWG process was proposed by Peng et al. [33], in which microalgae were gasified at 400 °C and 280 bar into methane-rich gas (55–60 vol%) using Ru/C catalyst. Long-term stability of 55 hours was achieved by effective salt separator and sulfur removal units. Salt separator employed supercritical water for precipitation of inorganic sulfur and minerals, and eventually, it produced nutrient-rich brine (N, K, S, P, and Na) for microalgae cultivation [34]. Subsequently, the organic sulfurs from salt-free feed stream were chemically absorbed by a zinc oxide bed.

The reaction mechanism of CSCWG is not fully revealed due to the complexity of biomass conversion chemistry where many reaction paths are present. Despite the active debate, free radical mechanisms are suggested to be plausible, and three reaction pathways are proposed to be responsible [35].

Steam reforming of biomass

$$C_xH_yO_z + (x - z)H_2O \rightarrow xCO + (y/2 + x - z)H_2 \tag{35.4}$$

$$C_xH_yO_z + (2x - z)H_2O \rightarrow xCO_2 + (y/2 + 2x - z)H_2 \tag{35.5}$$

Water–gas shift

$$CO + H_2O \rightleftharpoons CO_2 + H_2 \tag{35.6}$$

Methanation

$$CO + 3H_2 \rightleftharpoons CH_4 + H_2O \tag{35.7}$$

$$CO_2 + 4H_2 \rightleftharpoons CH_4 + 2H_2O \tag{35.8}$$

Thermodynamically, increasing the reaction temperature promotes hydrogen production via the endothermic steam reforming reactions (35.4) and (35.5) and suppresses water–gas shift (35.6) and methanation (35.7) reactions. On the other hand, increasing the reaction pressure is favorable for methanation reactions, even in the supercritical state. The increase in pressure can significantly compensate for high reaction temperature and help maintain the high rates for gasification and methane production [35].

35.2.3 A Greener High-Pressure System Using Microchannel Reactor

Steam reforming of methane ((35.9), the reverse reaction of (35.7)) is one of the most well-established technologies to produce hydrogen and synthesis gas for various industrial chemical processes such as ammonia synthesis, methanol synthesis, Fischer–Tropsch synthesis, and petroleum hydrotreating. In a steam reformer, an excess of steam is required to suppress carbon formation, which simultaneously promotes water–gas shift reaction (35.6), producing more hydrogen and carbon dioxide. By changing the steam-to-carbon ratio of the feed, the H_2/CO ratio of the product can be adjusted for various applications. As mentioned previously, steam reforming of methane is usually operated at high pressure to avoid compression step (or lowering the compression cost) of the syngas required by following process (e.g. ammonia or methanol synthesis), despite the thermodynamic disadvantage in terms of methane conversion and CO selectivity due to an increase in a number of moles of the reaction, as shown below:

$$CH_4 + H_2O \rightleftharpoons CO + 3H_2 \tag{35.9}$$

In order to compensate the decreased productivity due to elevated pressure, increased temperature up to 800–900 °C is required to drive this endothermic reaction forward [4]. Under such a harsh condition with high energy demands to operate the reaction, reactor efficiency becomes extremely important to achieve a stable running process. Among others, various types of microchannel reactors

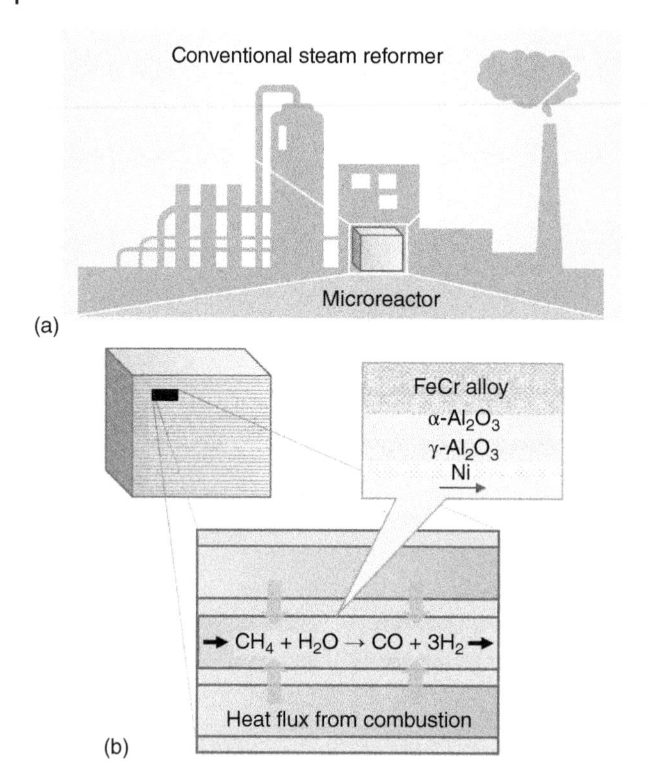

Figure 35.4 The depiction of process intensification by using microreactor (a) and the scheme of microreactor and catalyst layers for steam reforming of methane at high pressure (b).

such as coated-wall, packed-bed, structured-catalyst, and membrane microreactors were developed. A microreactor that provides a higher surface to volume because of its small passages (0.1–0.3 mm range) can enhance the heat and mass transfer compared with the conventional reactors that are large pipes and vessels. This accelerates the reactions rates up to 1000 times higher than using the conventional reactor, intensifying the efficiency of methane steam reforming. Figure 35.4 [36] shows an example of the commercialized Velocys methane steam reformer that achieves 60 times more intensive volumetric heat flux than the traditional steam reformer. Through the integration of multiple microreactor units, the space–time yield of such process is increased at the level of the industrial requirement [37]. The recent advances in the modular integration and process intensification of microreactors are expected not only to substantially increase the hydrogen production capacity but also to reduce capital and operating cost due to less constructing material and energy usage.

Performing methane steam reforming at high pressure can further intensify the process by decreasing the size of the reactor and increasing space–time yield. However, only a few studies have been reported on the catalytic behavior in the reaction using microreactors at high pressure. Zhang et al. performed high-pressure steam reforming up to 20 bar using a coated-wall microreactor [38], which is one of the most common types of microreactors. This type of microreactor is favorable for large-scale production than the others, especially at high pressure, because of excellent temperature control for a highly endothermic (and exothermic) reaction and negligible pressure drop. The coating of ceramic support onto the metal reactor wall is an essential step because it determines the strength of adherence between the reactor wall and catalyst coating. In this case, a strongly adhesive ceramic–metal interface was prepared by spraying half-melted α-Al_2O_3 onto FeCrAl alloy substrate. The γ-Al_2O_3 support layer was then coated on that substrate by dip coating followed by impregnation of the Ni active component on support. This microreactor performed very well even at a relatively higher gas hourly space velocity (GHSV $\sim$ 40 000 h^{-1}) than the commercial tubular reformer (2000–8000 h^{-1}) [39] and achieved nearly 90% CH_4 conversion close to the thermodynamic limit at industrial relevant conditions (steam to carbon of 3, 800–900 °C, and 5–20 bar). The high-pressure operation also increases the actual residence time of the reactants in the reactor by gas compression. This indicates that microreactor can be of great potential for industrial application when scale-up challenges such as scaling volume and production capacity without losing its key characteristics are overcome. Since thousands to tens of thousands of reactors can be required for high throughput, the uniform and precise fluid distribution and control in each reactor are required to retain the hydrodynamics and transfer characteristics, which may significantly impact the cost of scale-up [40].

35.2.4 Surpassing Chemical Equilibrium by *In situ* Product Separation Using a High-Pressure Membrane Reactor

In the industrial hydrogen production, especially for petroleum hydrotreating, a water–gas shift reaction (35.7) is usually combined with a steam reformer to achieve maximum hydrogen production efficiency by converting the remaining CO from synthesis gas (H_2, CO, and CO_2) [4]. However, the CO conversion is thermodynamically limited due to the exothermicity of the reaction, especially at kinetically favored high temperatures. In order to achieve a complete CO conversion, many efforts have been made to develop a membrane reactor, where H_2 can be removed directly from the reactor through the H_2-permselective membrane (Figure 35.5) [41]. High pressure plays an important role in mass transfer across the membrane as the driving force. Using such a membrane reactor, the reaction is driven more toward the product side, while H_2 and CO_2 can also be separated simultaneously. It should be noted that the conventional H_2 and CO_2 separation processes rely on CO_2 absorption using an organic solvent, which is not eco-friendly, energy efficient, or economical. Therefore, employing membrane reactors seems to be a promising emerging option for water–gas shift reaction.

For a membrane reactor, pressure difference and thus elevated pressure are essential for an effective H_2 transport through the membrane, which determines the equilibrium CO conversion and purity of CO_2 product [42]. Moreover, operating at high pressure drastically reduces the energy used for compression of CO_2 before the CO_2 sequestration or conversion process. The main challenges in implementing membrane reactor are the permselectivity and permeability for H_2 and thermochemical stability of the membrane itself. A good membrane must

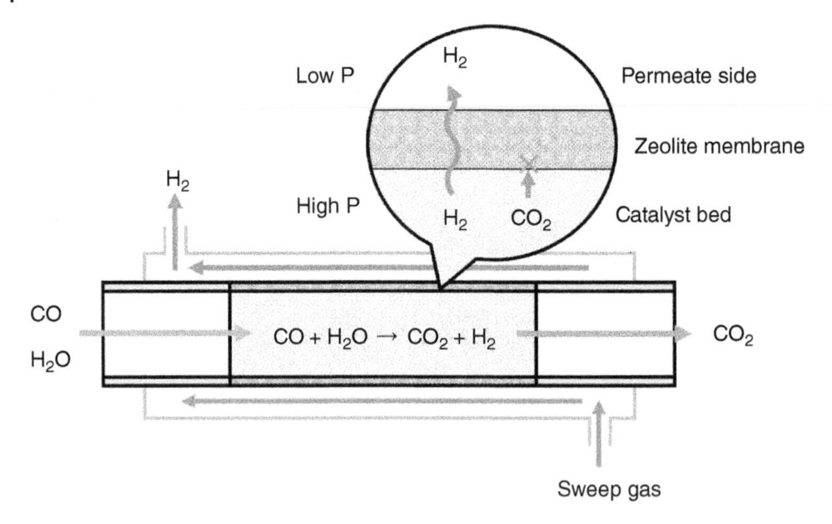

Figure 35.5 The schematic showing a shell-and-tube membrane with countercurrent flow configuration and showing the mass transfer of H_2 across the membrane.

provide high H_2 selectivity while withstanding harsh reaction conditions such as high temperature, pressure, and poisoning. For example, a dense Pd-alloy membrane is the most commonly used membrane for high-temperature water–gas shift reaction, but it is susceptible to sulfur poisoning and CO_2 embrittlement (also H_2 embrittlement in case of pure Pd membrane) [43]. MFI zeolite membrane is a strong candidate for high temperature and pressure conditions, thanks to the crystalline structure of the zeolite that provides thermal and chemical resistance. Typically, the intracrystalline nanoporous structure of zeolite (effective diameter ~0.56 nm) is large enough for small gas molecules involved in water–gas shift reaction to travel and offers high H_2 permeance, but with limited H_2 permselectivity over CO_2 [44]. The zeolitic channels near the membrane surface can be modified further with silica deposition to decrease the internal pore diameter down to <0.36 nm, which improves H_2 permselectivity and still maintains high H_2 permeance [45].

No studies have achieved CO conversion and H_2 recovery more than 99.5% using a single-stage membrane reactor until Arvanitis et al. [46] have succeeded in fabricating an MFI zeolite membrane with the aforementioned procedure. The water–gas shift reaction performed in the tubular zeolite membrane reactor packed with an iron oxide-based catalyst (Fe_2O_3–CeO_2–CrO_3) has set the first achievement for the nearly complete CO conversion (>99.9%) and total H_2 recovery (>99.9%) under the practical condition (500 °C and 20 bar) and reasonable gas hourly space velocity (15 000 h^{-1}). This CO conversion obtained from the membrane reactor exceeded the thermodynamic limit (92.5%). This is obviously impossible for a conventional fixed-bed reactor in the same condition. Although elevated pressure cannot alter equilibrium conversion due to the equal number of moles before and after the reaction, reaction pressure affects the contact time between gas and catalyst surface and resulted in enhanced CO conversion. Moreover, increasing reaction pressure within the membrane reactor will greatly enhance the driving force for H_2 diffusion across the membrane and result in much higher CO conversion. This membrane reactor was able to produce a high-pressure pure CO_2 retentate suitable for a carbon capture process, although there was a trade-off in H_2 purity due to the moderate H_2 permselectivity of zeolite membrane. However, that compromise is overcome by long-term stability up to 600 hours.

35.3 Concluding Remark

Despite the global trends to use milder pressure reaction conditions according to green chemistry principles, high-pressure technologies are still considered vital because of various unique advantages. The positive influences of high pressure on "thermodynamics" especially for "chemical equilibrium" and "phase behavior," as well as "reaction kinetics," in the bulk and on the catalyst surface are representative. High pressure also allows creating a driving force for molecular diffusion that can be used to enhance product separation and shift chemical equilibrium. Furthermore, a high-pressure condition facilitates the process intensification, reducing the size and increasing the efficiency of overall catalytic processes. Such prominent advantages that cannot be achieved at ambient pressure are expected to result in the innovative design of future greener catalytic processes.

References

1 Liu, H. (2014). Ammonia synthesis catalyst 100 years: practice, enlightenment and challenge. *Chin. J. Catal.* 35: 1619–1640. https://doi.org/10.1016/S1872-2067(14)60118-2.
2 Erisman, J.W., Sutton, M.A., Galloway, J. et al. (2008). How a century of ammonia synthesis changed the world. *Nat. Geosci.* 1: 636. https://doi.org/10.1038/ngeo325.
3 Sheldon, D. (2017). Methanol production – A technical history. *Johnson Matthey Technol. Rev.* 61: 172–182. https://doi.org/10.1595/205651317x695622.
4 Moulijn, J.A., Makkee, M., and van Diepen, A.E. (2013). *Chemical Process Technology*. Wiley https://books.google.es/books?id=7C0eQ1AiQRcC.
5 Demirors, M. (2011). The history of polyethylene. In: *100+ Years Plastics. Leo Baekeland and Beyond* (eds. E.T. Strom and S.C. Rasmussen), 115–145. https://doi.org/10.1021/bk-2011-1080.ch009.
6 Zhang, X., Kong, R.-M., Du, H. et al. (2018). Highly efficient electrochemical ammonia synthesis via nitrogen reduction reactions on a VN nanowire array under ambient conditions. *Chem. Commun.* 54: 5323–5325. https://doi.org/10.1039/C8CC00459E.
7 Wildfire, C., Abdelsayed, V., Shekhawat, D., and Spencer, M.J. (2018). Ambient pressure synthesis of ammonia using a microwave reactor. *Catal. Commun.* 115: 64–67. https://doi.org/10.1016/j.catcom.2018.07.010.

8 Van Tran, T., Le-Phuc, N., Nguyen, T.H. et al. (2018). Application of NaA membrane reactor for methanol synthesis in CO_2 hydrogenation at low pressure. *Int. J. Chem. Reactor Eng.* 16: 1–7. https://doi.org/10.1515/ijcre-2017-0046.

9 Corral-Pérez, J.J., Bansode, A., Praveen, C.S. et al. (2018). Decisive role of perimeter sites in silica-supported Ag nanoparticles in selective hydrogenation of CO_2 to methyl formate in the presence of methanol. *J. Am. Chem. Soc.* 140: 13884–13891. https://doi.org/10.1021/jacs.8b08505.

10 Bansode, A., Tidona, B., von Rohr, P.R., and Urakawa, A. (2013). Impact of K and Ba promoters on CO_2 hydrogenation over Cu/Al_2O_3 catalysts at high pressure. *Catal. Sci. Technol.* 3: 767–778. https://doi.org/10.1039/C2CY20604H.

11 Zhong, H., Iguchi, M., Song, F.-Z. et al. (2017). Automatic high-pressure hydrogen generation from formic acid in the presence of nano-Pd heterogeneous catalysts at mild temperatures. *Sustainable Energy Fuels* 1: 1049–1055. https://doi.org/10.1039/C7SE00131B.

12 Alazemi, J. and Andrews, J. (2015). Automotive hydrogen fuelling stations: an international review. *Renewable Sustainable Energy Rev.* 48: 483–499. https://doi.org/10.1016/j.rser.2015.03.085.

13 Olah, G.A., Goeppert, A., and Prakash, G.K.S. (2009). *Beyond Oil and Gas: The Methanol Economy*, 2e. Weinheim: Wiley-VCH Verlag GmbH & Co. KGaA https://doi.org/10.1002/9783527627806.

14 Bertau, M., Offermanns, H., Plass, L. et al. (2014). *Methanol: The Basic Chemical and Energy Feedstock of the Future: Asinger's Vision Today.* Heidelberg, New York, Dordrecht, London: Springer https://doi.org/10.1007/978-3-642-39709-7.

15 Tidona, B., Urakawa, A., and Rudolf von Rohr, P. (2013). High pressure plant for heterogeneous catalytic CO_2 hydrogenation reactions in a continuous flow microreactor. *Chem. Eng. Process. Process Intensif.* 65: 53–57. https://doi.org/10.1016/j.cep.2013.01.001.

16 Álvarez, A., Bansode, A., Urakawa, A. et al. (2017). Challenges in the greener production of formates/formic acid, methanol, and DME by heterogeneously catalyzed CO_2 hydrogenation processes. *Chem. Rev.* 117: 9804–9838. https://doi.org/10.1021/acs.chemrev.6b00816.

17 Kommoß, B., Klemenz, S., Schmitt, F. et al. (2017). Heterogeneously catalyzed hydrogenation of supercritical CO_2 to methanol. *Chem. Eng. Technol.* 40: 1907–1915. https://doi.org/10.1002/ceat.201600400.

18 van Bennekom, J.G., Venderbosch, R.H., Winkelman, J.G.M. et al. (2013). Methanol synthesis beyond chemical equilibrium. *Chem. Eng. Sci.* 87: 204–208. https://doi.org/10.1016/j.ces.2012.10.013.

19 Gaikwad, R., Bansode, A., and Urakawa, A. (2016). High-pressure advantages in stoichiometric hydrogenation of carbon dioxide to methanol. *J. Catal.* 343: 127–132. https://doi.org/10.1016/j.jcat.2016.02.005.

20 Sedunov, B. (2012). The analysis of the equilibrium cluster structure in supercritical carbon dioxide. *Am. J. Anal. Chem.* 03: 899–904. https://doi.org/10.4236/ajac.2012.312A119.

21 Kalinichev, A.G. and Churakov, S.V. (1999). Size and topology of molecular clusters in supercritical water: a molecular dynamics simulation. *Chem. Phys. Lett.* 302: 411–417. https://doi.org/10.1016/S0009-2614(99)00174-8.

22 Marrone, P.A. (2013). Supercritical water oxidation – current status of full-scale commercial activity for waste destruction. *J. Supercrit. Fluids* 79: 283–288. https://doi.org/10.1016/j.supflu.2012.12.020.

23 Savage, P.E. (1999). Organic chemical reactions in supercritical water. *Chem. Rev.* 99: 603–622. https://doi.org/10.1021/cr9700989.

24 Jensen, C.U., Rodriguez Guerrero, J.K., Karatzos, S. et al. (2017). Fundamentals of Hydrofaction™: renewable crude oil from woody biomass. *Biomass Convers. Biorefin.* 7: 495–509. https://doi.org/10.1007/s13399-017-0248-8.

25 Qian, L., Wang, S., Xu, D. et al. (2016). Treatment of municipal sewage sludge in supercritical water: a review. *Water Res.* 89: 118–131. https://doi.org/10.1016/j.watres.2015.11.047.

26 Gassner, M., Vogel, F., Heyen, G., and Maréchal, F. (2011). Optimal process design for the polygeneration of SNG, power and heat by hydrothermal gasification of waste biomass: thermo-economic process modelling and integration. *Energy Environ. Sci.* 4: 1726–1741. https://doi.org/10.1039/c0ee00629g.

27 Schubert, M., Aubert, J., Müller, J.B., and Vogel, F. (2012). Continuous salt precipitation and separation from supercritical water. Part 3: Interesting effects in processing type 2 salt mixtures. *J. Supercrit. Fluids* 61: 44–54. https://doi.org/10.1016/j.supflu.2011.08.011.

28 Kruse, A. (2008). Supercritical water gasification. *Biofuels, Bioprod. Biorefin.* 2: 415–437. https://doi.org/10.1002/bbb.93.

29 Guo, Y., Wang, S.Z., Xu, D.H. et al. (2010). Review of catalytic supercritical water gasification for hydrogen production from biomass. *Renewable Sustainable Energy Rev.* 14: 334–343. https://doi.org/10.1016/j.rser.2009.08.012.

30 Elliott, D.C. (2008). Catalytic hydrothermal gasification of biomass. *Biofuels, Bioprod. Biorefin.* 2: 254–265. https://doi.org/10.1002/bbb.74.

31 Nielsen, R.P., Olofsson, G., and Søgaard, E.G. (2012). CatLiq – high pressure and temperature catalytic conversion of biomass: the CatLiq technology in relation to other thermochemical conversion technologies. *Biomass Bioenergy* 39: 399–402. https://doi.org/10.1016/j.biombioe.2012.01.035.

32 Singh, J. and Gu, S. (2010). Commercialization potential of microalgae for biofuels production. *Renewable Sustainable Energy Rev.* 14: 2596–2610. https://doi.org/10.1016/j.rser.2010.06.014.

33 Peng, G., Vogel, F., Refardt, D., and Ludwig, C. (2017). Catalytic supercritical water gasification: continuous methanization of *Chlorella vulgaris*. *Ind. Eng. Chem. Res.* 56: 6256–6265. https://doi.org/10.1021/acs.iecr.7b00042.

34 Reimer, J., Peng, G., Viereck, S. et al. (2016). A novel salt separator for the supercritical water gasification of biomass. *J. Supercrit. Fluids* 117: 113–121. https://doi.org/10.1016/j.supflu.2016.06.009.

35 Duan, P.G., Li, S.C., Jiao, J.L. et al. (2018). Supercritical water gasification of microalgae over a two-component catalyst mixture. *Sci. Total Environ.* 630: 243–253. https://doi.org/10.1016/j.scitotenv.2018.02.226.

36 Lerou, J.J., Tonkovich, A.L., Silva, L. et al. (2010). Microchannel reactor architecture enables greener processes. *Chem. Eng. Sci.* 65: 380–385. https://doi.org/10.1016/j.ces.2009.07.020.

37 Tonkovich, A.Y., Perry, S., Wang, Y. et al. (2004). Microchannel process technology for compact methane steam reforming. *Chem. Eng. Sci.* 59: 4819–4824. https://doi.org/10.1016/j.ces.2004.07.098.

38 Zhang, N., Chen, X., Chu, B. et al. (2017). Catalytic performance of Ni catalyst for steam methane reforming in a micro-channel reactor at high pressure. *Chem. Eng. Process. Process Intensif.* 118: 19–25. https://doi.org/10.1016/j.cep.2017.04.015.

39 Murzin, D.Y. (2015). Chapter 5. Hydrogen and syngas generation. In: *Chemical Reaction Technology*. Berlin, München, Boston: De Gruyter https://doi.org/10.1515/9783110336443-007.

40 Marlin, D.S., Sarron, E., and Sigurbjörnsson, Ó. (2018). Process advantages of direct CO_2 to methanol synthesis. *Front. Chem.* 6: 1–8. https://doi.org/10.3389/fchem.2018.00446.

41 Voldsund, M., Jordal, K., and Anantharaman, R. (2016). Hydrogen production with CO_2 capture. *Int. J. Hydrogen Energy* 41: 4969–4992. https://doi.org/10.1016/j.ijhydene.2016.01.009.

42 Kim, S.J., Yang, S., Reddy, G.K. et al. (2013). Zeolite membrane reactor for high-temperature water-gas shift reaction: effects of membrane properties and operating conditions. *Energy Fuels* 27: 4471–4480. https://doi.org/10.1021/ef302014n.

43 Rahimpour, M.R., Samimi, F., Babapoor, A. et al. (2017). Palladium membranes applications in reaction systems for hydrogen separation and purification: a review. *Chem. Eng. Process. Process Intensif.* 121: 24–49. https://doi.org/10.1016/j.cep.2017.07.021.

44 Wang, H. and Lin, Y.S. (2012). Effects of water vapor on gas permeation and separation properties of MFI zeolite membranes at high temperatures. *AlChE J.* 58: 153–162. https://doi.org/10.1002/aic.12622.

45 Masuda, T., Fukumoto, N., Kitamura, M. et al. (2001). Modification of pore size of MFI-type zeolite by catalytic cracking of silane and application to preparation of H_2-separating zeolite membrane. *Microporous Mesoporous Mater.* 48: 239–245. https://doi.org/10.1016/S1387-1811(01)00358-4.

46 Arvanitis, A., Sun, X., Yang, S. et al. (2018). Approaching complete CO conversion and total H_2 recovery for water gas shift reaction in a high-temperature and high-pressure zeolite membrane reactor. *J. Membr. Sci.* 549: 575–580. https://doi.org/10.1016/j.memsci.2017.12.051.

36

Electro-, Photo-, and Photoelectro-chemical Reduction of CO_2

Jonathan Albo, Manuel Alvarez-Guerra, and Angel Irabien

University of Cantabria, School of Industrial Engineering and Telecommunications, Department of Chemical and Biomolecular Engineering, Avda. Los Castros s/n, 39005 Santander, Spain

36.1 Introduction

The large-scale consumption of fossil fuels for the production of energy has raised CO_2 levels from a preindustrial level of about 270 ppm to surpass the 400 ppm. According to International Energy Agency (IEA) statistics, oil, coal, and natural gas still represent 80.6% of the total energy supply in the world in 2015, while renewable energies only accounted for 10.2% [1]. The same report also indicated that CO_2 emissions have substantially grow accordingly, going from approximately 5500 $MTCO_2$ in 1973 to 14 500 $MTCO_2$ in 2015. The natural carbon cycle has been exceeded, and it is presently unable to keep up with the input of large amounts of anthropogenic CO_2 emitted, causing severe environmental problems such as the undesirable effects of global warming. Renewable energies, and, in particular, wind and solar energy, are becoming more technically and economically feasible to replace fossil fuels, but their implementation is being slower than expected, and the world energy supply will still rely on fossil fuels in the next decades. In this scenario, the development of methodologies to reduce CO_2 emissions needs to be a priority in the political agenda worldwide in order to advance toward a carbon-neutral energy cycle.

A variety of strategies may alleviate the CO_2 issue via improving the combustion efficiency of fossil fuels or exploring clean and renewable energy sources, but these strategies should be really hard pressed to replace fossil fuels, at least in the short term. A better efficiency in energy utilization would also help to save fossil fuels and reduce the emissions of CO_2. At the same time, great efforts are being undertaken to develop carbon capture and storage (CCS), as one of the technologies to handle large quantities of CO_2 emissions, where CO_2 capture (i.e. absorption, adsorption, or membrane separation) seems to be the bottleneck step where the efforts have to be applied [2]. Other methods include direct CO_2 capture from air, which could be important to address the dispersed emissions in the atmosphere from cars, domestic heating, or trains [3]. Limitations to CCS technologies, however, include high costs and energy requirements, the uncertainty on the permanence of stored CO_2 in storage sites or the impact of

Heterogeneous Catalysts: Advanced Design, Characterization and Applications, First Edition.
Edited by Wey Yang Teoh, Atsushi Urakawa, Yun Hau Ng, and Patrick Sit.

large volumes of CO_2 on natural systems. Additionally, processes for converting captured CO_2 into useful and valuable products, in the so-called carbon capture and utilization (CCU), are being developed in parallel. In view of the vastness of CO_2 supplied in the atmosphere, it makes sense to consider CO_2 a resource rather than a waste for its direct use or its transformation into new products. It is also being recognized that exploiting part of the captured CO_2 to generate value from it can complement its storage via CCS, thus contributing to meet the carbon reduction targets.

The utilization of CO_2 can be grouped into three main categories: (i) technological utilization, which may not require the conversion of CO_2; (ii) enhanced biological utilization; and (iii) production of chemicals and fuels [4]. Firstly, CO_2 can be directly used in numerous technological applications. For several decades, CO_2 has been used for enhanced oil recovery (EOR). In this process, a percentage of the CO_2 stays in the reservoir, and EOR is therefore considered as a form of long-term storage. CO_2 can be also used as a solvent for supercritical extraction of components such as pigments, oils, or caffeine. CO_2 finds use in the food industry, including carbonated beverages and food packaging, or in water treatment to reduce pH levels in alkaline water. Secondly, CO_2 is a key component of photosynthesis, and by adding additional CO_2 to greenhouses, the growth of plants can be enhanced. Lastly, CO_2 can be chemically converted into a variety of value-added chemicals, including salicylic acid (an important active metabolite of aspirin), urea, and different polymers. In fact, CO_2 can be activated and chemically converted through different approaches such as thermochemical and biochemical methods and more innovative technologies such as the electrochemical, photochemical, or photoelectrochemical reduction [5]. These latter innovative technologies are appealing since they could enable an economically competitive industrial production of CO_2-based chemicals by using renewable energy, contributing to rebalance the carbon cycle.

The final oxidation state of the carbon atom in the value-added chemicals synthesized with these technological solutions is determined by the specific reaction pathway and the number and rates of electrons exchanged. Table 36.1 shows the sequence of reactions and the corresponding redox potentials for CO_2 to commonly reported products (e.g. CO, HCOOH, CH_3OH, and CH_4) [5].

Since CO_2 is a thermodynamically stable molecule, its multistep reduction via electro-, photo-, and photoelectro-chemistry is challenging and confronts

Table 36.1 Redox potentials for CO_2 reduction.

Reaction	E^0 (V vs. NHE) at pH = 7
$CO_2 + e^- \rightarrow CO_2^{\cdot-}$	−1.9
$CO_2 + 2H^+ + 2e^- \rightarrow HCOOH$	−0.61
$CO_2 + 2H^+ + 2e^- \rightarrow CO + H_2O$	−0.52
$CO_2 + 4H^+ + 4e^- \rightarrow HCHO + H_2O$	−0.51
$CO_2 + 6H^+ + 6e^- \rightarrow CH_3OH + H_2O$	−0.38
$CO_2 + 8H^+ + 8e^- \rightarrow CH_4 + 2H_2O$	−0.24
$2H^+ + 2e^- \rightarrow H_2$	−0.42

many fundamental technical hurdles. The kinetics for CO_2 reduction are also, in general, more sluggish than the thermodynamically favorable two-electron H_2 evolution reaction (HER), which competes with CO_2 reduction. In any case, the achievements so far are exciting, and the authors believe that with continued research efforts, these innovative technologies for CO_2 conversion might become technically and economically feasible in the near future.

This chapter provides the basic principles of the electro-, photo-, and photoelectro-chemical processes for CO_2 conversion, taking mainly into consideration the applied catalytic materials and cell geometries. The key challenges to be faced are envisaged, proposing future steps to be undertaken to advance as fast as possible toward realistic solutions for CO_2 utilization.

36.2 Fundamentals

36.2.1 Redox Processes

There are many reactions in which it is convenient to regard electrons as participants in the transformation of reactants into products. Reactions of this kind are known generally as electron transfer reactions, or more commonly as oxidation–reduction reactions [6]. In order to convert CO_2 into useful products, CO_2 must be "reduced," i.e. it must gain electrodes from another species (which is "oxidized"), in a transactional process of "reduction–oxidation," commonly known as "redox."

The equipment for carrying out a redox reaction is normally called electrochemical reactor or electrochemical cell. These systems are most generally defined as two electrodes separated by at least one electrolyte phase [7]. When a net reaction proceeds in an electrochemical cell, oxidation occurs at one electrode (the anode), and reduction takes place at the other electrode (the cathode). We can think of the cell as consisting of two half-cells joined together by an external circuit through which electrons flow and an internal pathway that allows ions to migrate between them so as to preserve electroneutrality [6]. The redox process occurs at the same active surface in photocatalysis as explained later in the chapter, but for the sake of clarity, we focus at this point in electrochemical cells.

The thermodynamics of electrochemical cells is treated in all textbooks of physical chemistry as are the conventions. The equilibrium (or reversible) cell potential can be calculated by subtracting the equilibrium potential of the anode (E_e^A) from that of the cathode (E_e^C), and this is related to the free energy of the overall cell reaction (ΔG) by the well-known equation [8]:

$$\Delta G = -nFE_{\mathrm{Cell}}^{e} = -nF(E_e^C - E_e^A) \tag{36.1}$$

where n is the number of moles of electrons per mole reacted and F is the Faraday constant (i.e. the charge on a mole of electrons, 96 485 C·mol^{-1}).

Therefore, obtaining the equilibrium cell potential for the redox reaction, we can then calculate the free energy change associated with the redox reaction. If ΔG is negative, this tells us that the reaction is favorable, while if it is positive, it is thermodynamically unfavorable and can only occur when (electrical) energy is supplied. The equilibrium potentials for some CO_2 reduction reactions are listed in Table 36.1, showing us that several different products can be obtained as reduced forms of CO_2.

In any case, this is a thermodynamic discussion, but it does not consider the rate of chemical change (i.e. the current that will flow). The reactivity of CO_2 is low because it is the most stable carbon-based structure in nature, and an input of energy must be supplied to reduce CO_2 and convert it into reduced, useful products. Much larger voltages than the equilibrium cell potentials have to be applied in practice across the two electrodes to drive the conversion. The rate of chemical conversion will depend on the kinetics of the two electrode reactions, and an overpotential (η) is necessary to carry out the reaction at a required rate (measured by the current density, j, the current per unit area of electrode surface, as explained later). Moreover, an input of energy is also necessary to drive the migration of ions between the electrodes through the cell, which leads to a potential drop iR_{Cell} (where R_{Cell} represents the internal resistance of the cell, which depends on the electrolyte properties, the configuration of the electrodes, and the design of the cell). Therefore, the cell voltage required to observe a current i in a real cell can be expressed by

$$E_{\mathrm{Cell}} = E_e^C - E_e^A - \sum|\eta| - iR_{\mathrm{Cell}} \tag{36.2}$$

Both the η and iR_{Cell} terms increase with current density and may be regarded as inefficiencies whereby electrical energy is converted into heat [8].

36.2.2 Assessment of Reaction Performance: Figures of Merit

Several different figures of merit are commonly used to assess the performance of electrocatalytic CO_2 reduction processes:

(a) *Rate of product formation*: The amount of the desired product from CO_2 reduction obtained per unit of cathode area and unit of time (e.g. expressed in units of $\mathrm{mol{\cdot}m^{-2}{\cdot}min^{-1}}$) is sometimes used as a useful figure for quantifying the rate at which the product can be obtained in the reduction process.

(b) *Current density (j, usually expressed in $mA{\cdot}cm^{-2}$)*: The current density is defined as the electric current flow divided by the area of the electrode (usually the geometric area). It measures the rate of the conversion, and a high j as possible is desirable to minimize the area of electrode (and hence the size of the reactor) needed for a desired production rate.

(c) *Faradaic efficiency (FE, %) for a certain product*: It is the yield based on the electrical charge passed, i.e. the % of the total charge supplied that is used in forming the desired CO_2 reduction product, so ideally it should be as close as possible to 100%. Also known as "current efficiency," FE measures the selectivity of the process of reducing CO_2 to a certain product, and it can be calculated using the equation:

$$\mathrm{FE} = \frac{zFm}{Q} \times 100 \tag{36.3}$$

where z is the number of electrons exchanged (e.g. $z = 2$ for the reduction of CO_2 to HCOOH or CO and $z = 6$ for the reduction to CH_3OH; see Table 36.1), F is the Faraday constant, m is the number of moles of the certain product, and Q is the total charge passed.

(d) *Energetic efficiency (EE, %)*: It is a measure of the overall energy utilization toward the desired CO_2 reduction products, which can be defined by Eq. (36.4):

$$EE = \frac{E^{\circ}}{E} \times 100 \quad (36.4)$$

where E° is the theoretical equilibrium cell potential and E represents the real potential applied in the process (which includes all the cell overpotentials). The EE describes the ratio between energy stored in the desired CO_2 reduction product and the input energy needed to produce it, so a high EE is desirable since this means a small energy penalty of the process.

36.2.3 Role of Heterogeneous Catalysts

The electrochemical reaction occurs at the interface between the electrode (an electronic conductor) and the electrolyte (an ionic conductor). It is composed of a series of steps, including (i) the approach of the reactant species to the electrode surface, (ii) the reaction via heterogeneous electron transfer across the interface (actual electrochemical step), and (iii) the transfer of the reduced CO_2 product species away from the reaction area into bulk solution [9].

Most of the proposed mechanisms for direct CO_2 reduction on an electrode involve the initial formation of intermediate species. That species is usually called $CO_2^{\cdot-}$, although the exact structure of the species is unknown [10]. Hence, $CO_2^{\cdot-}$ does not necessarily represent a bare $CO_2^{\cdot-}$ anion, but it denotes whatever intermediate species is formed when an electron is transferred to a CO_2 molecule. The detailed mechanistic pathways for each product are not clear at present, and in many cases, several different schemes have been proposed [9].

We already mentioned that the actual potentials for CO_2 reduction are much more negative than the values estimated from the thermodynamic data. Now we may understand the reason: it costs energy to create the $CO_2^{\cdot-}$ intermediate species (which requires highly negative potential for the formation), and extra energy has to be provided to get the reaction to occur. Figure 36.1 represents the qualitative reaction scheme for CO_2 conversion. As shown in Figure 36.1, the overpotential required for the CO_2 reduction reaction to proceed is much lower when catalysts are present ("η_{cat}") than when no catalyst is involved in the reaction ("$\eta_{no\ cat}$"). Accordingly, catalysts play a key role by lowering the energy of the intermediates and therefore improving the EE of the CO_2 conversion.

Catalytic strategies thus aim to avoid or at least reduce the penalty of the high-energy intermediate $CO_2^{\cdot-}$ for obtaining the various products from reduction of CO_2 at lesser energetic costs. Energetically more favorable strategies

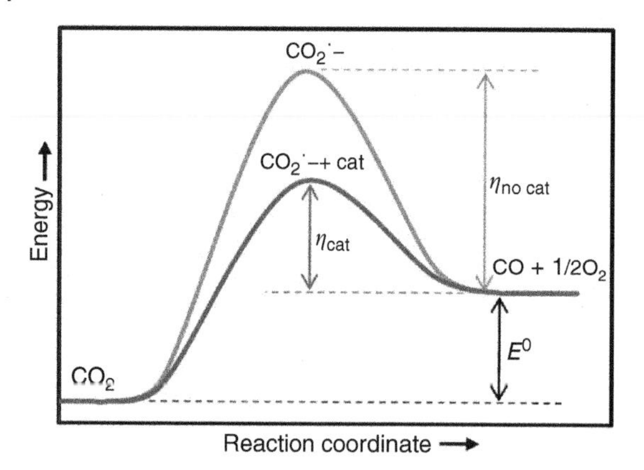

Figure 36.1 Qualitative reaction scheme for CO_2 conversion. Source: Whipple and Kenis 2010 [11]. Reproduced with permission of American Chemical Society.

consist of catalyzing the target reaction by means of a homogeneous or a heterogeneous catalyst. Homogeneous catalysis has essentially, but not exclusively, involved reduced states of transition metal complexes [12]. The Section 36.3 will briefly summarize the importance of heterogeneous catalysts on the electro-, photo-, and photoelectro-catalytic reduction of CO_2.

36.3 Innovative Technologies for CO_2 Reduction

36.3.1 Electrochemical Reduction

36.3.1.1 How Does the Technology Work?

Figure 36.2 schematically shows the process that takes place in an electrochemical reactor when electrical energy is supplied to establish a potential between a cathode and an anode to allow CO_2 to be electrochemically transformed into reduced forms. One of the advantages of this technology for CO_2 valorization is that the conversion of CO_2 to useful materials can be carried out under mild

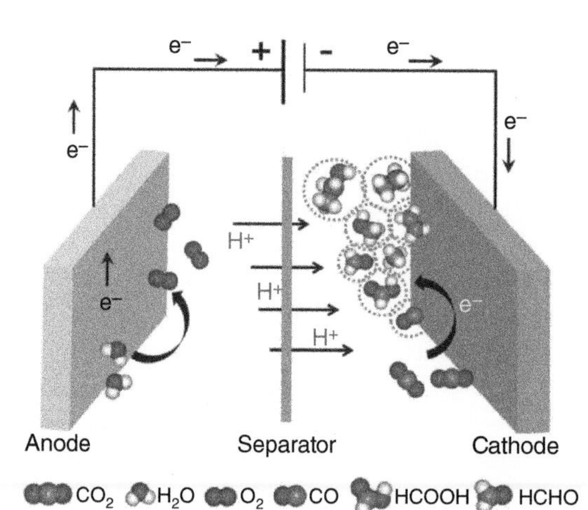

Figure 36.2 Scheme of the CO_2 electrochemical reduction process and different possible products generated in an electrochemical cell. Source: Zhang et al. 2018 [13]. https://onlinelibrary.wiley.com/doi/10.1002/advs.201700275. Licensed under CCBY 4.0.

conditions of temperature and pressure (normally ambient conditions). As the main goal is the reduction of CO_2, research efforts have obviously been focused on the cathodic compartment where CO_2 can be converted to different products, simply leaving that in the anodic compartment a reaction like oxygen evolution (i.e. oxidation of water to give molecular oxygen) can take place without adversely affecting the reduction of CO_2.

Different types of electrochemical reactors can be used. Undivided cells (where both the anode and the cathode are in the same compartment) can be useful for studies that aim at gaining fundamental understanding of the CO_2 electroreduction process. Nevertheless, the most typical configuration is shown in Figure 36.2: an electrochemical cell in which cathode and anode are placed in different compartments separated by an ion conducting membrane, usually a cation-exchange membrane like Nafion. The separated two-compartment configuration prevents the oxidation in the anode of the desired products obtained from CO_2 reduction in the cathode. The "H-type cells" (name derived from their typical "H" shape) have been popular two-compartment cells for studying CO_2 electroreduction, but flow or filter-press-type reactors (in which the electrolytes flow through their own compartment) allow continuous operation and appear as a more suitable cell configuration for an industrial real application. Recent studies to evaluate the techno-economic feasibility of producing different chemicals through CO_2 electroreduction indicated that CO and formic acid appear as the most economically viable products [14]. Important research has also been carried out to other products with more potential such as methanol or gaseous hydrocarbons, although their performance is still further from practical implementation [15].

36.3.1.2 Key Factors Influencing Reaction Performance

The electrochemical reduction of CO_2 is a promising and challenging process in which many different variables can have an influence on its performance. Undoubtedly, the catalyst used has a crucial influence on the process. As explained in the Section 36.2, catalysts are needed to reduce the high overpotentials, and they can induce the selective formation of desired products. In the last few decades, research efforts on the direct heterogeneous reduction of CO_2 have mostly focused on different metal catalysts, showing that different products can be obtained depending on the metal used. For example, in aqueous media at ambient conditions, CO is the main product with metal electrodes like Au, Ag, or Zn [16]; Cu mainly yields mixtures of hydrocarbons (e.g. methane and ethylene [17]) and alcohols (e.g. methanol [18]); and other metals such as In, Sn, Hg, or Pb are selective for the production of formic acid/formate [19]. Over the last few years, researchers have also started to study other materials beyond bulk monometals, including metal oxides, bimetallic materials, metal–organic frameworks (MOFs), and organometallic or nonmetallic (e.g. nitrogen- or boron-doped diamond) catalysts, whose promising results represent new trends and opportunities in CO_2 electroreduction [20, 21].

In catalysis, the Sabatier principle suggests that the interactions between a catalyst and the reactant should be neither too strong nor too weak. According to this principle, if the reaction rate is represented vs. a catalyst property, such as the energy of adsorption (which can be calculated by density functional theory

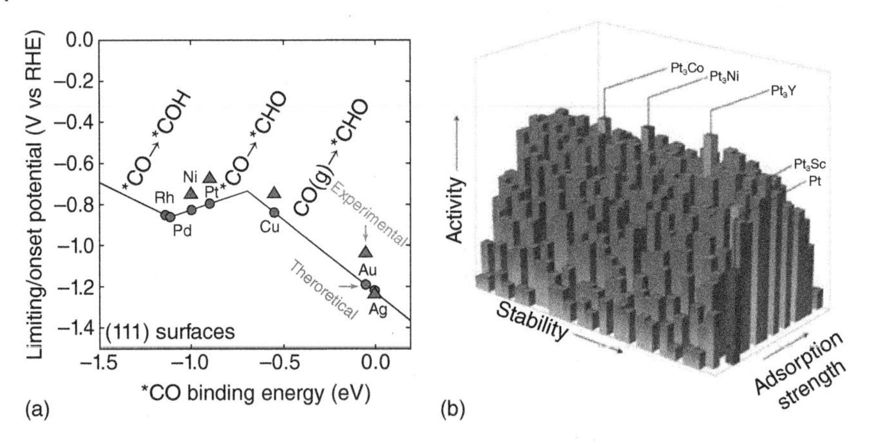

Figure 36.3 Volcano plots: (a) volcano plot for carbon dioxide reduction on metals. (b) A three-dimensional volcano scheme of metal alloy catalysts. Source: (a) Seh et al. 2017 [22]. Reproduced with permission of Royal Society of Chemistry; (b) Mayrhofer and Arenz 2009 [23]. Reproduced with permission of Springer Nature.

[DFT]), plots with the shape of a volcano should be obtained, such as those shown in Figure 36.3. Understanding how to control binding energies of reactive intermediates on a surface is the key to designing catalytic materials with improved performance. Therefore, the further development of volcano plots, such as those pioneered by Norskov's group [22, 24], will help in the understanding of trends in electrocatalytic activity for CO_2 reduction over different catalysts, and hence, they will be useful in the design of advanced, more active catalysts.

Moreover, apart from the nature of the material used as catalyst, the configuration of the electrode has also a great influence on the performance of the CO_2 electroreduction. For example, a certain metal can be used in electrodes with different forms, such as plates, granules, powders, or nanoparticles deposited on porous supports. Particularly, the configuration of gas diffusion electrodes (GDEs) has resulted to be especially successful. GDEs are usually consist of a carbonaceous support, a microporous layer, and a catalytic layer formed by metal nanoparticles, as represented in Figure 36.4.

The improved performance of GDEs can be attributed to an enhancement of the three-phase boundary area between the solid catalyst, the gaseous reactants, and the electrolyte, which avoids mass transfer limitations, and therefore, allows working at higher current densities with high FEs [26].

In addition to the catalyst, the electrolyte used has also a great influence on CO_2 electroreduction. The vast majority of studies on electrochemical reduction of CO_2 have been carried out in aqueous media. However, due to the low solubility of CO_2 in water and the presence of the competing side reaction of hydrogen evolution, attempts in other nonaqueous media have also been reported. Propylene carbonate, acetonitrile, dimethylformamide (DMF), or dimethyl sulfoxide (DMSO) are aprotic solvents, and therefore hydrogen evolution is greatly suppressed in them [10]. Using the same catalyst, the major products from CO_2 reduction are different in a nonaqueous electrolyte than in an aqueous electrolyte (for more details, see, e.g. [27]). Using aprotic solvents, typical products obtained

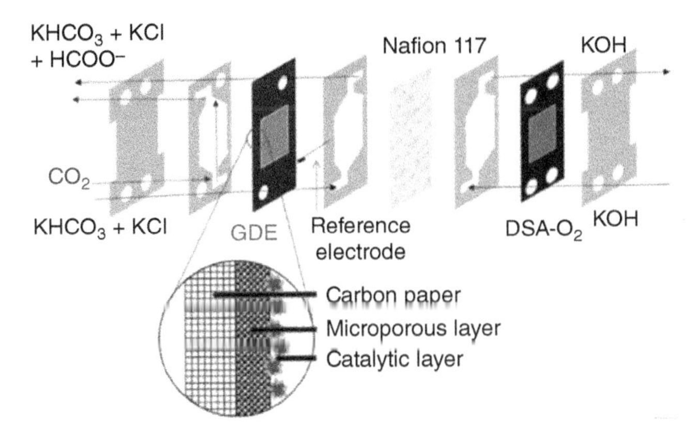

Figure 36.4 Filter-press electrochemical cell with a GDE configuration. Source: Del Castillo et al. 2017 [25]. Reproduced with permission of Elsevier.

do not need protons to take part in CO_2 reduction (e.g. CO or oxalates), but when protons are needed for the reaction (e.g. for formic acid, alcohols, etc.), small amounts of water may be added. Interestingly, in recent years the use of ionic liquids (a family of compounds – organic salts that are liquid below 100 °C – with unique properties) has allowed remarkable improvements in CO_2 electroreduction [28, 29].

The electrochemical reduction of CO_2 in gas phase can be carried out using a solid polymer electrolyte instead of a liquid electrolyte. Ion-exchange membranes (e.g. Nafion cation-exchange membranes or Selemion anion-exchange membranes) coated with metal catalyst have been mainly used for the reduction of CO_2 to gaseous products like CO, CH_4, or C_2H_6 [30].

36.3.1.3 Main Challenges

Significant progress has been achieved in recent years, as denoted by the rapidly increasing number of research publications, but the maturity of CO_2 electrochemical technology has yet to reach the requirements for commercialization. Key challenges that will have to be tackled in this exciting area of research include:

(a) *Improved and/or novel electrocatalytic materials*: The great research efforts on research on the catalysis of the electroreduction of CO_2 during the last decade have allowed significant advances, although improvements in electrocatalysts are definitely necessary. An ideal electrocatalyst should combine:
   (i) high FE to the desired CO_2 reduction product (which means that a high % of the electrical charge supplied is used in the desired product, and therefore it implies that the selectivity of the process is high);
   (ii) high formation rates of the product (measured by the current density, the electric current flow divided by the area of the electrode, at which the process can be operated); and
   (iii) a low overpotential for CO_2 reduction to allow an energy efficient process (a lower overpotential means that the real potential applied in the process will be closer to the equilibrium potential, so the energy penalty of the process will be lower).

Future research will definitely continue tackling the challenge of optimizing all these figures of merit (i.e. FE, *j*, EE, etc.) at the same time, so improving one is not at the expense of making other worse. The still insufficient stability and durability of the materials used as electrodes can also be highlighted as another important challenge to be addressed for economically viable industrial processes.

Moreover, further understanding of reaction mechanisms will indeed help in the design of better electrocatalytic materials. Due to the complexity of the reaction environment and multiple bond-forming and bond-breaking processes, there is a lack of understanding of the CO_2 electroreduction mechanism, pathways, and intermediates. The determination of the transient catalytic intermediates is a real challenge due to their very short lifetimes and low concentrations. Coupled to advanced methods of experimental analysis (e.g. scanning tunneling microscope [STM], electron paramagnetic resonance [EPR], Fourier transform infrared spectroscopy [FTIR], etc.), the use of first principles computational techniques, such as DFT, will be helpful for improving the understanding of how the CO_2 electroreduction reaction proceeds on catalytic materials. Such theoretical modeling techniques could also guide the development of new catalysts with improved performance. As explained before, further development of Norskov's volcano plot will be useful as a rational platform to understand trends in activity of different catalysts, helping to design better and more active electrocatalytic materials for CO_2 electroreduction.

(b) *Improved/novel electrolytes*: An interesting way of improvement is related to the nature of the solvent used as liquid electrolyte, which, together with the electrocatalyst, plays a crucial role in the CO_2 electroreduction. The use of aqueous media presents limitations, especially due to the concomitant side reaction of hydrogen evolution that takes place at similar thermodynamic potential. Therefore, relevant research efforts point to the use of new solvents, like ionic liquids, as a promising way of enhancing the performance. Extending the application of solid polymer electrolytes to avoid the use of liquid electrolytes that dilute liquid products of CO_2 electroreduction may also open opportunities for improvement.

(c) *More efficient electrocatalytic reactors*: As in any electrochemical process, the design of the electrochemical reactor for CO_2 conversion is crucial for a successful performance at a large scale. Undoubtedly, future work will have to bridge the gap between laboratory-scale reactors and industrial reality. Engineering research should focus on the design and scale-up of electrochemical reactors that are able to operate in a continuous mode for the industrial implementation of economically viable CO_2 electroreduction processes.

36.3.2 Photochemical Reduction

36.3.2.1 How Does the Technology Work?

Since the first report on photocatalysis by Fujishima and Honda in 1972 [31], the research community has been attempting to mimic nature and pursue the spontaneous transformation of atmospheric CO_2 and H_2O to chemicals using sunlight as the sole energy input and over the same active surface, in contrast

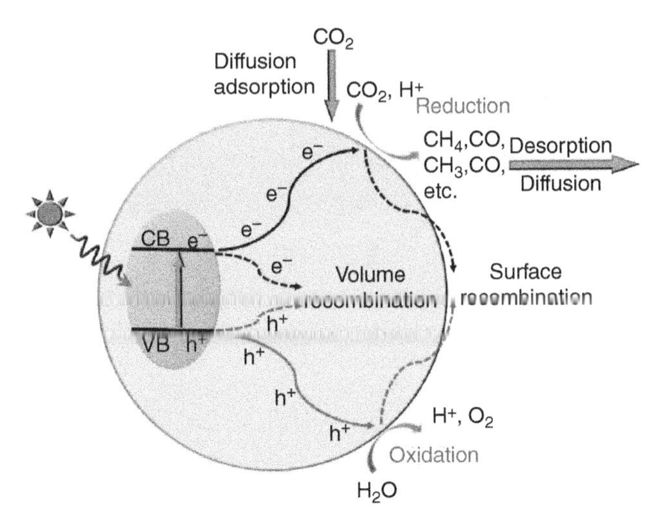

Figure 36.5 Photoinduced generation of electron–hole pairs in CO_2 reduction with H_2O. Source: Wu et al. 2017 [5]. https://onlinelibrary.wiley.com/doi/10.1002/advs.201700194. Licensed under CCBY 4.0.

with the electroreduction process. Although the efficiencies reported are still low [32], the artificial photosynthesis is believed to have a great potential to enable a shift to a sustainable energy economy and chemical industry.

The process starts with the adsorption of CO_2 in the photocatalyst surface, where the molecule is susceptible to reduction by photogenerated electrons. Under the irradiation of light source, the electrons are excited from the valence band (VB) to the conduction band (CB), and, simultaneously, an equal number of holes (which serve as a positive charge carrier) are formed in the CB due to the absorption of photons from the incident light. These electron–hole pairs separate from each other, producing the reduction of CO_2 molecule and the oxidation of H_2O (releasing H^+) at the surface of a semiconductor material, as represented in Figure 36.5.

In order for these photogenerated electrons and holes to be energetically favorable to reduce CO_2 and oxidize H_2O, photocatalysts should have a suitable band structure. Their CB edge must be more negative than the redox potential of CO_2 reduction, and the VB edge should be more positive than the redox potential of water oxidation (+0.82 V vs. normal hydrogen electrode [NHE] in pH = 7). After reduction, the products need to desorb and diffuse to accomplish the whole conversion process. The redox potential levels of the adsorbate species and the band gap energy determine the likelihood and rate of the charge transfer processes for electrons and holes [33], resulting in different products (Table 36.1).

The general criterion to evaluate the performance for the photocatalytic conversion of CO_2 is to measure the concentration of a certain product, n, produced within a period, t, under light irradiation per gram of catalyst, m, according to Eq. (36.5). A common unit for this formation rate, r, is $\mu mol \cdot g_{cat.}{}^{-1} \cdot h^{-1}$:

$$r = \frac{n}{t \cdot m} \tag{36.5}$$

Another relevant parameter to evaluate the photocatalytic performance is the effectivity of the catalyst to convert light into chemical energy. The efficiency of light absorption, charge separation, and surface redox reaction can be expressed by the apparent quantum yield, AQY (%), defined as the rate of electrons transferred toward a certain product per rate of incident photons on the photoelectrode surface. The AQY can be comparable with the FEs used in electrocatalysis and is defined as:

$$\text{AQY} = \frac{n_{e^-} R}{I} \times 100 \tag{36.6}$$

where n_{e^-}, R, and I denote the number of electrons involved in the photocatalytic reaction, the molecular production rate, and the rate of incident photons, respectively.

36.3.2.2 Key Factors

The performance of a photocatalysis system for CO_2 conversion is mainly related to the photoactive material and photoreactor design applied. An ideal semiconductor photocatalyst may include an adjusted band gap and proper position of CB and VB edges to utilize visible-light range (which correspond to about 50% of the energy received at the Earth's surface), together with an enhanced light harvesting, suitable electron–hole separation, large surface area, favorable surface chemistry/morphology, and sufficient stability. For reasons that include affordability and stability, titanium dioxide (TiO_2), mainly in its anatase and rutile forms, has been the most widely used semiconductor in photocatalysis. Unfortunately, after an intensive research in this area, it appears that TiO_2 possesses important drawbacks as photoactive material for CO_2 conversion since it can mainly absorb in the UV part of the spectrum ($\lambda < 390$ nm) and it presents a poor electron–hole pair separation [34]. Various strategies have been followed by the research community to manipulate TiO_2 band gap and enhance light utilization [35], including doping with metal/nonmetal catalysts, surface decoration, dye sensitization, heterojunction formation, or structuralization, among others, as described in detail in other chapters in this book. These techniques may promote electron–hole pair separation and CO_2 adsorption and activation, enhancing CO_2 photoconversion efficiency to desired products. Particular interest has been lately directed toward the use of low-dimensional structured materials, composed of ultrafine particles, as photocatalysts primarily for their tunable optical and electronic properties due to quantum size effects as well as their enhanced photoelectric performance compared with bulk materials [36].

Nonetheless, the charge separation and selectivity of TiO_2-based materials are still low, even after applying in combination several of the aforementioned strategies. Therefore, it is of crucial importance to explore the performance of alternative materials to titanium-based semiconductors. For example, silicon carbide (SiC), copper oxides (Cu_2O), gallium (Ga), or cadmium sulfide (CdS)-based compounds may show better light harvesting properties than TiO_2 and may be good candidates to substitute it [37]. Figure 36.6 shows common semiconductor materials that may be applicable for the photoreduction of CO_2, attending to their CB and VB edge positions.

Besides, efficient photoreactor designs for real-world applications need to consider various factors, including (i) light source and geometrical configuration, to

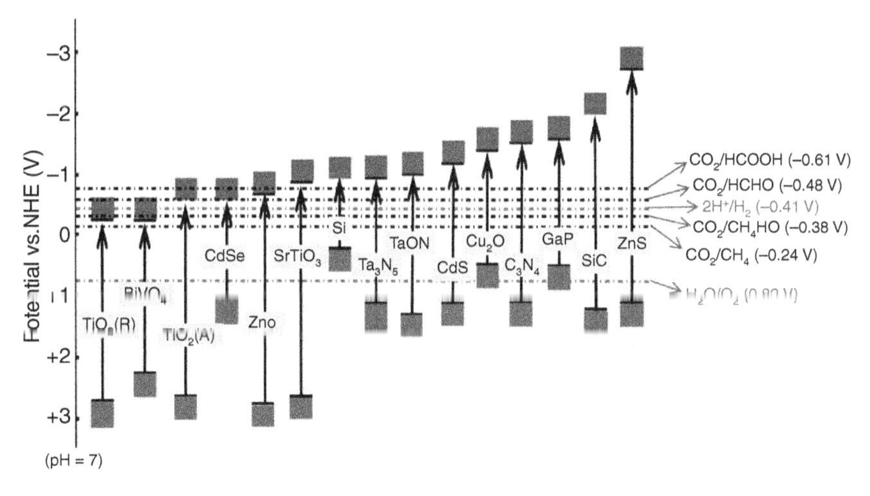

Figure 36.6 Band gap energies for common semiconductors materials relative to the redox potentials of CO_2 reduction products at pH = 7. Source: Li et al. 2014 [38]. Reproduced with permission of Springer Nature.

have uniform light distribution throughout the entire system; (ii) construction material, limited by requirements of light transmission; (iii) heat exchange, to remove the heat generated by the lamp; (iv) mixing and flow characteristics, to ensure an appropriate contact between reactants, electrons/holes, and catalysts; and (v) phases involved and mode of operation [39].

A number of different laboratory photoreactor designs have been assessed for the photocatalytic reduction of CO_2. Normally, batch-type reactors (with the catalyst introduced as powder on the bottom of the cell) and slurry reactors (in which light scattering occurs) have been applied. These simple configurations have been demonstrated to be inefficient to induce the CO_2 conversion to more reduced species, due to a low surface area-to-volume ratio produced by particle agglomeration, and the required separation of the photocatalyst material from the products obtained. Fixed reactors are also applied, although they present low specific area and light scattering, leading to poor light utilization [40]. The agglomeration phenomena occurring in the aforementioned configurations may be partially solved by the use of reactors equipped with coated catalysts (e.g. optical fiber technology and internally illuminated reactor) [41], although their optical properties vary for different coating strategies that lead to different particles sizes and porosities. In any case, the application of these reactor designs is still scarce, and further developments are needed. Figure 36.7 shows a representation of slurry, optical fiber, and internally illuminated reactors.

36.3.2.3 Main Challenges

The most relevant issues that need to be tackled, together with a further exploration of CO_2 photoreduction mechanisms, are:

(a) *Innovative photoactive materials*: The research efforts will require a multidisciplinary approach to develop novel photocatalytic materials with higher stability, selectivity, and efficiency toward products under visible-light irradiation. It should take the fullest advantage of new nanoscale structures

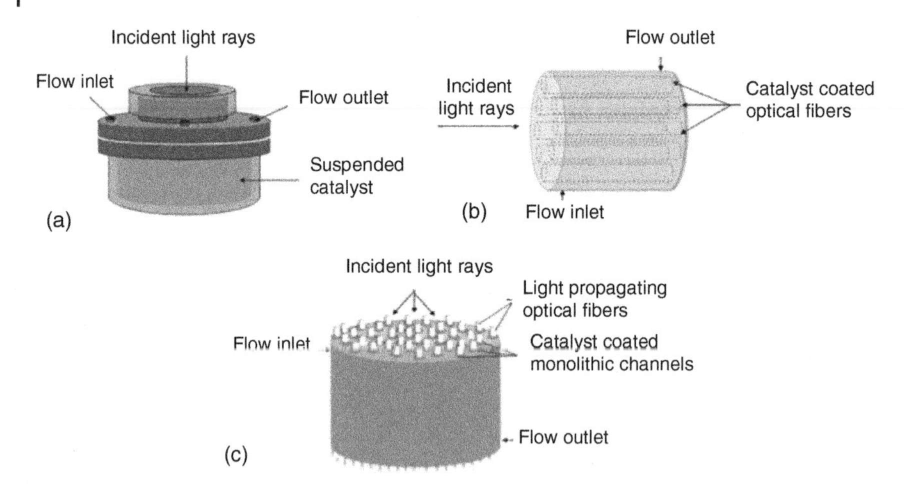

Figure 36.7 Scheme for a (a) slurry reactor with top illumination, (b) optical fiber reactor with side illumination, and (c) internally illuminated reactor with top illumination. Source: Oluwafunmilola and Maroto-Valer 2015 [39]. https://www.sciencedirect.com/science/article/pii/S1389556715000271. Licensed under CCBY 4.0.

with large surface area, mixed crystal phase, defect disorders, material modifications, improved charge separation, and directional electron transfers. The breakthrough seems to be in the synthesis of non-TiO_2 materials with improved physicochemical properties. Metal-free cocatalysts are also under development to help structuring robust artificial photocatalysts. In particular, carbonaceous materials including carbon nitride, graphitic carbon nitride (g-C_3N_4), graphene, and carbon nanotubes (CNTs) show great potential to replace some of the existing metal catalysts, providing an enhanced electrical and thermal conductivity, light harvesting, surface CO_2 adsorption capacity, and low cost. MOFs may also represent an opportunity in the photocatalytic reduction of CO_2. The integration of active semiconductors with bio-based materials presents also promise for an enhanced transformation of CO_2.

(b) *Efficient photocatalytic reactors*: The research efforts must consider important parameters such as reactor geometry, flow rate, irradiation source, or catalyst mass. Modeling of the effect of operation parameters on CO_2 reduction is also required in order to extrapolate results and design a pilot-scale system. Besides, considering the limited rates of CO_2 photocatalytic reduction to products, the recirculation of unconverted products is also advised. These aspects may have an important implication on both economy and efficiency of the overall process.

36.3.3 Photoelectrochemical Reduction

36.3.3.1 How Does the Technology Work?

The photoelectrocatalytic reduction of CO_2 was initially reported by Halmann in 1978 using a GaP semiconductor as photoelectrode in aqueous solution [42]. Since then, photoelectrochemistry and the development of efficient photoelectrocatalysts for an efficient reduction of CO_2 have received much attention [43].

Basically, a photoelectrochemical system is similar to a standard electrochemical cell, except that light is explicitly involved in providing energy to cause the redox reactions to proceed. In principle, such integration reduces the system capital cost and enables higher efficiency by reducing loses in transporting electricity to the electrolysis cell, eliminating current collectors, and interconnections between devices. Compared with photocatalysis, the applied bias can cause band bending and help the oriented transfer of the photogenerated electrons, decreasing the recombination of the photogenerated electron–hole pairs. Besides, photocatalytic materials with unfavorable band positions for CO_2 reduction and H_2O oxidation can be still used in photoelectrocatalytic systems when applying an external bias. In addition, the obtained redox products can be readily separable in two-compartment photoelectrochemical cells. Overall, the options of photoelectrocatalysis may be broader than photocatalysis for CO_2 transformation. At the photoelectrodes, light is absorbed by a semiconductor material, exciting electrons to a higher energy level. These electrons and the corresponding holes generated are capable of carrying out redox reactions at the semiconductor/liquid interface. The photoelectrochemistry is therefore a multidisciplinary field involving surface science, electrochemistry, solid-state physics, and optics.

The efficiency of a CO_2 photoelectrochemical reduction system is commonly expressed as the product formation rate, r (Eq. (36.5)), and the percentage of light energy input converted to chemical energy output, AQY (Eq. (36.6)) or FE (Eq. (36.3)), since generally an external bias potential is required.

36.3.3.2 Key Factors

The development of high-efficiency photoelectrodes that satisfy the requirements of real applications entails the processing of materials with enhanced solar energy conversion efficiency, durability, and low cost. Photoelectrocatalysis normally exploits semiconductor materials for CO_2 conversion including inorganic binary compounds, such as TiO_2, ZnO, GaP CdS, and SiC [44]. As it is for photocatalysis, TiO_2-based materials with various structures (e.g. nanocrystals, nanocomposites, nanotube arrays, etc.) are the most investigated photoelectrocatalysts. The same strategies followed in photocatalysis to overcome the important limitations of TiO_2 (see Section 36.3.2.2) are of application in the development of materials for photoelectrocatalysis. Research efforts have also been focused on multielement metal oxides (e.g. Bi_2WO_6, ZnGaNO, $Ca_2Nb_2TaO_{10}$, or $MnCo_2O_4$), as their band gaps and band edge positions can be rationally tuned and also metal complexes, which may exhibit excellent CO_2 adsorption ability. Moreover, metal-free compounds (e.g. graphitic g-C_3N_4 or MOFs) give good hopes to improve the photoelectrochemical conversion of CO_2.

For a practical use of photoelectrochemical cells, anode and cathode should be separated by an ion-exchange membrane in a dual-compartment cell in order to allow an efficient separation/transport of charges, dealing with a reduction–oxidation reaction in each compartment and enhancing the separation of products. Gas-phase operation on the cathode compartment is also preferable in order to avoid the limitations of CO_2 solubility [45]. Research on CO_2 reduction to liquid products has mostly focused on batch photoelectrochemical reactors, but production of valuable chemicals on an industrial scale will likely require continuous flow reactors to minimize capital costs and maximize product consistency.

Moreover, there are different electrode configurations for photoelectrochemical systems depending on which electrode (i.e. anode, cathode, or both) acts as photoelectrode [44] as shown in Figure 36.8.

(a) *Photocathode–dark anode*: The most common configuration usually employs a photocathode made of a p-type semiconductor such as those based on Cu (e.g. Cu_2O, CuO, Cu_2ZnSnS_4, etc.), Co_3O_4, InP, or CdSeTe, among others, due

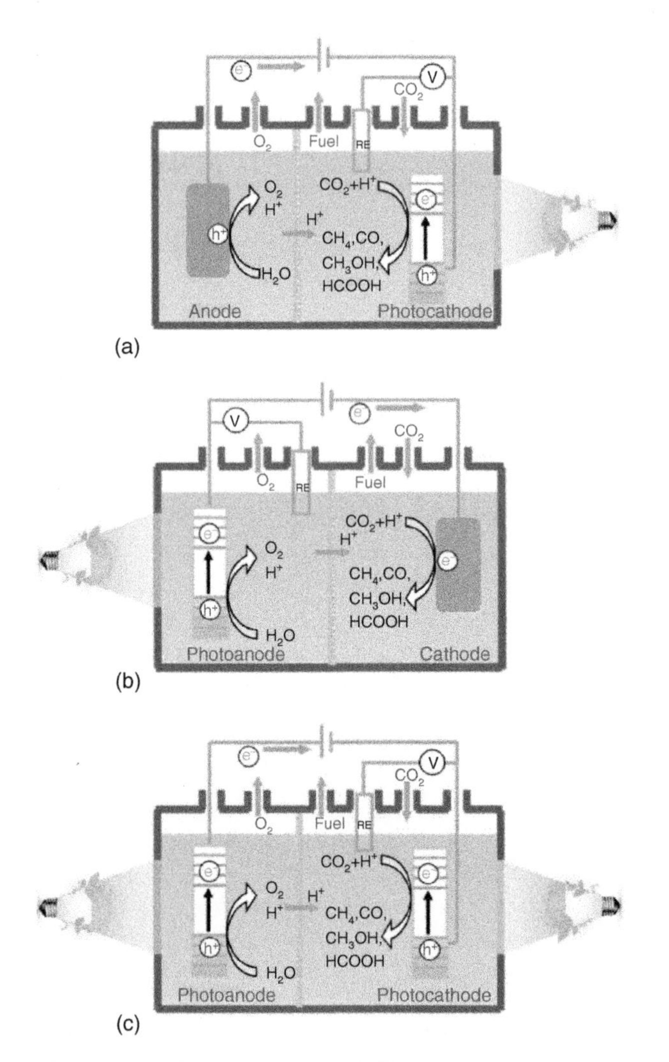

Figure 36.8 Representation of a (a) photocathode–dark anode, (b) photoanode–dark cathode, and (c) photocathode–photoanode electrode configurations. Source: Xie et al. 2016 [44]. Reproduced with permission of Royal Society of Chemistry.

to their high CB energy suitable for CO_2 reduction and a metallic anode for H_2O oxidation. However, CO_2 reduction on p-type semiconductors requires a high bias potential since their VB are not sufficiently positive to oxidize H_2O. These materials are also expensive, toxic, or unstable during the reaction, and two-electron compounds such as CO and formic acid are usually the main products of CO_2 reduction. Moreover, combination of a photocathode with a cocatalyst able to activate CO_2 molecules is needed since p-type semiconductors do not act as a true catalyst for the activation of CO_2 molecules, but just as light harvester to generate electrons and holes.

(b) *Photoanode–dark cathode*: The use of n-type semiconductors such as TiO_2, ZnO, $BiVO_4$, or WO_3, which are earth abundant, cheap, and stable, as photoanode for H_2O oxidation and a metallic electrocatalyst that is active for CO_2 reduction as the cathode is also an attractive alternative. This configuration allows reducing the external electric bias over an electrochemical cell configuration, since the voltage generated by the light in the anode supplies an extra negative potential to the cathode for CO_2 reduction. It also provides active protons through a proton-exchange membrane toward the cathode side, which plays an important role for CO_2 reduction.

(c) *Photocathode–photoanode*: The combination of a photocathode made of a p-type semiconductor for CO_2 reduction with a photoanode made of a n-type semiconductor for H_2O oxidation can also be seen. Ideally, this configuration allows an efficient CO_2 reduction without external voltage by using two appropriate n-type and p-type semiconductors to form multi-semiconductor systems, in which the materials have matched band gap positions to transfer electrons and holes. In some cases, the system may need an extra electrical bias to facilitate charge separation and overcome parasitic loses and reaction overpotentials.

Some recent reports also consider photovoltaic (PV) cell tandem devices, where PV panels are coupled to photoelectrochemical cells to supply the voltage required for CO_2 reduction/H_2O splitting [46].

36.3.3.3 Main Challenges

Photoelectrocatalysis combines the advantages of both electrocatalysis and photocatalysis. In spite of that, the efficiencies of the systems presented to date are modest. The challenges are basically those mentioned above for electro- and photo-catalysis for CO_2 transformation, with some particularities.

(a) *Photoactive materials*: Most of the n-type semiconductors applied suffer photocorrosion during the process, and modification of photoelectrodes using surface coatings (e.g. TiO_2, Al_2O_3, CNT, etc.) could be an effective way to improve the stability for CO_2 photoelectroreduction. In addition, Z-scheme junction systems with semiconductors with matched CB and VB edge positions (as both photocathodes and photoanodes) can lead to systems with facilitated charge transfer to promote its reduction efficiency and selectivity and conduct the reduction of CO_2 and oxidation of H_2O simultaneously without (or with reduced) external applied bias. Besides, the development of more efficient cocatalysts with strong adsorption and activation capacity of CO_2 will lead to the selective production of desirable chemicals in the photocathode. The development of materials with high conductivity for proton-exchange membrane is also key for the process.

(b) *Photocatalytic reactors*: The systems reported have evolved from photocathode to photoanode-driven cells, but more promising strategies such as two-compartment bias-free photocathode–photoanode systems and photoelectrochemical cells coupled with PV panels in stand-alone tandem cells should receive more attention to demonstrate their feasibility in the practical application of CO_2 reduction processes under sunlight illumination.

36.4 Concluding Remarks

The conversion of CO_2 into useful products is challenging because CO_2 is the most stable carbon-based structure in nature. Electro- and photoelectro-chemical approaches appear as attractive options for CO_2 reduction, especially when they are coupled to intermittent renewable energy, where there is a very interesting future possibility to use the excess electric energy from renewable sources, like solar or wind energy, to reduce CO_2 to value-added products and/or fuels. The products could then be used in fuel cells or in industrial processes, when and where needed. More potential for a shift toward a low-carbon economy presents the photochemical approach, where CO_2 reduction and H_2O oxidation occurs at the same catalytic surface using sunlight as the sole energy input.

The number of interesting studies, especially in the field of photo- and photoelectro-chemical reduction of CO_2, has grown rapidly. As highlighted in this chapter, promising results on electrocatalytic CO_2 conversion to different chemicals and fuels have already been achieved, but further research efforts are still needed before these technologies can reach practical commercialization. The challenges discussed in this chapter, including the development of catalytic materials for improved performance of CO_2 reduction, can be seen as exciting opportunities of research in this vibrant field of CO_2 conversion.

Acknowledgments

The authors gratefully acknowledge the financial support from the Spanish Ministry of Economy and Competitiveness (MINECO), under the projects CTQ2013-48280-C3-1-R and CTQ2016- 76231-C2-1-R, and PID2019-104050RA-I00, as well as Ramón y Cajal programme (RYC-2015-17080).

References

1 International Energy Agency (IEA). (2017). Key World Energy Statistics 2017.
2 Boot-Handford, M.E., Abanades, J.C., Anthony, E.J. et al. (2014). Carbon capture and storage update. *Energy Environ. Sci.* 7: 130–189.

3 Sanz-Perez, E.S., Murdock, C.R., Didas, S.A., and Jones, C.W. (2016). Direct capture of CO_2 from ambient air. *Chem. Rev.* 116: 11840–11876.
4 Aresta, M., Dibenedetto, A., and Angelini, A. (2014). Catalysis for the valorization of exhaust carbon: from CO_2 to chemicals, materials, and fuels. Technological use of CO_2. *Chem. Rev.* 114 (3): 1709–1742.
5 Wu, J., Huang, Y., Ye, W., and Li, Y. (2017). CO_2 reduction: from the electrochemical to photochemical approach. *Adv. Sci.* 4: 1700194.
6 Lower, S. (2018). General Chemistry Virtual Textbook. Chem1 Introduction to Electrochemistry. http://www.chem1.com/acad/webtext//elchem/ec-1.html (accessed 29 October 2018).
7 Bard, A.J. and Faulkner, L.R. (2001). *Electrochemical methods: Fundamentals and Applications*, 2e. New York, NY: John Wiley & Sons.
8 Pletcher, D. and Walsh, F.C. (1993). *Industrial Electrochemistry*, 2e. London, UK: Chapman and Hall.
9 Qiao, J., Liu, Y., and Zhang, J. (eds.) (2016). *Electrochemical Reduction of Carbon Dioxide: Fundamentals and Technologies*. Boca Raton, FL: CRC Press.
10 Masel, R., Liu, Z., Zhao, D. et al. (2016). CO_2 conversion to chemicals with emphasis on using renewable energy/resources to drive the conversion. In: *Commercializing Biobased Products: Opportunities, Challenges, Benefits, and Risks*, RSC Green Chemistry No. 43 (ed. S.W. Snyder), 215–257. Royal Society of Chemistry (RSC).
11 Whipple, D.T. and Kenis, P.J.A. (2010). Prospects of CO_2 utilization via direct heterogeneous electrochemical reduction. *J. Phys. Chem. Lett.* 1: 3451–3458.
12 Costentin, C., Robert, M., and Savéant, J.M. (2013). Catalysis of the electrochemical reduction of carbon dioxide. *Chem. Soc. Rev.* 42: 2423–2436.
13 Zhang, W., Hu, Y., Ma, L. et al. (2018). Progress and perspective of electrocatalytic CO_2 reduction for renewable carbonaceous fuels and chemicals. *Adv. Sci.* 5: 1700275.
14 Verma, S., Kim, B., Jhong, H.R.M. et al. (2016). A gross-margin model for defining technoeconomic benchmarks in the electroreduction of CO_2. *ChemSusChem* 9: 1972–1979.
15 Merino-Garcia, I., Alvarez-Guerra, E., Albo, J., and Irabien, A. (2016). Electrochemical membrane reactors for the utilisation of carbon dioxide. *Chem. Eng. J.* 305: 104–120.
16 Zhu, D.D., Liu, J.L., and Qiao, S.Z. (2016). Recent advances in inorganic heterogeneous electrocatalysts for reduction of carbon dioxide. *Adv. Mater.* 28: 3423–3452.
17 Merino-Garcia, I., Albo, J., and Irabien, A. (2018). Tailoring gas-phase CO_2 electroreduction selectivity to hydrocarbons at Cu nanoparticles. *Nanotechnology* 29: 014001.
18 Albo, J., Alvarez-Guerra, M., Castaño, P., and Irabien, A. (2015). Towards the electrochemical conversion of carbon dioxide into methanol. *Green Chem.* 17: 2304–2324.
19 Du, D., Lan, R., Humphreys, J., and Tao, S. (2017). Progress in inorganic cathode catalysts for electrochemical conversion of carbon dioxide into formate or formic acid. *J. Appl. Electrochem.* 47: 661–678.

20 Jhong, H., Ma, S., and Kenis, P.J. (2013). Electrochemical conversion of CO_2 to useful chemicals: current status, remaining challenges, and future opportunities. *Curr. Opin. Chem. Eng.* 2: 191–199.

21 Lu, Q. and Jiao, F. (2016). Electrochemical CO_2 reduction: electrocatalyst, reaction mechanism, and process engineering. *Nano Energy* 29: 439–456.

22 Seh, Z.W., Kibsgaard, J., Dickens, C.F. et al. (2017). Combining theory and experiment in electrocatalysis: insights into materials design. *Science* 355: eaad4998.

23 Mayrhofer, K.J.J. and Arenz, M. (2009). Log on for new catalysts. *Nat. Chem.* 1: 518–519.

24 Liu, X., Xiao, J., Peng, H. et al. (2017). Understanding trends in electrochemical carbon dioxide reduction rates. *Nat. Commun.* 8: 15438.

25 Del Castillo, A., Alvarez-Guerra, M., Solla-Gullón, J. et al. (2017). Sn nanoparticles on gas diffusion electrodes: synthesis, characterization and use for continuous CO_2 electroreduction to formate. *J. CO2 Util.* 18: 222–228.

26 Irabien, A., Alvarez-Guerra, M., Albo, J., and Domínguez-Ramos, A. (2018). Electrochemical conversion of CO_2 to value-added products. In: *Electrochemical Water and Wastewater* (eds. C.A. Martínez Huitle, M.A. Rodrigo and O. Scialdone). Elsevier. ISBN: 9780128131602.

27 Sánchez-Sánchez, C.M., Montiel, V., Tryk, D.A. et al. (2001). Electrochemical approaches to alleviation of the problem of carbon dioxide accumulation. *Pure Appl. Chem.* 73: 1917–1927.

28 Rosen, B.A., Salehi-Khojin, A., Thorson, M.R. et al. (2011). Ionic liquid-mediated selective conversion of CO_2 to CO at low overpotentials. *Science* 334: 643–644.

29 Alvarez-Guerra, M., Albo, J., Alvarez-Guerra, E., and Irabien, A. (2015). Ionic liquids in the electrochemical valorisation of CO_2. *Energy Environ. Sci.* 8: 2574–2599.

30 Hori, Y. (2016). CO_2 reduction using electrochemical approach. In: *Solar to Chemical Energy Conversion*, Lecture Notes in Energy 32 (eds. M. Sugiyama, K. Fujii and S. Nakamura). Cham, Switzerland: Springer International Publishing. ISBN: 978-3-319-25400-5.

31 Fujishima, A. and Honda, K. (1972). Photolysis of water at a semiconductor electrode. *Nature* 238: 37–38.

32 Sohn, Y., Huang, W., and Taghipour, F. (2017). Recent progress and perspectives in the photocatalytic CO_2 reduction of Ti-oxide-based nanomaterials. *Appl. Surf. Sci.* 396: 1696–1711.

33 Yuan, L. and Xu, Y.-J. (2015). Photocatalytic conversion of CO_2 into value-added and renewable fuels. *Appl. Surf. Sci.* 342: 154–167.

34 Chen, X. and Mao, S.S. (2017). Titanium dioxide nanomaterials: synthesis, properties, modifications, and applications. *Chem. Rev.* 107 (7): 2891–2959.

35 Tu, W., Zhou, Y., and Zou, Z. (2014). Photocatalytic conversion of CO_2 into renewable hydrocarbon fuels: state-of-the-art accomplishment, challenges, and prospects. *Adv. Mater.* 26 (27): 4607–4626.

36 Voiry, D., Suk Shin, H., Ping Loh, K., and Chhowalla, M. (2018). Low-dimensional catalysts for hydrogen evolution and CO_2 reduction. *Nat. Rev. Chem.* 2: 0105.

37 Navalón, S., Dhakshinamoorthy, A., Alvaro, M., and Garcia, H. (2013). Photocatalytic CO_2 reduction using non-titanium metal oxides and sulfides. *ChemSusChem* 6: 562–577.

38 Li, X., Wen, J., Low, J. et al. (2014). Design and fabrication of semiconductor photocatalyst for photocatalytic reduction of CO_2 to solar fuel. *Sci. China Mater.* 57: 70–100.

39 Oluwafunmilola, O. and Maroto-Valer, M.M. (2015). Review of material design and reactor engineering on TiO_2 photocatalysis for CO_2 reduction. *J. Photochem. Photobiol. C: Photochem. Rev.* 24: 16–42.

40 Roupp, G.B., Nico, J.A., Annangi, S. et al. (1997). Two-flux radiation-field model for an annular packed-bed photocatalytic oxidation reactor. *AlChE J.* 43: 792–801.

41 Tahir, M. and Amin, N.A.S. (2013). Photocatalytic CO_2 reduction with H_2O vapors using montmorillonite/TiO_2 supported microchannel monolith photoreactor. *Chem. Eng. J.* 230: 314–327.

42 Halmann, M. (1978). Photoelectrochemical reduction of aqueous carbon dioxide on p-type gallium phosphide in liquid junction solar cells. *Nature* 275: 115–116.

43 Wang, P., Wang, S., Wang, H. et al. (2018). Recent progress on photo-electrocatalytic reduction of carbon dioxide. *Part. Part. Syst. Char.* 35: 1700371.

44 Xie, S., Zhang, Q., Liu, G., and Wang, Y. (2016). Photocatalytic and photoelectrocatalytic reduction of CO_2 using heterogeneous catalysts with controlled nanostructures. *Chem. Commun.* 52: 35–59.

45 Ampelli, C., Centi, G., Passalacqua, R., and Perathoner, S. (2010). Synthesis of solar fuels by novel photoelectrocatalytic approach. *Energy Environ. Sci.* 3: 292–301.

46 Jang, Y., Jeong, I., Lee, J. et al. (2016). Unbiased sunlight-driven artificial photosynthesis of carbon monoxide from CO_2 using a ZnTe-based photocathode and a perovskite solar cell in tandem. *ACS Nano* 10: 6980–6987.

37

Photocatalytic Abatement of Emerging Micropollutants in Water and Wastewater

Lan Yuan[1,2], *Zi-Rong Tang*[2], *and Yi-Jun Xu*[1,2]

[1] *Fuzhou University, State Key Laboratory of Photocatalysis on Energy and Environment, College of Chemistry, Xueyuan Road No. 2, University New Area, Fuzhou, 350116, PR China*
[2] *Fuzhou University, College of Chemistry, New Campus, Xueyuan Road No. 2, University New Area, Fuzhou, 350116, PR China*

37.1 Introduction

The rapid development of manufacturing technology after the industrial revolution has improved the standards of living significantly but also threatened human health and the environment. Micropollutants discharged by human activities into the aquatic environment, including wastewater, surface water, groundwater, and drinking water, has become a serious global issue [1, 2]. It has been estimated that around 4 billion people have no or little access to clean and sanitized water supply worldwide, and millions of people die due to severe waterborne diseases every year. These statistical figures are expected to continue to grow in the near future [3, 4]. Micropollutants are commonly present in waters at trace concentrations. The "low concentration" and diversity of micropollutants not only complicate the associated detection and analysis procedures but also create challenges for water and wastewater treatment processes [5, 6]. It is an imperative task to develop newer eco-friendly methods to treat micropollutants in water and wastewater at a lower cost and with less energy [7, 8].

For treating micropollutants, traditional physical techniques (adsorption on activated carbon, ultrafiltration, reverse osmosis, coagulation by chemical agents, ion exchange on synthetic adsorbent resins, etc.) can generally be used efficiently [1, 2, 6]. Nevertheless, they are nondestructive, since they just transfer organic compounds from water to another phase, thus potentially causing secondary pollution. Consequently, regeneration of the adsorbent materials and posttreatment of solid wastes are needed, which are expensive operations [8]. In contrast, recently, heterogeneous photocatalysis using semiconductor photocatalysts is emerging as a destructive and green technology, which has received intense attention and been widely applied to the abatement of water micropollutants [4, 7]. The unique advantages of heterogeneous photocatalysis for the aquatic environment treatment include the following: (i) it is driven by sunlight, a completely renewable source of energy; (ii) it can occur under

Heterogeneous Catalysts: Advanced Design, Characterization and Applications, First Edition.
Edited by Wey Yang Teoh, Atsushi Urakawa, Yun Hau Ng, and Patrick Sit.

mild conditions (room temperature and atmospheric pressure); (iii) it does not involve mass transfer and thus not causing secondary pollution; and (iv) semiconductor photocatalysts such as TiO_2 are abundant, cheap, and nontoxic and show relatively high chemical stability [4, 7].

The utilization of combined photocatalysis and other solar technologies may be developed to reduce water pollution. This chapter first provides an overview of the basic principles of heterogeneous photocatalysis for micropollutant abatement in water and wastewater, followed by the introduction of different kinds of micropollutants and examples of those degradable by photocatalysis. For the low efficiency of traditional photocatalysts in micropollutant abatement, available strategies for the efficiency enhancement of the reactions are outlined. Finally, the future research challenges and prospects are concisely discussed.

37.2 Main Processes for Photocatalytic Abatement of Micropollutants in Water and Wastewater

In general, micropollutants consist of a vast and expanding array of anthropogenic as well as natural substances, which include pharmaceuticals, personal care products, surfactants, industrial chemicals, pesticides, heavy metals, and many other emerging compounds [1, 2, 6]. The categories and typical examples that are degradable by photocatalysis are listed in Table 37.1.

As for classical heterogeneous catalysis, the overall process can be divided into five independent steps: (i) transfer of the reactants in the fluid phase to the surface of photocatalyst, (ii) adsorption of the reactants, (iii) reaction in the adsorbed phase, (iv) desorption of the products, and (v) removal of the products from

Table 37.1 Nonexhaustive examples of micropollutants degradable by photocatalysis.

Category	Important subclasses	Examples degradable by photocatalysis
Pharmaceuticals	Lipid regulator, anti-inflammatory drugs, anticonvulsants, antibiotics	Tetracycline, lincomycin, salicylic, ciprofloxacin, naproxen, paracetamol, caffeine
Personal care products	Fragrances, preservatives, antimicrobials, insect repellents	Benzene, naphthalene, chloroxylenol
Surfactants	Nonionic surfactants	Sodium dodecyl sulfate, sodium dodecyl benzene sulfonate, trimethyl phosphate, tetrabutylammonium phosphate
Industrial chemicals	Plasticizers and plastic additives	Phthalic, bisphenol A
Pesticides	Insecticides, herbicides, and fungicides	Parathion, lindane, tetrachlorvinphos, fenitrothion
Heavy metals	Al, Cd, Cr, Cu, Fe, Pb, Hg, Ni, Zn	Al, Cu, Zn, Cr, Fe

A rather complete list of the micropollutants in water and wastewater and more examples of photocatalytic degradation can be found in [2, 7], respectively.

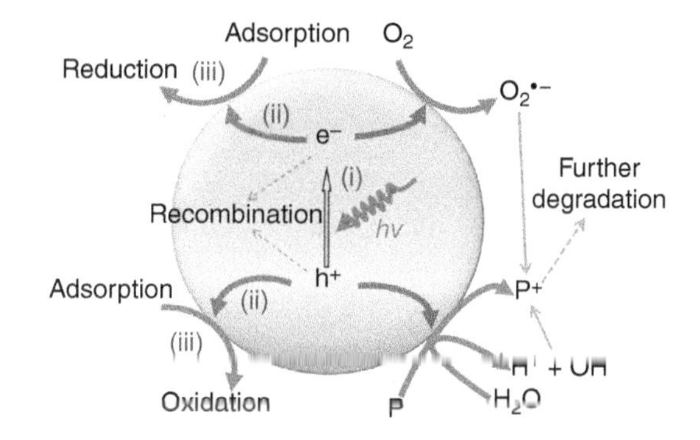

Figure 37.1 Schematic illustration of semiconductor-based photocatalytic processes. P stands for micropollutants.

the interface region [9]. Photocatalytic reactions differ from conventional catalysis only by the mode of activation of the catalyst in which thermal activation is replaced by photonic activation, which occurs in the adsorbed phase in step (iii). In general, this photonic activation process based on semiconductors involves three main steps (Figure 37.1). Specifically, (i) upon light irradiation, the photocatalyst absorbs supra-bandgap photons ($\geq E_g$, Eq. (37.1)), and photoexcited electron (e^-) and hole (h^+) pairs are produced in the conduction band (CB) and the valence band (VB), respectively (Eq. (37.2)). (ii) The photogenerated electrons and holes are either separated and migrated to catalytically active sites at the semiconductor surface or recombined. (iii) An efficient charge utilization of the electron–hole pairs allows the respective oxidation and reduction reactions with the adsorbates on the particle surface (Eqs. (37.3) and (37.4)).

$$\text{Bandgap (eV)} = \frac{1240}{\lambda(\text{nm})} \tag{37.1}$$

$$\text{PC} + h\nu(\geq E_g) \longrightarrow \text{h}^+ + \text{e}^- \tag{37.2}$$

$$A(\text{ads}) + \text{e}^- \longrightarrow A^-(\text{ads}) \tag{37.3}$$

$$D(\text{ads}) + \text{h}^+ \longrightarrow D^+(\text{ads}) \tag{37.4}$$

$$\text{e}^- + \text{O}_2 \rightarrow \text{O}_2^{\cdot -} \tag{37.5}$$

$$\text{h}^+ + \text{OH}^-(\text{or HO}) \longrightarrow {}^{\cdot}\text{OH} \tag{37.6}$$

$$\text{O}_2 + 2\text{H}^+ + 2\text{e}^- \longrightarrow \text{H}_2\text{O}_2 \tag{37.7}$$

The first reaction type for water pollutant abatement is oxidative degradation. It is well known that O_2 and water are essential for photooxidation. Molecular oxygen usually functions as an electron acceptor by interacting with the photogenerated electrons on the CB of the photocatalyst, thus forming reactive oxygen species (ROS), superoxide radicals ($O_2^{\cdot -}$), an oxidizing agent (Eq. (37.5)). The photogenerated holes can oxidize the micropollutants (P) to form P^+ or react with OH^- or oxidize H_2O into $OH^{\cdot}$ radicals, a different type of ROS acting as a strong oxidizing agent (Eq. (37.6)). Together with other highly oxidant species, peroxide radicals (Eq. (37.7)) are reported to be responsible for heterogeneous photocatalytic oxidation [4, 5]. Another reaction type is the direct reduction of micropollutants such as heavy metal ions in solution by photogenerated electrons [9, 10]. Heavy metals are generally toxic and can be removed from industrial waste effluents as small crystallites deposited on the photocatalyst according to the redox process, provided the redox potential of the metal cation is more positive than the flat band potential of the semiconductor.

37.3 Advancements in Photocatalysts for Photocatalytic Abatement of Micropollutants in Water and Wastewater

Various semiconductors including TiO_2 [11], ZnO [12], CeO_2 [13], WO_3 [14], C_3N_4 [15], $BiVO_4$ [16], Bi_2WO_6 [17] can be used as photocatalysts for micropollutant abatement in the aqueous environment [5, 18]. Among them, TiO_2 has been the most commonly used due to its well-known advantages such as cost-effectiveness, nontoxicity, regeneration ability, photocatalytic efficiency, and high chemical stability [8, 19]. However, a major drawback of pure TiO_2 is its relatively large bandgap, as it can only absorb a small portion of the ultraviolet (UV) radiation spectrum [20]. In addition, most of the photocatalysts encounter a common problem of fast charge carrier recombination, which results in low photocatalytic conversion efficiency. Therefore, it is highly desired to carry out proper modifications to enhance the photolysis potential of semiconductors.

37.3.1 Photocatalysts Components Optimization

An efficient approach is the modification of catalyst structure and composition, which is able to largely influence each reaction process, including light harvesting, charge carrier separation, and surface reaction kinetics, and ultimately results in photocatalytic performance regulation. The strategies for optimization of semiconductor-based photocatalyst components include semiconductor doping, metal deposition, quantum dots sensitization, plasmon-based photocatalysts, semiconductor combinations, and so on.

37.3.1.1 Semiconductor Doping

In most cases, proper dopants not only enhance the absorption of visible light due to the formation of localized or delocalized electronic states but also enhance the separation efficiency of photogenerated electron–hole pairs [5]. Metal ion doping with a lower dopant content can introduce localized electronic states, such a donor level above the VB (Figure 37.2a) or an acceptor level below the CB (Figure 37.2b) in the forbidden band of wide bandgap photocatalysts, such as TiO_2 and ZnO, which can narrow their bandgaps and thus enhance their activity under visible-light irradiation [20, 21]. Metal ion doping with a high dopant content often leads to the formation of the band of delocalized states in the middle of the bandgap (Figure 37.2c). On the other hand, some ions, especially nonmetal ions, such as N, I, B, Cl, F, S, P, and C, can contribute to new valence

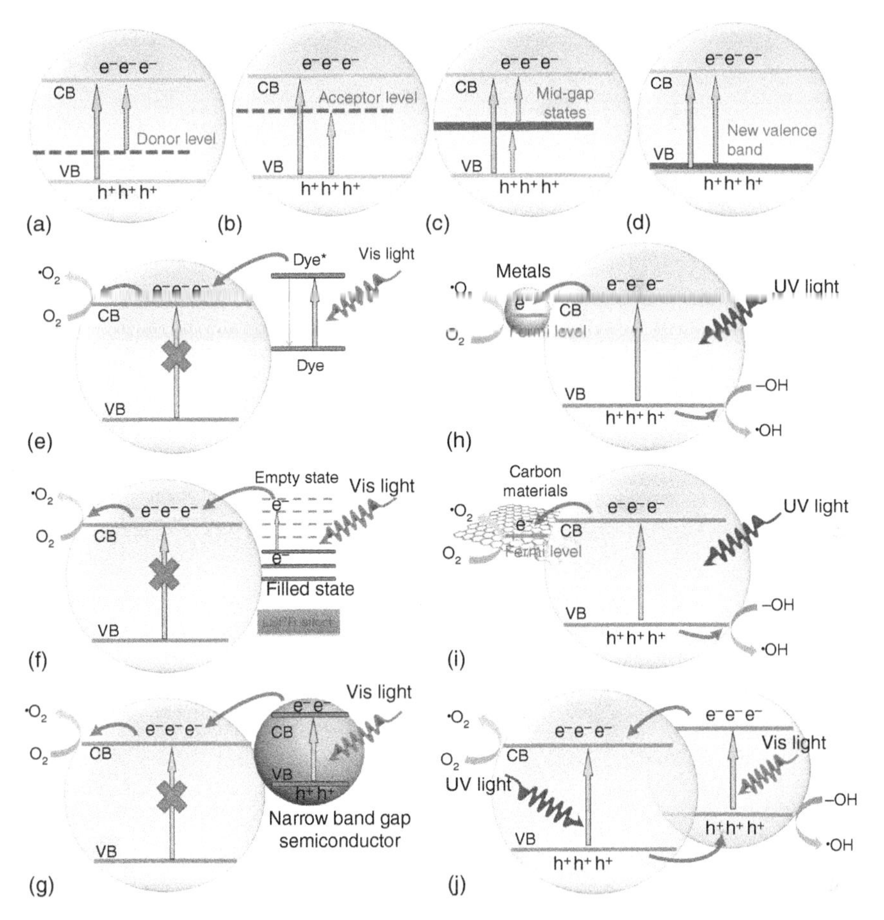

Figure 37.2 (a) Donor level, (b) acceptor level, (c) mid-gap states formed by metal ion doping; (d) new valence band formation by nonmetal ions doping; schematic basic principle of (e) dye-sensitized, (f) Localized surface plasmon resonance (LSPR)-sensitized, and (g) narrow bandgap semiconductor-sensitized activity under visible-light irradiation; schematic illustration of charge transfer for (h) metal deposition, (i) carbon materials combination, and (j) type II semiconductors combination. The * in Figure 37.2e stands for the excited state of the dye.

band formations and shift valence band edge upward to narrow the bandgap (Figure 37.2d) [20, 22, 23].

For example, N-doped TiO_2 has been reported to show excellent visible-driven photocatalytic activity for the degradation of pharmaceutical pollutants, including ciprofloxacin, naproxen, and paracetamol [24]. C-, N-, and S-doped mesoporous anatase–brookite nanoheterojunction titania have been synthesized through a simple sol–gel method in the presence of triblock copolymer pluronic P123. The photocatalytic efficiency of the catalyst under visible-light illumination dramatically increased with the addition of the C, N, and S nonmetals as dopants, achieving complete degradation (close to 100%) of cyanotoxin microcystin [25]. The results demonstrate the great potential of the visible-light-activated C-, N-, and S-doped titania photocatalysts for the treatment of organic micropollutants in contaminated waters under visible light.

37.3.1.2 Surface Sensitization

The surface modifications of wide bandgap photocatalysts using a combination of dye sensitization, metals with surface plasmon resonance (SPR) effects or other visible-light-responsive photocatalysts provide another way for visible-light activity enhancement. In dye sensitization, the dye is excited by absorbing visible light, causing charge injection into the CB of the semiconductor at sub-bandgap excitation, which is followed by catalytic processes through interfacial electron transfer (Figure 37.2e) [26]. The metallic plasmonic nanoparticles (such as Au, Ag, and Cu), anchored to a semiconductor, acting similarly to a dye sensitizer, which absorbs resonant photons and transfers the energetic electrons formed in the process of SPR excitation to the nearby semiconductor to initiate the photocatalytic reactions (Figure 37.2f) [20, 27]. In addition, combining wide bandgap semiconductors with narrow bandgap semiconductors has also been reported to result in visible-light activity (Figure 37.2g) [21, 28].

For example, nanocrystalline TiO_2 has been combined with D35 organic dye to fabricate a visible-light photocatalyst, which exhibits excellent visible-light degradation of bisphenol A (BPA). Under the experimental conditions with initial BPA concentration of 5 mg/L, initial pH of 7, external bias of 0.25 V, and sensitizer concentration of 0.1 mM, BPA was almost completely degraded in 300 minutes, and the four intermediates were gradually mineralized [29]. Interestingly, using a photosensitizing material tris(4,7-diphenyl-1,10-phenanthroline)ruthenium(II) chloride immobilized in a porous poly(dimethylsiloxane) inert support, Diez-Mato et al. have investigated the elimination of three common micropollutants such as ibuprofen, paracetamol, and BPA. The process effectively removed ibuprofen and BPA in ultrapure water with conversion rates of 100% and 80%, respectively [30]. Graphitic carbon nitride (g-C_3N_4), possessing a suitable bandgap (2.7 eV), has been employed in the modification of TiO_2 to form g-C_3N_4/TiO_2 nanocomposite, which showed superior photocatalytic performance for the degradation of ciprofloxacin (CIP) under visible-light irradiation, compared with pure TiO_2 and commercial P25 [31].

37.3.1.3 Metal Deposition

Considering the practical applications, a higher reaction rate is still required since the quantum efficiency under visible light is still lower than that under UV light [18, 32]. One way to improve the photocatalytic activity is to deposit metallic nanoparticles of noble metals (such as Pt, Pd, Au, Ag, Ru, and Fe) onto semiconductors to enhance the photocatalytic activity by suppressing the electron–hole pair recombination [28]. The photoinduced electrons migrate to the metal due to the relatively low Fermi level of metals (Figure 37.2h), which makes the photoinduced holes more stable by increasing the lifetime of the charge carrier. Therefore, more $OH^{\cdot}$ radicals and superoxide radicals are generated and enhanced the redox reaction.

For example, Xu's group has reported an efficient visible-light ($\lambda > 420$ nm)-driven photoreduction of Cr(VI) to Cr(III) over noble metal nanoparticles loaded on TiO_2 (P25) with oxygen vacancies (OV). The results show that the noble metal deposition not only decreases the concentration of OV, the charge recombination sites on the defective P25-OV, but also serve as electron sinks to promote the separation of electron–hole pairs, thereby enhancing the photoactivity [33]. Modifying TiO_2 surface with Pt and SiOx has accelerated the degradation of all the tested pharmaceuticals (i.e. caffeine, cimetidine, propranolol, and sulfamethoxazole) to a certain degree [14]. Moreover, Ag and Au cocatalysts decorated on $BiVO_4$ nanofibers have been developed to show greater photocatalytic performance for phenol degradation. Ag–$BiVO_4$ showed enhancement due to increased carrier traps, while Au–$BiVO_4$ showed enhancement due to both carrier traps and SPR. It is considered that $BiVO_4$ nanofibers have the potential to become efficient photocatalysts alternative to TiO_2 for the removal of emerging organic contaminants [34].

37.3.1.4 Carbon Materials Combination

Recently, combining semiconductors with various carbon materials, such as activated carbon nanofibers, carbon nanotubes (CNTs), and graphene, has been investigated as promising materials toward enhanced photocatalytic activities through suppression of e^-/h^+ recombination (the electron-sink effect of carbon materials benefited from their low Fermi level and high conductivity) (Figure 37.2i). In addition, the carbon materials provide high-surface-area scaffolds for semiconductor particles distribution and immobilization, resulting in greater photocatalytic efficiency [35, 36]. For example, enhanced photocatalytic activity of graphene-modified TiO_2 in comparison to bare TiO_2 has been demonstrated in the degradation of different types of micropollutants, including diuron, alachlor, isoproturon, microcystin-LA, BPA, and diphenhydramine. The enhanced activity was attributed to the good assembly and interfacial coupling between TiO_2 and graphene sheets as well as to the respective quenching of photoluminescence [37, 38].

37.3.1.5 Combining Semiconductors

Another promising approach is to combine different types of semiconductors to attain simultaneously a sufficient redox potential of photoexcited electron–hole pairs and efficient absorption of a large proportion of sunlight. Figure 37.2j illustrates the type II band structure (the most effective type for charge carrier transfer among three types of semiconductor heterojunctions) [39] of a composite photocatalyst prepared by a mixture of wide and narrow bandgap photocatalysts. Upon light irradiation, the photogenerated electrons can flow from the semiconductor with a higher CB minimum to the other with a lower CB minimum. Moreover, the formation of heterojunction leads to a more efficient inter-electron transfer between the two catalysts, thus enhancing photocatalytic activity. For example, using sunlight-driven g-C_3N_4/P25 photocatalyst, the degradation of clofibric acid (CA) was investigated. A very low g-C_3N_4 content of 8.0 wt% resulted in a 3.36 and a 2.29 times faster reaction rate for CA photodegradation compared to pristine g-C_3N_4 and P25, respectively [40]. A broad-spectrum N-doped carbon quantum dot/TiO_2 nanosheet composites with significantly improved broad-spectrum utilization exhibited higher photocatalytic activity toward the degradation of diclofenac. This excellent photocatalytic performance was largely attributed to the efficient charge separation induced by the fabricated heterostructures [41]. Semiconductors combination is a primary strategy to develop broad-spectrum photocatalysts for photocatalytic pollutant elimination.

37.3.2 Photocatalysts Configuration Optimization

37.3.2.1 Freestanding Particulate

In general, photocatalysts can be used either as freestanding particulates or as thin film upon coating on a substrate. As for particulate photocatalysts, the structure, size, and shape can be optimized to maximize the photocatalytic activities during the water treatment process, since the morphology or scale modification improves the adsorption capacity of the contaminants and the active surface area with nanoscales [5]. Figure 37.3a shows the various dimensions of the structure of photocatalysts: 0D spheres, 1D fibers, rods, and tubes, 2D nanosheets, and 3D interconnected architectures, as well as their property advantages. Typically, it is suggested that with the decreasing size of the nanoparticles, the e^-/h^+ recombination behaviors decrease due to the enhanced interfacial charge carrier transfer on the semiconductor surfaces [5]. 1D structure was reported to be efficient in

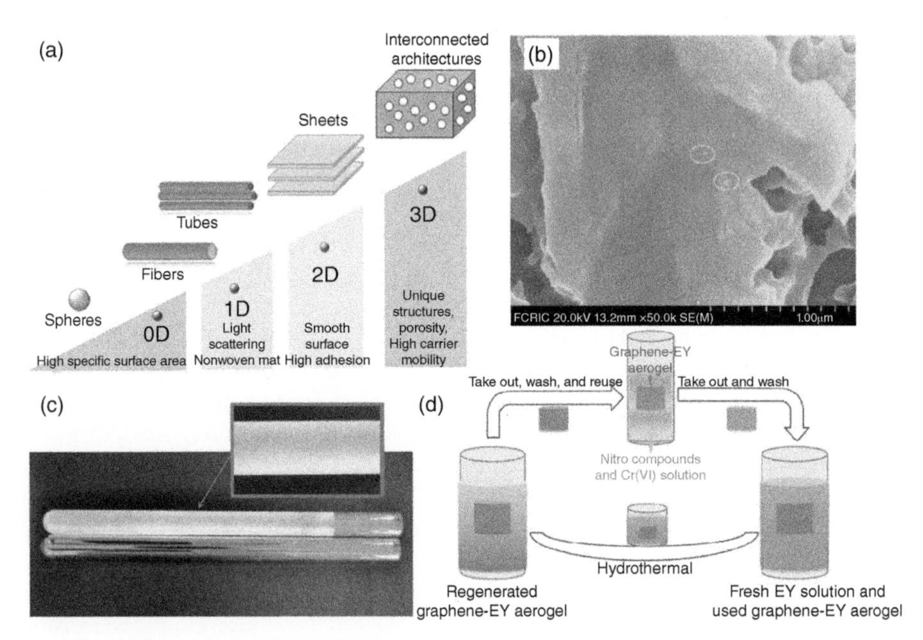

Figure 37.3 (a) Semiconductor structures according to the structural dimensionality and expected property. (b) SEM image of the nanocomposite Au NPs/TiO_2 films. NP, nanoparticles. (c) Images of borosilicate tube before and after abrasive blasting treatment. (d) Schematic illustration of the regeneration of 3D graphene–EY aerogel. Source: (a) Reprinted with permission from Lee and Park [5]. Copyright 2013, Elsevier; (b) Reprinted with permission from Nil et al. [46]. Copyright 2018, Springer-Verlag; (c) Reprinted with permission from Espino-Estévez et al. [47]. Copyright 2015, Elsevier; (d) Reprinted with permission from Yang et al. [48]. Copyright 2017, Elsevier.

the adsorption and decomposition of nonbiodegradable organic compounds by exhibiting a shorter carrier diffusion pathway and faster mass transfer of organic compounds [34]. Because of the unique mesoporous channel for facilitating the intraparticle molecular transfers of ions, the well-defined 3D architectures with a large specific surface area could further help increase their photocatalytic efficiency [12].

37.3.2.2 Film, Immobilized, and Aerogel-Based Catalysts

A great challenge for powder photocatalysts to be used in the treatment of wastewater lies in the difficulty of phase separation and recovery of the catalysts. In this context, coated catalyst configurations eliminate the need for catalyst filtration or centrifugation, thus possessing greater applicability in the repeated catalytic operations [42]. Therefore, more coated photocatalysts and immobilization techniques have been investigated, such as preparation of film materials and reactive photocatalysts immobilized on inert supports [43–45].

Recently, Au NPs/TiO_2 has been coated onto borosilicate glass disk to obtain thin-film nanocomposite materials and used in the photocatalytic degradation of emerging micropollutants [46]. The scanning electron microscopy (SEM) image (Figure 37.3b) of Au NPs/TiO_2 thin film showed excellent distribution of fine grains of TiO_2 on the surface of borosilicate glass, while Au nanoparticles were evenly distributed on the TiO_2 structure. The film exhibited efficient photocatalytic removal of sulfamethoxazole and triclosan from aqueous solutions using UV-A light (λ_{max} = 330 nm). The stability of the film was reassessed with the repeated use of the catalyst through simple wash and dry, which showed no significant decrease in photocatalytic degradation efficiency. The preparation of stable and highly photoactive titania coatings on inert supports is necessary for realistic applications of TiO_2 photocatalysis in water purification treatments. The coatings of TiO_2 on the outer wall of the inner tube of a glass tubular reactor by dip-coating method (Figure 37.3c) have been reported [47]. The results of the adhesion tape test showed that either milling of aggregate material with a planetary mill or chemical stabilization of the particles was necessary to obtain TiO_2 coatings on the glass tube with acceptable quality to be used in water treatment applications. The photoactivity results showed that the coatings prepared with 30 minutes wet milling of the catalyst followed by chemical disaggregation were more efficient in degradation and mineralization of phenol, diclofenac, and isoproturon. The reusability of the TiO_2 coatings was evaluated, and a promising photocatalytic performance was observed with a very low variation in the decay rate after five consecutive usages.

3D graphene aerogel materials are another viable option of using the photocatalysts that make the phase separation easy. The 3D graphene aerogel architectures possess not only the inherent properties of graphene (GR) of high electrical conductivity but also unique hierarchically porous structures for reactant adsorption, both of which are beneficial for the degradation of pollutants. Typically, organic dyes Eosin Y (EY) as photosensitizers have been spatially confined and distributed in the graphene framework to form the 3D macroscopic graphene–organic aerogel photocatalysts, which display efficient visible-light-driven activity toward photoreduction of Cr(VI) to Cr(III) [48].

In addition, such a bulk aerogel manifests excellent regenerability via simple replenishment of fresh dyes, which guarantees the long-term photoactivity of the 3D aerogel (Figure 37.3d).

37.4 Reaction System Optimization

37.4.1 Reaction Conditions

The rate and efficiency of a photocatalytic reaction depend on a number of factors that govern the kinetics of photocatalysis. An extensive review of the influential variables showed that dissolved oxygen, light intensity distribution, loading of the photocatalyst, air flow rate, temperature, and hydrogen peroxide concentration had constructive impacts on the process performance, while initial concentration of the reactant, light wavelength, feed flow rate, irradiation time, and pH showed detrimental effects [3, 9]. For more details on the effect of variables on the process performance, readers are referred to Ref. [18].

37.4.2 Solar Reactors

Depending on the reaction considered, various photoreactors can be chosen, including fixed-bed photoreactors and slurry batch photoreactors, either mechanically or magnetically stirred [9, 49]. In laboratory experiments, near-UV light is provided by a lamp placed in front of an optical window of the photoreactor, with wavelength adjustable using optical filters. For different types of photocatalytic reactors tested, refer to Ref. [18]. For pilot experiments using sunlight, extensive effort has been devoted to the design of efficient solar light collectors. The original solar photoreactor designs for photochemical were based on line-focusing parabolic trough concentrator (PTC) (Figure 37.4a), which is based on conventional solar thermal collector designs. Subsequently, it has been found that PTCs were unsuitable for photocatalytic applications since water was heated and radiation flux was too high. Most of the photons were not used efficiently, and their cost was high. Attempts were made to use nonconcentrating solar collectors (NCCs) (Figure 37.4b) as an alternative to PTCs. Despite important advantages, the design of NCC is not trivial, due to the need for weather-resistant and chemically inert UV-transmitting reactors. Besides, NCC designs possess other disadvantages of low mass transfer and reactant vaporization and contamination. In this regard, compound parabolic concentrators (CPCs) (Figure 37.4c), a type of low-concentration collector used in thermal applications, becomes an option of interest. Having the advantages of both nonconcentrating and concentrating and testing of small nontracking systems without their original disadvantages, CPCs seem to be the best option for solar photocatalytic processes. Figure 37.4d shows the photographs of a commercial nonconcentrating solar detoxification system using the CPC technology, consisting of a TiO_2 separation system (left) and a CPC (right). For more details of the development of the photoreactors, see the reviews from the group of S. Malato [3, 50].

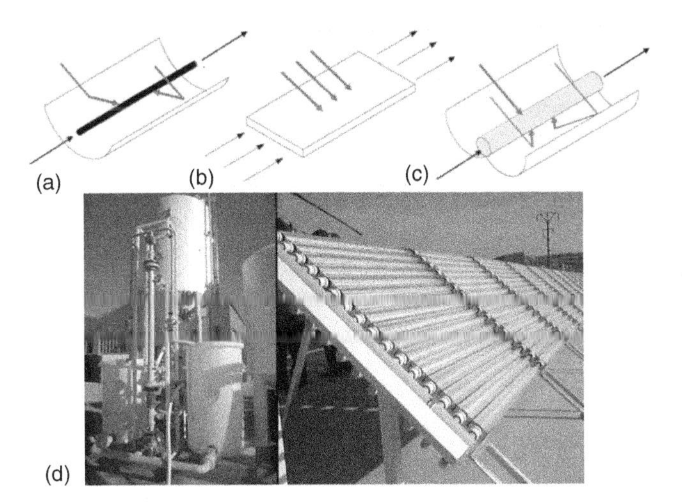

Figure 37.4 Design concepts for solar water photocatalytic reactors: (a) concentrating (parabolic trough), (b) nonconcentrating (one-sun reactor), and (c) compound parabolic collector. (d) Photographs of solar detoxification demonstration plant constructed in "SOLARDETOX" project at HIDROCEN (Madrid, Spain). Left: TiO_2 separation system. Right: compound parabolic collector. Source: Reprinted with permission from Malato et al. [3]. Copyright 2009, Elsevier.

In summary, the design procedure for a solar photocatalytic system requires the selection of a reactor, catalyst operating mode (slurry or fixed matrix), reactor-field configuration (series or parallel), treatment system mode (once-through or batch), flow rate, pressure drop, pretreatment, catalyst and oxidant loading method, pH control, and so on. The optimization of these reaction conditions from a system level is of great importance to achieve a desirable photocatalytic efficiency.

37.5 Future Challenges and Prospects

Heterogeneous photocatalysis appears to be one of the most promising potential applications for micropollutant abatement in water and wastewater, since many toxic micropollutants, either organic or inorganic, can be oxidized or reduced into harmless final compounds under ambient conditions. Despite the attractive advantages and great progresses made in the design of efficient materials and photocatalytic systems, it still remains a big challenge to achieve efficiency and practicality potentially applied in industrial operations. Thus, it is more realistic to address the package of challenges as a whole in this research field.

First, more efficient photocatalytic engineering should be developed in terms of photocatalysts and reaction systems optimization. In particular, for practical applications, it is necessary for the water to be transparent before the photocatalytic treatment. In this regard, conventional wastewater treatment plants (WWTPs), including physical and chemical adsorption, coagulation, precipitation, and biotransformation, as well as advanced oxidation processes (AOPs) such as UV radiation, UV/Cl_2, UV/O_3, and UV/H_2O_2 treatment, should be combined to pave the way for novel integrated water treatment technologies, thereby overcoming high treatment costs. Second, monitoring the disappearance rate of the target micropollutants is not the most appropriate parameter to classify the efficiency. More efforts should be devoted to understanding the thorough mechanisms with detailed reaction steps of the different pathways, the quantification of various intermediates, and by-product evaluation, which are key factors to optimize each treatment and to maximize the overall process. Third, toxicity tests of the treated water will also gather useful information about the practical application of the photocatalytic process, particularly when incomplete degradation is planned. A better understanding of the photocatalytic process and the operative conditions could give great opportunities for its application in the abatement of emerging micropollutants in water and wastewater.

Acknowledgments

The support from the National Natural Science Foundation of China (NSFC) (U1463204, 21872029, and 21173045), the Award Program for Minjiang Scholar Professorship, the Natural Science Foundation (NSF) of Fujian Province for Distinguished Young Investigator Rolling Grant (2017J07002), the Independent Research Project of State Key Laboratory of Photocatalysis on Energy and Environment (No. 2014A05), the 1st Program of Fujian Province for Top Creative Young Talents, and the Program for Returned High-Level Overseas Chinese Scholars of Fujian Province is gratefully acknowledged.

References

1 Kim, M.-K. and Zoh, K.-D. (2016). *Environ. Eng. Res.* 21: 319–332.
2 Margot, J., Rossi, L., Barry, D.A., and Holliger, C. (2015). *WIREs Water* 2: 457–487.
3 Malato, S., Fernández-Ibáñez, P., Maldonado, M.I. et al. (2009). *Catal. Today* 147: 1–59.
4 Chong, M.N., Jin, B., Chow, C.W.K., and Saint, C. (2010). *Water Res.* 44: 2997–3027.
5 Lee, S.-Y. and Park, S.-J. (2013). *J. Ind. Eng. Chem.* 19: 1761–1769.
6 Luo, Y., Guo, W., Ngo, H.H. et al. (2014). *Sci. Total Environ.* 473–474: 619–641.
7 Gaya, U.I. and Abdullah, A.H. (2008). *J. Photochem. Photobiol. C: Photochem. Rev.* 9: 1–12.
8 Akpan, U.G. and Hameed, B.H. (2009). *J. Hazard. Mater.* 170: 520–529.
9 Herrmann, J.-M. (1999). *Catal. Today* 53: 115–129.
10 Amin, M.T., Alazba, A.A., and Manzoor, U. (2014). *Adv. Mater. Sci. Eng.* 2014: 24.
11 Mahmoud, W.M.M., Rastogi, T., and Kummerer, K. (2017). *Curr. Opin. Green Sustainable Chem.* 6: 1–10.

12 Chang, J.S., Tan, J.K., Shah, S.N. et al. (2017). *J. Taiwan Inst. Chem. Eng.* 81: 206–217.

13 Sponza, D.T. and Guney, G. (2017). *Water Sci. Technol.* 76: 2603–2622.

14 Choi, J., Lee, H., Choi, Y. et al. (2014). *Appl. Catal., B* 147: 8–16.

15 Zheng, Q., Durkin, D.P., Elenewski, J.E. et al. (2016). *Environ. Sci. Technol.* 50: 12938–12948.

16 Chen, S.J., Zhai, X.N., Li, Y.J. et al. (2018). *J. Nanosci. Nanotechnol.* 18: 2472–2480.

17 Chen, S.J., Li, Y.J., Lu, R.J., and Wang, P. (2013). *J. Nanosci. Nanotechnol.* 13: 5624–5630.

18 Zangeneh, H., Zinatizadeh, A.A.L., Habibi, M. et al. (2015). *J. Ind. Eng. Chem.* 26: 1–36.

19 Konstantinou, I.K. and Albanis, T.A. (2004). *Appl. Catal., B* 49: 1–14.

20 Dong, H., Zeng, G., Tang, L. et al. (2015). *Water Res.* 79: 128–146.

21 Li, X., Yu, J., Low, J. et al. (2015). *J. Mater. Chem. A* 3: 2485–2534.

22 Serpone, N. and Emeline, A.V. (2012). *J. Phys. Chem. Lett.* 3: 673–677.

23 Lee, K.M., Lai, C.W., Ngai, K.S., and Juan, J.C. (2016). *Water Res.* 88: 428–448.

24 Shetty, R., Chavan, V.B., Kulkarni, P.S. et al. (2017). *Indian Chem. Eng.* 59: 177–199.

25 El-Sheikh, S.M., Zhang, G.S., El-Hosainy, H.M. et al. (2014). *J. Hazard. Mater.* 280: 723–733.

26 Chatterjee, D. and Mahata, A. (2001). *Appl. Catal., B* 33: 119–125.

27 Zhang, N., Han, C., Fu, X., and Xu, Y.-J. (2018). *Chem* 4: 1832–1861.

28 Pelaez, M., Nolan, N.T., Pillai, S.C. et al. (2012). *Appl. Catal., B* 125: 331–349.

29 Bai, X., Yang, L., Hagfeldt, A. et al. (2019). *Chem. Eng. J.* 355: 999–1010.

30 Diez-Mato, E., Cortezon-Tamarit, F.C., Bogialli, S. et al. (2014). *Appl. Catal., B* 160: 445–455.

31 Yang, Z., Yan, J., Lian, J.B. et al. (2016). *ChemistrySelect* 1: 5679–5685.

32 Hu, Y., Zhang, X., and Wei, C. (2010). *Res. Chem. Intermed.* 36: 95–101.

33 Pan, X. and Xu, Y.-J. (2013). *J. Phys. Chem. C* 117: 17996–18005.

34 Nalbandian, M.J., Zhang, M.L., Sanchez, J. et al. (2015). *J. Mol. Catal. A: Chem.* 404: 18–26.

35 Zhang, N., Yang, M.-Q., Liu, S. et al. (2015). *Chem. Rev.* 115: 10307–10377.

36 Yang, M.-Q., Zhang, N., Pagliaro, M., and Xu, Y.-J. (2014). *Chem. Soc. Rev.* 43: 8240–8254.

37 Moreira, N.F.F., Narciso-da-Rocha, C., Polo-Lopez, M.I. et al. (2018). *Water Res.* 135: 195–206.

38 Martins, P.M., Ferreira, C.G., Silva, A.R. et al. (2018). *Composites Part B* 145: 39–46.

39 Wang, Y., Wang, Q., Zhan, X. et al. (2013). *Nanoscale* 5: 8326–8339.

40 Chen, P., Wang, F.L., Zhang, Q.X. et al. (2017). *Chemosphere* 172: 193–200.

41 Wang, F.L., Wu, Y.L., Wang, Y.F. et al. (2019). *Chem. Eng. J.* 356: 857–868.

42 Luna, R., Solis, C., Ortiz, N. et al. (2018). *Int. J. Chem. Reactor Eng.* 16.

43 Bellobono, I.R. and Groppi, F. (2017). *Curr. Opin. Green Sustainable Chem.* 6: 69–77.

44 Sanches, S., Nunes, C., Passarinho, P.C. et al. (2017). *J. Chem. Technol. Biotechnol.* 92: 1727–1737.
45 Arlos, M.J., Hatat-Fraile, M.M., Liang, R. et al. (2016). *Water Res.* 101: 351–361.
46 Nil, L., Tiwari, A., Shukla, A. et al. (2018). *Environ. Sci. Pollut. Res.* 25: 20125–20140.
47 Espino-Estévez, M.R., Fernández-Rodríguez, C., González-Díaz, O.M. et al. (2015). *Chem. Eng. J.* 279: 488–497.
48 Yang, M.-Q., Zhang, N., Wang, Y., and Xu, Y.-J. (2017). *J. Catal.* 346: 21–29.
49 Pronina, N., Klauson, D., Rudenko, T. et al. (2016). *Photochem. Photobiol. Sci.* 15: 1492–1502.
50 Malato, S., Blanco, J., Alarcón, D.C. et al. (2007). *Catal. Today* 122: 137–149.

38

Catalytic Abatement of NO_x Emissions over the Zeolite Catalysts

Runduo Zhang, Peixin Li, and Hao Wang

College of Chemical Engineering, Beijing University of Chemical Technology, State Key Laboratory of Chemical Resource Engineering, No.15 North 3rd Ring Road, Chaoyang District, Beijing, 100029, PR China

For the past 50 years, there has been a growing focus on NO_x emission control from both stationary and mobile sources, leading to worldwide research on deNO_x technologies. Over 40% of NO_x emissions are from stationary sources, such as power plants and industrial boilers using fossil fuels, and over 50% come from automotive sources, such as gasoline and diesel engine cars. Currently, three commercial catalytic systems are available: noble-metal-based three-way catalysts (TWCs) for the purification of gasoline engine exhausts, transition-metal–zeolite catalysts for the purification of diesel engine exhausts, and vanadium–titania catalysts for the elimination of power plant effluent gases. However, due to stricter environmental legislation and the demand to achieve energy savings, there is an increasing desire to develop more efficient deNO_x catalysts.

For the stationary deNO_x technology, especially for that used in power plants, it is required that the industrial catalyst possesses excellent low-temperature activity as well as high N_2 selectivity. In addition to these criteria, the automobile deNO_x demands a wide temperature window for ideal performance due to the large temperature swings in the exhaust gases. Selective catalytic reduction (SCR) was first discovered by Engelhard and patented in 1957, and it was mainly summarized into the following four parts:

(1) *Noble-metal-supported alumina (Pt, Ph, Pd, Ag)*: These catalysts were developed as catalytic materials for emission control in the early 1970s and became the standard for SCR reaction, particularly for diesel engine emissions. The reaction uses CO and H_2 or hydrocarbons as the reducing agents while requiring temperature above 300 °C to achieve decent conversions [1, 2]. The noble metal catalyst is not only expensive but is also prone to sulfur poisoning [3–5] and restricted to a narrow operation temperature range [6, 7]. These factors limit their wider implementations.
(2) *Metal oxide catalysts*: Japan developed the SO_2-resistant V_2O_5/TiO_2 catalyst in the 1970s [8, 9], to satisfy the request according to country's environmental protection policy. The catalyst exhibited high activity for NO_x reduction in the temperature range of 300–400 °C [10, 11] and for oxidative

Heterogeneous Catalysts: Advanced Design, Characterization and Applications, First Edition.
Edited by Wey Yang Teoh, Atsushi Urakawa, Yun Hau Ng, and Patrick Sit.

desulfurization of flue gas [12, 13]. The reaction used ammonia or urea as the reducing agent. Despite being commercialized since 1978, the vanadium catalyst remains widely used today. However, the temperature of the plant obviously declined after desulfurization and dedusting, which reduces the corresponding catalytic activity. Simultaneously, due to the highly biological toxicity and relatively high temperature necessary for the reaction, the use of the vanadium catalyst has been largely limited in the United States and Japan. To overcome this drawback, an effort on designing an efficient and environmentally friendly low-temperature catalyst for NH_3-SCR has been made.

(3) *Carbon-based catalysts*: These catalysts were used as a catalyst carrier for SCR reaction with the large surface area and high adsorption capacity [14, 15]. The currently studied carbon-based catalysts mainly include activated carbon (AC) [16], activated carbon fibers (ACFs) [17], and AC moldings [18]. Due to complicated preparation process, high cost, and instability under an oxidative atmosphere, they lack practical value and are difficult to popularize. At present, the carbon-based catalysts are still in the state of laboratory development.

(4) *Zeolite catalysts*: Zeolites are water-containing aluminosilicates of natural or synthetic origin with a highly ordered crystal structure, generally formulated as $M^IM^{II}_{0.5}[(AlO_2)_x{\cdot}(SiO_2)_y{\cdot}(H_2O)_z]$, where M^I and M^{II} are preferentially alkali and alkaline-earth metals. They consist of SiO_4^- and AlO_4^- tetrahedra, which are interlinked by common oxygen atoms to form a three-dimensional network. The tetrahedra are the smallest structural units by which zeolites can be classified. Linking these primary building units together leads to 23 possible secondary building blocks (polygons). Zeolites are predominantly distinguished on the basis of the geometries of the cages (α, β, γ, and super) and channels (straight and sinusoidal) formed by the rigid tetrahedral frameworks. To date, 229 types of zeolite framework structures have been recognized by the Structure Commission of the International Zeolite Association [19].

With the high catalytic activity for NH_3-SCR and a wide range of active temperature, zeolites have been attracting attention in the research. The types of zeolites used mainly include Y (FAU), ZSM-5 (MFI), β (BEA), SSZ-13 (CHA), mordenite (MOR), etc. The zeolite catalysts for the SCR process are mainly prepared by ion-exchange method, and the metal elements that can be used for ion exchange usually include Mn, Cu, Co, Pd, V, Ir, Fe, Ce, etc. More details are given in the following sections. Summarized above, zeolite catalysts are the most promising deNO_x catalysts. The research direction has thus been pursued: designing an efficient low-temperature zeolite catalyst for the NH_3-SCR system.

38.1 Zeolite Catalysts with Different Topologies

Metal-exchanged zeolites represent one kind of SCR catalysts, which exhibit considerable performances across a wide temperature range. Fe- and Cu-modified zeolite catalysts were widely investigated for NH_3-SCR, due to their higher activities and thermal stabilities with respect to the commercialized vanadium catalysts [20]. Cu-based zeolite catalysts (Cu–ZSM-5, Cu-β, Cu–mordenite) commonly exhibited higher activities at lower temperature ranges (<400 °C) [21, 22], because Cu catalysts are capable of high NO_x conversions at $T < 400$ °C; however, the overconsumption of NH_3 would suppress the related $deNO_x$ activities as the temperature increased ($T > 400$ °C). Activity evaluation suggested that the simultaneous presence of Cu and Fe species in β-type zeolites significantly improved the low-temperature NO conversion. Although a large amount of research have been done over the conventional zeolite catalysts with medium micropores, such as β (BEA), ZSM-5 (MFI), mordenite (MOR), and ferrierite (FER) [23, 24], several practical challenges arise when these metal-exchanged zeolites are used in lean-burn or diesel vehicles. Two of the most serious problems are hydrothermal deactivation and chemisorption of impurities (hydrocarbon, phosphorus, and potassium) on the active sites of catalysts [25, 26]. Recently, many researchers concentrate on the newly discovered small-pore Cu-exchanged zeolite catalysts with a CHA topology, such as Cu–SAPO-34 and Cu–SSZ-13, which were frequently reported to be extremely active for NH_3-SCR with high hydrothermal stability [27, 28]. NH_3-SCR performance was gradually improved along with the zeolite framework structures changing from straight channel (ZSM-5, β, Y) to cage type (SSZ-13, SSZ-16, SSZ-17), bridged by hybrid structures (OFF-ERI series: offretite, ZSM-34, UZM-12) [29]. The catalytic activities for NH_3-SCR of NO_x over Cu-exchanged zeolites with diverse topologies are illustrated in Figure 38.1. Cage-type zeolites exhibit a superior activity (NO_x conversion >95% at a relatively wide temperature range of 150–400 °C) compared to Cu–ZSM-5 and Cu-β at either low temperatures (<200 °C) or high temperatures (>400 °C). For the structural stability, cage-type zeolites exhibit the best resistance to dealumination and have long life,

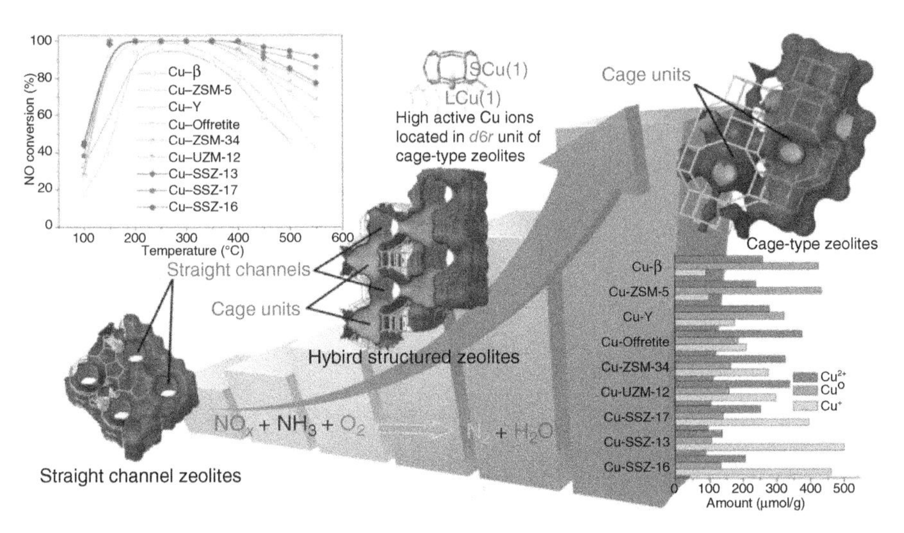

Figure 38.1 Catalytic activity profiles for NH_3-SCR of NO_x over Cu zeolites. Source: Reprinted with permission from Wang et al. [29]. Copyright 2018, Elsevier B.V.

due to their small pore openings. For the exchanged Cu sites, straight channels favor the formation of dimeric $[Cu–O–Cu]^{2+}$, while eight-membered ring enclosed cages favor the formation of isolated monomers Cu^{2+} and $Cu(OH)^{+}$. It was realized that cage-type zeolites favored the formation of copper cations, especially generating much more Cu^{+} ions than the others, rather than CuO [30].

38.2 Essential Nature of Novel Cu–CHA catalyst

38.2.1 Shape Selectivity

CHA-type SSZ-13 has a three-dimensional tetrahedral framework composed of double six-membered rings (*d6r*) in an AABBCC sequence, forming a "cage" per unit cell as depicted in Figure 38.2. It has been pointed out that SSZ-13 exhibits outstanding SCR performance (~100% NO conversion) and excellent N_2 yields (>95%) across a wide temperature range of 150–400 °C after copper ion exchange [30, 32]. This remarkable performance is certainly attributable to the special topology of the SSZ-13 zeolite.

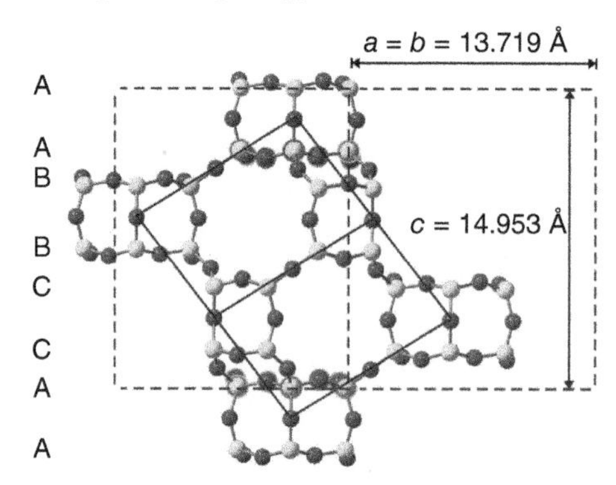

Figure 38.2 Hexagonal unit cell of an SSZ-13 zeolite (dashed lines) illustrating the AABBCCAA stacking sequence and equivalent rhombohedral unit cell (solid lines). The zeolite cage (delimited by 12 4-MR, 6 8-MR, and 2 highlighted 6-MR) is depicted as well. Light gray spheres (yellow line) are Si atoms and dark gray spheres (red line) are O atoms. Source: Reprinted with permission from McEwen et al. [31]. Copyright 2012, Elsevier. (See online version for color figure).

An attempt has been made to correlate the SCR behavior with the unique topology of SSZ-13 on the basis of the "shape selectivity" concept [33], which means that only the reactant, intermediate, and product of a certain size and shape can penetrate into the interior of the zeolite pores and undergo reaction at catalytically active sites. The supersmall pores (~3.80 Å, 8-membered rings [MR]) of SSZ-13, which serve as the exits of the reaction channels, are smaller than the dimensions of the undesired by-products (NO_2 [>3.83 Å] and N_2O [3.83 Å]); however, they are passable by the target product (N_2 [3.64 Å]). Accordingly, the pore size effect may provide an explanation for the lesser amounts of harmful NO_2 and N_2O by-products and the high N_2 yields achieved over Cu–SSZ-13. In addition, it has been widely accepted that both C_3H_6 and SO_2 often cause deactivation during the catalytic elimination of NO by Cu zeolites. Although C_3H_6 can participate in the SCR reaction as a reductant, it has the disadvantage of giving rise to carbonaceous deposition on the reactive centers, especially at low temperatures [28, 34]. Moreover, SO_2 poisoning is frequently observed, primarily because of its reaction with Cu active sites to form $CuSO_3$ or $CuSO_4$ species at $T < 200$ °C. Fortunately, the pore openings (3.8 Å) of Cu–SSZ-13 are smaller than the kinetic diameters of both C_3H_6 (4.68 Å) and SO_2 (4.11 Å). As a result, the pore size restriction can greatly inhibit the diffusion of C_3H_6 and SO_2 molecules into the main channels of SSZ-13, thereby preserving a majority of the active sites from poisoning. In other words, the shape selectivity exhibited by SSZ-13 due to its supersmall pores is confirmed to be responsible for its high resistance to C_3H_6 and SO_2 poisoning.

38.2.2 Cation Location

Due to its unique structural topologies with supersmall pores, the chemical status of active sites constituted one of the hottest topics during NH_3-SCR investigation on the CHA-type zeolite catalysts. The location of Cu^{2+} cations inside the zeolite pores was investigated by using Rietveld refinement of synchrotron-based powder X-ray diffraction (XRD) [35]. The results suggested that Cu^{2+} ions were mostly isolated and exclusively occupied in the plane of 6-MR of SSZ-13. Moreover, XRD patterns further suggested that the thermal stability of SSZ-13 was improved significantly after copper introduction compared with the parent zeolite with an acidic form. The characterization methods of electron paramagnetic resonance (EPR) and temperature-programmed reduction with hydrogen (H_2–TPR) were used to study the distribution of Cu species for Cu–SSZ-13 prepared by the wet ion-exchange (WIE) method [27]. Five kinds of locations for Cu^{2+} cations (A–E) were proposed, as depicted in Figure 38.3, showing one unit cell of CHA topology. The following conclusions were accordingly reached:

(i) Only under dehydrated conditions and low Cu loading amount (ion-exchange level <30%), Cu^{2+} ions can be located within the face of 6-MR (position A′ in Figure 38.3).
(ii) H_2O and NH_3 molecules can strongly bond to the Cu^{2+} ions during NH_3-SCR process, which induces a slight migration of Cu^{2+} ions to site A.
(iii) As the Cu loadings increase to a certain extent, two Cu^{2+} ions are located in one unit cell (ion-exchange level >80%). The second Cu^{2+} could be located

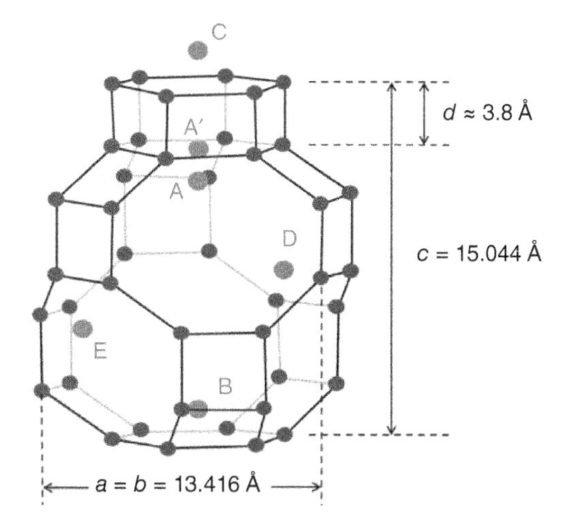

Figure 38.3 Schematic of the SSZ-13 hexagonal unit cell structure and possible Cu^{2+} locations. Source: Reprinted with permission from Gao et al. [27]. Copyright 2013, Elsevier.

in position B or C if it stays close to a 6-MR; or instead, it might be located inside the large cage if it is close to an 8-MR (position D or E).

The density functional theory (DFT) method was used to check the location and energy of Cu ions of Cu–SSZ-13 zeolite [36]. The isolated Cu^{2+} cations were confirmed to be more favorably located at 6-MR, which is consistent with those reported by experimental approaches [37]. However, in the presence of various adsorbates with –OH ligand, for example, $[Cu^{II}(OH)]^+$, the extra-framework site in the 8-MR was found to be more energetically stable than that of 6-MR for Cu^{2+} ions. Furthermore, the effect of Si/Al and Cu/Al ratios was also taken into account during DFT calculations. It was revealed that the Cu ions could be three- and fourfold coordinated to the lattice O atoms of Cu–SSZ-13 with different Si/Al ratios, as shown in Figure 38.4 for ZCu and Z_2Cu models. Examination of the partial density of state (PDOS) further verified that the Cu ions were, respectively, in +1 and +2 oxidation states in ZCu and Z_2Cu. Additionally, the infrared (IR) vibration frequencies of NO adsorption on Cu^+, Cu^{2+}, and $[Cu^{II}(OH)]^+$ sites were DFT calculated, being located in the range of 1770–1808, 1850–1950, and 1870–1915 cm^{-1}, respectively, and also consistent with those derived from the IR experimental results. This finding verified the correctness of DFT simulations [31]. Based on the DFT, the oxidation state and coordination environment of Cu active sites in Cu–SSZ-13 were also investigated during NH_3-SCR. This indicated that the fourfold coordinated Cu(II) was the dominant Cu species under the "fast SCR" and "slow SCR" conditions, wherein the NO_2/NO_x ratios were 0.5 and 1, respectively. Under the standard SCR conditions, containing no NO_2 in the feed, the mixed Cu(I) and Cu(II) oxidation states could be both observed. As

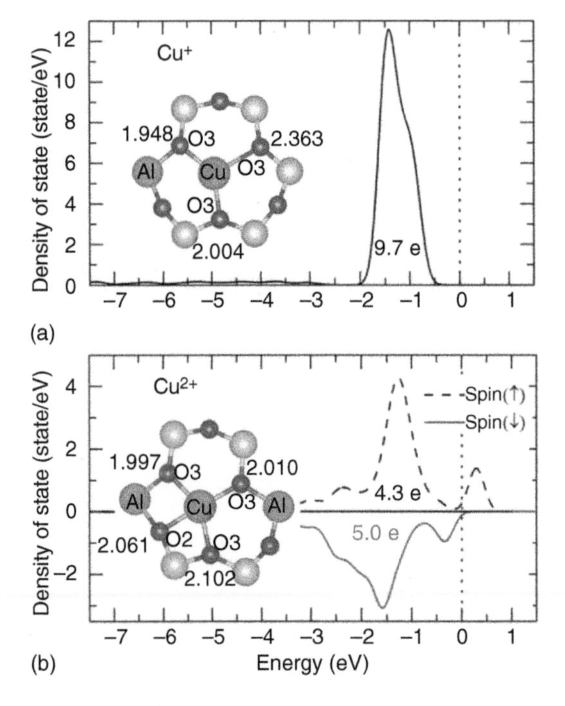

Figure 38.4 Partial density of state (PDOS) of Cu 3d states in (a) ZCu and (b) Z_2Cu. Insets are the local structures of Cu in the 6-MR sites. Cu—O bond lengths are indicated in units of Å. The numbers of integrated electrons from PDOS plots are shown. In (b), the dashed and solid lines show spin-up and spin-down states for Cu 3d, respectively, and the dotted vertical line highlights the location of the Fermi level. Source: Reprinted with permission from Zhang et al. [36]. Copyright 2014, American Chemical Society.

also reported, twofold Cu(I) and fourfold Cu(II) bound with H_2O or OH were found to be the most stable species over a wide range of $deNO_x$ conditions [36].

38.2.3 Copper Status

As stated earlier, the Si/Al ratio could influence the dispersion as well as the detailed chemical status of the introduced Cu active species. Combination of the experimental and theoretical (DFT) approaches could investigate the chemical status of Cu species doped on SSZ-13 [38]. This indicates that at least two kinds of Cu species existed over this zeolite catalyst, which was closely dependent on the ratio of copper species to total Al atoms in zeolite (Cu/Al_{total}). As $Cu/Al_{total} < 0.2$, Cu ions were dominant in the status of isolated Cu^{2+} cations located near the 6-MR, while Cu_xO_y species could be formed as $Cu/Al_{total} > 0.2$ being located in the 8-MR. A series of Cu-exchanged SSZ-13 catalysts ($Cu/Al_{total} = 0–0.41$) for the standard NH_3-SCR were investigated to verify the effect of Cu/Al_{total} ratio on copper status [39]. The isolated Cu^{2+} ions, acting as the active centers, were confirmed to be located at the 6-MR of SSZ-13 during NH_3-SCR as $Cu/Al_{total} < 0.2$, and the standard SCR reaction rate increased linearly up to $Cu/Al_{total} = 0.2$, with a maximum of 3.8×10^{-6} mol NO/g cat/s. Because NH_3 acting as the actual reducing agent plays a key role in SCR reaction, investigations to identify the species formed by NH_3 adsorption upon Cu–SSZ-13 and their involvement were put forward based on *in situ* EPR, solid-state nuclear magnetic resonance (NMR), and DFT calculations [40]. Five kinds of NH_3 adsorption modes were observed under different conditions: $[Cu(NH_3)_5]^{2+}$, $[Cu(O_f)_2(NH_3)_2]^{2+}$, $[Cu(O_f)_3NH_3]^{2+}$, $[Cu(NH_3)_2]^+$, and $[CuO_f(NH_3)]^+$ (O_f representing the framework oxygen). The adsorbed NH_3 was demonstrated to be able to reduce Cu^{2+} ions into Cu^+ ions.

38.2.4 CHA-Type Silicoaluminophosphate

Silicoaluminophosphate-34 (SAPO-34), being generated from the incorporation of an Si atom into neutral $AlPO_4$, has the same spatial topology as that of SSZ-13. According to the literature, Cu–SAPO-34 catalysts exhibit outstanding activities and durability for the NH_3-SCR process, and their active site status has attracted an especially large amount of attention. In an early study, a series of Cu–SAPO-34 samples were prepared by a WIE and precipitation method for NO reduction [41]. Various techniques (XRD, H_2–TPR, scanning transmission electron microscopy [STEM], and diffused reflectance infrared Fourier transform spectroscopy [DRIFTS]) were used to identify the location and status of Cu species in these samples. The results consistently indicated that the Cu species existed predominantly as isolated ions at the exchange sites in the ion-exchanged samples; however, as for the precipitated sample, CuO on the external surface was the dominant species. Superior NH_3-SCR activity was observed for the ion-exchanged samples, suggesting that isolated Cu cations at the exchange sites constituted the active centers. A series of Cu–SAPO-34 samples with various Cu loadings (0.7–3.0 wt%) through the solid-state ion-exchange (SSIE) method were also prepared [41]. The chemical statuses of the loaded Cu species were characterized by *in situ* DRIFTS, XRD,

H_2–TPR, and ultraviolet and visible spectroscopy (UV/Vis), which suggest that two different Cu species existed on the prepared Cu–SAPO-34 samples: Cu^{2+} ions and Cu_xO_y clusters (dimeric or oligomeric Cu species). Cu^{2+} ions were verified to be the active centers for NO abatement, while Cu_xO_y could promote the NH_3 oxidation, leading to the observed decrease in the standard SCR process at high temperatures. The acidity–activity correlation for Cu–SAPO-34 during NH_3-SCR was subsequently studied by the same group, wherein a series of Cu–SAPO-34 samples with varying numbers of Brönsted acid sites were prepared by WIE with potassium to investigate the role of the Brönsted acidity. Along with an increase in the potassium loading, the SCR activity of the Cu/K–SAPO-34 catalysts diminished in accordance with the decreasing Brönsted acidity. The reaction rate was further found to be dependent on the NH_3 coverage on Brönsted acid sites at 180–280 °C. At elevated temperatures, the acidic sites could act as a source of NH_3 for the SCR reaction because the NH_4^+ was initially adsorbed on the Brönsted acid sites and could then gradually migrate to the copper sites to finally react with the NO_x species.

38.3 SCR Reaction Mechanism

NH_3-SCR process generally occurs via three types of reaction paths according to the literature [42], which depends on the fraction of NO_2 attended in reaction as listed in Eqs. (38.1)–(38.4):

$$4NO + 4NH_3 + O_2 \rightarrow 4N_2 + 6H_2O \text{ (standard SCR)} \tag{38.1}$$

$$NO + NO_2 + 2NH_3 \rightarrow 2N_2 + 3H_2O \text{ (fast SCR)} \tag{38.2}$$

$$6NO_2 + 8NH_3 \rightarrow 7N_2 + 12H_2O \text{ (slow SCR to } N_2\text{)} \tag{38.3}$$

$$2NO_2 + 2NH_3 \rightarrow N_2 + N_2O + 3H_2O \text{ (slow SCR to } N_2O\text{)} \tag{38.4}$$

As for the standard SCR, many authors suggest that the reaction starts with an oxidation of NO to NO_2 on the active sites. Subsequently, NO_2 can react with the adsorbed NH_3, yielding NH_4NO_2 or NH_4NO_3, which can be further decomposed into the final products of N_2 and H_2O as well as the undesired pollutants of N_2O and NO_x [43]. The NO oxidation into NO_2 is reported to be the rate-determining step. Most researchers believed that the standard SCR could take place over the zeolite catalysts, fulfilling the Langmuir–Hinshelwood mechanism. However, strange SCR behavior was observed in the case of Cu–SSZ-13 prepared through one-pot synthesis [44]. Although the "fast SCR mechanism," the first step of which being a rapid oxidation of partial NO resulting in a $NO + NO_2$ mixture, is typical in effect for the catalytic reduction of NO by NH_3 at low temperatures, it has been proposed that the "standard SCR mechanism" is more relevant for Cu–SSZ-13 because of the lesser NO_2 generation caused by a transition state constraint imposed by the small window of the CHA cage. However, the standard SCR mechanism is still a subject of debate, which is mainly because the generated intermediates are so active that they cannot be readily identified. An intermediate of nitrite/HONO during a standard SCR mechanism study of Fe–ZSM-5 was found to exist by a novel method [45]. During the research, BaO/Al_2O_3, known as an LNT (lean NO_x trap) catalyst, was mixed with the SCR catalyst of Fe–ZSM-5. A trap experiment was thereafter conducted by exposing this mixed catalyst under an atmosphere of NO and O_2 at a low temperature (120 °C) [46]. As reported, the inclusion of BaO/Al_2O_3 in a physical mixture with Fe–ZSM-5 resulted in stabilizing nitrite species adsorbed on Fe sites after exposure to $NO + O_2$ and captured upon BaO via gas-phase equilibrium with HONO. Finally, an alternative mechanism for the standard SCR process was also proposed: (i) the nitrite species (NO_2^-) in equilibrium with gaseous nitrous acid (HONO) could be initially generated in the presence of NO and O_2; (ii) the formed nitrite/HONO species were subsequently decomposed into gaseous NO_2 to a certain degree. However, in the presence of NH_3, the nitrite/HONO could quickly react with NH_3 to form the final product of N_2. It was also reported that NO oxidation to NO_2 hardly was the rate-determining step of the standard SCR reaction, whereas the nitrite/HONO served as the most important intermediate.

Many researchers suggest that most of typical microporous zeolites (such as ZSM-5, Y, and β) favor "fast SCR mechanism," due to which the pore size of zeolites are larger than the dimensions of the undesired by-products. Different copper species have different effects on the SCR mechanism of Cu zeolite catalysts [28]. The CuO generated at high Cu loadings can catalyze the oxidation of NO into NO_2, which is favorable for low-temperature deNO_x activity because the $NH_3 + NO + NO_2$ (fast SCR) and $NH_3 + NO_2$ (slow SCR) reactions are known to be faster than the $NH_3 + NO$ reaction (standard SCR) at low temperatures. An overall SCR scheme as a function of the NO_2/NO_x ratio as well as reaction temperature was thereafter profiled, as shown in Figure 38.5. It is seen that a

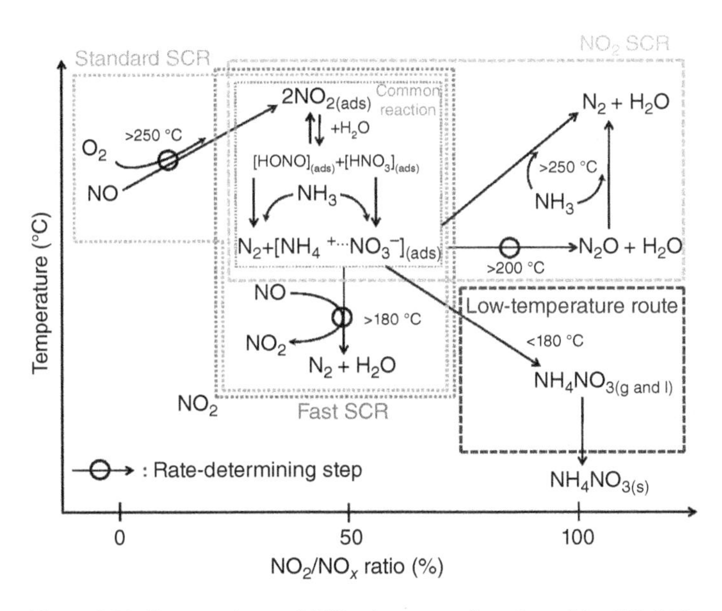

Figure 38.5 Proposed overall SCR scheme as a function of the NO_2/NO_x ratio and temperature. Source: Reprinted with permission from Iwasaki and Shinjoh [46]. Copyright 2010, Elsevier.

common step in each kind of SCR mechanism is the formation of surface species of adsorbed ammonium nitrate. These mechanism studies are beneficial for clarifying the structure–performance relationship and give us ideas on the design of SCR catalyst with considerable activity.

The main differences among the three pathways are reaction sequence and reaction rate. NO is easy to be oxidized to NO_2, and for "fast SCR" and "slow SCR," NO_2 is used as a reactant to combine with NH_3 to form NH_4NO_3 and further react with NO to form NH_4NO_2 (rate-determining step) to decompose into N_2 and H_2O. In the standard SCR, NO as a reactant combines with NH_3 to directly form NH_4NO_2 and decompose into N_2 and H_2O. The reaction sequences of the three pathways are distinguishable, and since the reaction of NH_4NO_3 to NH_4NO_2 is the rate-determining step, the reaction rates are accordingly different.

38.4 Conclusions and Perspectives

In brief, zeolites are so far the superior catalysts for the reduction of NO_x. Zeolites are highly porous materials having special pore structures with diverse topologies, associated with relatively high surface areas [47]. Their ion-exchange ability is beneficial for the better dispersion of active components such as cations [48, 49]. Simultaneously, static electrical field of the zeolite framework facilitates the activation of reactants [50]. The aforementioned special characteristics of zeolites make them promising materials served as SCR catalysts. Furthermore, acidity will help in stabilizing NH_3 at high temperatures [51, 52], which is the reducing agent for NO elimination. The low cost and the convenience of industrialization have promoted the rapid development of zeolite catalysts.

Perspectives in deNO_x reaction system are listed as follows:

(a) NH_3-SCR has become one of the most promising deNO_x methods under lean conditions. However, due to some drawbacks of commercialized V–W–Ti catalysts, more attention is now paid to the development of new superiorly active materials for NO_x removal. Fortunately, Cu–SSZ-13 zeolite has recently been proposed to exhibit extremely high low-temperature activity and N_2 selectivity. The related structure–activity relationship is also illustrated on the basis of both experimental and theoretical approaches. However, using the very expensive and toxic template of *N,N,N*-trimethyl-1-adamantylammonium hydroxide (TMA-daOH) for Cu–SSZ-13 synthesis leads to a significant increase in the catalyst cost, which seriously prevents industrial realization. Therefore, developing an economical way to synthesize SSZ-13 by using a cheap template has become an important direction for the NH_3-SCR technique.

(b) On the basis of the concept of "quasi shape selectivity" illustrated for SSZ-13, investigations of other microporous zeolite catalysts with supersmall apertures, such as OFF (3.6×4.9 Å^2), ERI (3.6×5.1 Å^2), LEV (3.6×4.8 Å^2), and AFX (3.6×3.4 Å^2), should also be taken into account for the NH_3-SCR study to obtain a common principle for the design of highly efficient zeolite catalysts.

(c) Developing deNO_x technologies other than the traditional NH_3-SCR, such as H_2-SCR and NO direct decomposition. Although some research has already been done on these topics, they still need a lot of efforts.
(d) The hot topic of special structures in zeolites, such as "hierarchical," "core–shell," "yolk–shell," may play a role in the deNO_x reaction.
(e) Theoretical simulation based on DFT is believed to be promising to describe the structure and local environment of active sites, as well as to determine the depollution mechanism at the molecular or atomic level. However, due to the complications of some depollution systems, such as NO-SCR involving many molecules (NO, O_2, and reductant), causing a huge amount of calculation, and being more uncertain, related DFT studies on the deNO_x mechanism were scarcely conducted. Hence, these mechanism simulations might become prevalent in the future with the development of computational technology.

References

1 Vaccaro, A.R., Mul, G., Pérez-RamíRez, J. et al. (2003). On the activation of Pt/Al_2O_3 catalysts in HC–SCR by sintering: determination of redox-active sites using multitrack. *Appl. Catal., B* 46: 687–702.
2 Bion, N., Saussey, J., Haneda, M. et al. (2003). Study by in situ FTIR spectroscopy of the SCR of NO_x by ethanol on Ag/Al_2O_3 – evidence of the role of isocyanate species. *J. Catal.* 217: 47–58.
3 Guenin, M., Breysse, M., Frety, R. et al. (1987). Resistance to sulfur poisoning of metal catalysts: Dehydrogenation of cyclohexane on Pt/Al_2O_3 catalysts. *J. Catal.* 105: 144–154.
4 Corro, G., Cano, C., and Fierro, J.L.G. (2010). A study of Pt–Pd/γ-Al_2O_3 catalysts for methane oxidation resistant to deactivation by sulfur poisoning. *J. Mol. Catal. A: Chem* 315: 35–42.
5 Angelidis, T.N., Christoforou, S., Bongiovanni, A. et al. (2002). On the promotion by SO_2 of the SCR process over Ag/Al_2O_3: influence of SO_2 concentration with C_3H_6 versus C_3H_8 as reductant. *Appl. Catal., B* 39: 197–204.
6 Jagtap, N., Umbarkar, S.B., Miquel, P. et al. (2009). Support modification to improve the sulphur tolerance of Ag/Al_2O_3 for SCR of NO_x with propene under lean-burn conditions. *Appl. Catal., B* 90: 416–425.
7 Wang, J., He, H., Feng, Q. et al. (2004). Selective catalytic reduction of NO_x with C_3H_6 over an Ag/Al_2O_3 catalyst with a small quantity of noble metal. *Catal. Today* 93: 783–789.
8 Kasaoka, S., Sasaoka, E., and Senda, K. (1977). Catalysts of V_2O_5-TiO_2 & system for reduction of nitrogen oxide with ammonia. *J. Fuel Soc. Jpn.* 56: 818–823.
9 Phil, H.H., Reddy, M.P., Kumar, P.A. et al. (2008). SO_2 resistant antimony promoted V_2O_5/TiO_2 catalyst for NH_3-SCR of NO_x at low temperatures. *Appl. Catal., B* 78: 301–308.

10 Wang, X., Wang, J., and Zhu, T.Y. (2013). Coupled control of chlorobenzene and NO over V_2O_5/TiO_2 catalyst in NH_3-SCR reaction. *Adv. Mater. Res.* 811: 83–86.

11 Jung, S.M. and Grange, P. (2001). Characterization and reactivity of $V_2O_5-WO_3$ supported on $TiO_2-SO_4^{2-}$ catalyst for the SCR reaction. *Appl. Catal., B* 32: 123–131.

12 Morikawa, S., Takahashi, K., Yoshida, H. et al. (1982). Life of $V_2O_5-TiO_2$ catalyst for the NO_x reduction with NH_3 in flue gas from oil fired boiler. *J. Fuel Soc. Jpn.* 61: 1024–1030.

13 Wan, Q., Duan, L., He, K.B. et al. (2011). Removal of gaseous elemental mercury over cerium doped low vanadium loading $V_2O_5-WO_3/TiO_2$ in simulated coal-fired flue gas. *Environ. Sci.* 32: 2800–2804.

14 Shen, B., Chen, J., Yue, S., and Li, G. (2015). A comparative study of modified cotton biochar and activated carbon based catalysts in low temperature SCR. *Fuel* 156: 47–53.

15 Wei, L., Shan, T., Yun, S. et al. (2015). Utilization of Sargassum based activated carbon as a potential waste derived catalyst for low temperature selective catalytic reduction of nitric oxides. *Fuel* 160: 35–42.

16 Boyano, A., Gálvez, M.E., Moliner, R. et al. (2008). Carbon-based catalytic briquettes for the reduction of NO: effect of H_2SO_4 and HNO_3 carbon support treatment. *Fuel* 87: 2058–2068.

17 Shen, B.X. and Liu, T. (2010). Deactivation of MnO_x-CeO_x/ACF catalysts for low-temperature NH_3-SCR in the presence of SO_2. *Acta Phys. Chim. Sin.* 26: 3009–3016.

18 Li, Y., Guo, Y., Xiong, J. et al. (2016). The roles of sulfur-containing species in selective catalytic reduction of NO with NH_3 over activated carbon. *Ind. Eng. Chem. Res.* 55: 12341–12349.

19 Baerlocher, Ch. and McCusker, L.B. Database of Zeolite Structures: http://www.iza-structure.org/databases/.

20 Delahay, R., Kieger, S., Tanchoux, N. et al. (2004). Kinetics of the selective catalytic reduction of NO by NH_3 on a Cu-faujasite catalyst. *Appl. Catal., B* 52: 251–257.

21 Komatsu, T., Nunokawa, M., Moon, I.S. et al. (1994). Kinetic studies of reduction of nitric oxide with ammonia on Cu^{2+}-exchanged zeolites. *J. Catal.* 148: 427–437.

22 Rahkamaa-Tolonen, K., Maunula, T., Lomma, M. et al. (2005). The effect of NO_2 on the activity of fresh and aged zeolite catalysts in the NH_3-SCR reaction. *Catal. Today* 100: 217–222.

23 Kröcher, O., Devadas, M., Elsener, M. et al. (2006). Investigation of the selective catalytic reduction of NO by NH_3 on Fe–ZSM5 monolith catalysts. *Appl. Catal., B* 66: 208–216.

24 Long, R.Q. and Yang, R.T. (2001). Selective catalytic oxidation (SCO) of ammonia to nitrogen over Fe-exchanged zeolites. *J. Catal.* 201: 145–152.

25 Brandenberger, S., Krocher, O., Tissler, A., and Althoff, R. (2008). The state of the art in selective catalytic reduction of NO_x by ammonia using metal-exchanged zeolite catalysts. *Catal. Rev.* 50: 492–531.

26 Bartholomew, C.H. (2001). Mechanisms of catalyst deactivation. *Appl. Catal., A* 212: 17–60.

27 Gao, F., Walter, E.D., Karp, E.M. et al. (2013). Structure–activity relationships in NH_3-SCR over Cu–SSZ-13 as probed by reaction kinetics and EPR studies. *J. Catal.* 300: 20–29.

28 Pereda-Ayo, B., De La Torre, U., Illán-Gómez, M.J. et al. (2014). Role of the different copper species on the activity of Cu/zeolite catalysts for SCR of NO_x with NH_3. *Appl. Catal., B* 147 (5): 420–428.

29 Wang, H., Xu, R., Yi, J., and Zhang, R.D. (2019). Zeolite structure effects on Cu active center, SCR performance and stability of Cu-zeolite catalysts. *Catal. Today* 327: 295–307.

30 Chen, B., Xu, R., Zhang, R., and Liu, N. (2014). Economical way to synthesize SSZ-13 with abundant ion-exchanged Cu^+ for an extraordinary performance in selective catalytic reduction (SCR) of NO_x by ammonia. *Environ. Sci. Technol.* 48: 13909–13916.

31 McEwen, J.S., Anggara, T., Schneider, W.F. et al. (2012). Integrated operando X-ray absorption and DFT characterization of Cu–SSZ-13 exchange sites during the selective catalytic reduction of NO_x with NH_3. *Catal. Today* 184: 129–144.

32 Kwak, J.H., Tonkyn, R.G., Kim, D.H. et al. (2010). Excellent activity and selectivity of Cu–SSZ-13 in the selective catalytic reduction of NO_x with NH_3. *J. Catal.* 275 (2): 187–190.

33 Hussain, M., Fino, D., and Russo, N. (2014). Development of modified KIT-6 and SBA-15-spherical supported Rh catalysts for N_2O abatement: from powder to monolith supported catalysts. *Chem. Eng. J.* 238: 198–205.

34 Ye, Q., Wang, L., and Yang, R.T. (2012). Activity, propene poisoning resistance and hydrothermal stability of copper exchanged chabazite-like zeolite catalysts for SCR of NO with ammonia in comparison to Cu/ZSM-5. *Appl. Catal., A* 427–428: 24–34.

35 Fickel, D.W. and Lobo, R.F. (2010). Copper coordination in Cu–SSZ-13 and Cu–SSZ-16 investigated by variable-temperature XRD. *J. Phys. Chem. C* 114: 1633–1640.

36 Zhang, R.Q., McEwen, J.S., Kollár, M. et al. (2014). NO chemisorption on Cu/SSZ-13: a comparative study from infrared spectroscopy and DFT calculations. *ACS Catal.* 4: 4093–4105.

37 Deka, U., Juhin, A., Eilertsen, E.A. et al. (2012). Confirmation of isolated Cu^{2+} ions in SSZ-13 zeolite as active sites in NH_3-selective catalytic reduction. *J. Phys. Chem. C* 116: 4809–4818.

38 Verma, A.A., Bates, S.A., Anggara, T. et al. (2014). NO oxidation: a probe reaction on Cu–SSZ-13. *J. Catal.* 312: 179–190.

39 Bates, S.A., Verma, A., Paolucci, C. et al. (2014). Identification of the active Cu site in standard selective catalytic reduction with ammonia on Cu–SSZ-13. *J. Catal.* 312: 87–97.

40 Moreno-González, M., Hueso, B., Boronat, M. et al. (2015). Ammonia-containing species formed in Cu-chabazite as per in situ EPR, solid-state NMR, and DFT calculations. *J. Phys. Chem. Lett.* 6: 1011–1017.

41 Wang, L., Li, W., Schmieg, S.J., and Weng, D. (2015). Role of Brönsted acidity in NH_3 selective catalytic reduction reaction on Cu/SAPO-34 catalysts. *J. Catal.* 324: 98–106.
42 Zhang, R.D., Liu, N., and Lei, Z. (2016). Selective transformation of various nitrogen-containing exhaust gases toward N_2 over zeolite catalysts. *Chem. Rev.* 116 (6): 3658.
43 Wang, D., Zhang, L., Kamasamudram, K., and Epling, W.S. (2013). In situ-DRIFTS study of selective catalytic reduction of NO_x by NH_3 over Cu-exchanged SAPO-34. *ACS Catal.* 3: 871–881.
44 Xie, L., Liu, F., Ren, L. et al. (2014). Excellent performance of one-pot synthesized Cu–SSZ-13 catalyst for the selective catalytic reduction of NO_x with NH_3. *Environ. Sci. Technol.* 48: 566–572.
45 Ruggeri, M.P., Selleri, T., Colombo, M. et al. (2014). Identification of nitrites/HONO as primary products of NO oxidation over Fe–ZSM-5 and their role in the standard SCR mechanism: a chemical trapping study. *J. Catal.* 311: 266–270.
46 Iwasaki, M. and Shinjoh, H. (2010). A comparative study of "standard", "fast" and "NO_2" SCR reactions over Fe/zeolite catalyst. *Appl. Catal., A* 390: 71–77.
47 Puertolas, B., Garcia-Andujar, L., Garcia, T. et al. (2014). Bifunctional Cu/H–ZSM-5 zeolite with hierarchical porosity for hydrocarbon abatement under cold-start conditions. *Appl. Catal., B* 154–155: 161–170.
48 Hunger, B., Heuchel, M., Clark, L.A., and Snurr, R.Q. (2002). Characterization of acidic OH groups in zeolites of different types: an interpretation of NH_3–TPD results in the light of confinement effects. *J. Phys. Chem. B* 106: 3882–3889.
49 Weitkamp, J. (2000). Zeolites and catalysis. *Solid State Ionics* 131: 175–188.
50 Liu, N., Zhang, R., Li, Y., and Chen, B. (2014). Local electric field effect of TMI (Fe, Co, Cu)-BEA on N_2O direct dissociation. *J. Phys. Chem. C* 118: 10944–109560.
51 Miyamoto, Y., Katada, N., and Niwa, M. (2000). Acidity of β zeolite with different Si/Al ratio as measured by temperature programmed desorption of ammonia. *Microporous Mesoporous Mater.* 40: 271–281.
52 Zhang, W., Burckle, E.C., and Smirniotis, P.G. (1999). Characterization of the acidity of ultrastable Y, mordenite, and ZSM-12 via NH_3-stepwise temperature programmed desorption and Fourier transform infrared spectroscopy. *Microporous Mesoporous Mater.* 33: 173–185.

Index

Heterogeneous Catalysts: Advanced Design, Characterization and Applications, First Edition.
Edited by Wey Yang Teoh, Atsushi Urakawa, Yun Hau Ng, and Patrick Sit.

b

c

d

g

o

p

t

u